Harnessing AI for Human-Animal Communication

Walid Ben Ameur
https://orcid.org/0009-0000-5046-7408
University of Gabes, Tunisia

Adnène Arbi
https://orcid.org/0000-0002-2834-0248
University of Carthage, Tunisia

IGI Global
Scientific Publishing
Publishing Tomorrow's Research Today

Published in the United States of America by
IGI Global Scientific Publishing
701 East Chocolate Avenue
Hershey, PA, 17033, USA
Tel: 717-533-8845 | Fax: 717-533-7115
Website: https://www.igi-global.com E-mail: cust@igi-global.com

Library of Congress Cataloging-in-Publication Data

LCCN: 2025033244 (CIP Data Pending)
ISBN13: 9798337354835
Isbn13Softcover: 9798337354842
EISBN13: 9798337354859

British Cataloguing in Publication Data
A Cataloguing in Publication record for this book is available from the British Library.

Table of Contents

Detailed Table of Contents

and modifies achievement standards. A combination of theoretical models and real-world study transforms into actual insights about AI systems' twofold effects within the education field. AI creates individualized student learning along with active classroom participation because it delivers responsive guidance and support systems.

This chapter explores the emerging intersection of artificial intelligence (AI) and animal communication, focusing on the challenges and opportunities involved in interpreting non-human signals. It presents cutting-edge AI methodologies applied to analyze animal vocalizations, gestures, and behaviours, highlighting their applications in veterinary medicine, wildlife conservation, and companion animal care. By fostering interdisciplinary collaboration among AI specialists, ethologists, veterinarians, and ecologists, the chapter offers a holistic view of the technological advancements shaping this field. It also addresses critical ethical and philosophical questions surrounding the use of AI in understanding and influencing human-animal relationships. Serving as a foundational reference, this work aims to inspire future research and responsible innovation to bridge species communication and enhance animal welfare in an AI-driven world.

Knowledge of animal behavior is paramount to develop studies not only in the field of ecology, ethology, veterinary, and animal welfare. Conventional behavior observation techniques are highly dependent on manual observation which is cumbersome, subjective and can carry out its own errors as done by human beings. Computer vision has recently become an important unsolicited, scalable, non-invasive method to analyse animal behavior. The following paper gives an overview of methods of computer vision that were used in the study of animal behavior, namely object detection, pose estimation, tracking and activity recognition. It talks about the application of deep learning algorithms and data fitting of diverse organisms and environments, ranging between laboratories and wild. Computer vision incorporated into this field can increase accuracy and raise the opportunity of real-time surveillance, early detection of the disease, and behavioral phenotyping.

This chapter explores how Small and Medium Enterprises (SMEs) navigate digital transformation by developing dynamic capabilities—technological, relational, financial, and environmental—that enhance their strategic resilience in volatile markets. Recognizing the unique challenges faced by resource-limited SMEs, the volume updates current knowledge on managing digital maturity throughout the SME lifecycle. It suggests customized management frameworks and performance indicators aimed at enhancing innovation and competitiveness. The chapter also offers practical guidance for policymakers, development institutions, and entrepreneurship ecosystems focused on fostering sustainable growth and digital adoption in SMEs. Integrating perspectives from senior management, institutional economics, and sustainable entrepreneurship, this work broadens traditional views centered on large corporations, providing actionable insights for driving SME innovation and long-term economic impact.

IOT devices are often in the vicinity of animals. IOT devices on animals are used for experimentation and monitoring purposes. This chapter discusses the effect of different radiations emitted from IoT devices on animals. The specific animals in focus are rats, mice and chickens. There are different short-range and long-range frequency-based radiations emitted from IoT devices. The current literature discusses the experiments conducted on these animals and the duration of exposure to radiation. The literature review provides evidence regarding their effect on different classes of animals. These effects also depend upon the type of animal and the duration of exposure to radiations. The rats showed memory, blood, brain, cell loss, learning related effects. The mouse or mice showed reduced calcium absorption, memory loss, behavioral, anxiety-related,carcinogens-related changes. The chickens showed effect on neurons and blood-brain integrity changes. There were also studies which did not point to any drastic changes in these animals.

This chapter explores the implications of AI technologies used in decoding animal vocalisations, emotion recognition, behavioural prediction, and robotic interaction. Employing a qualitative, interdisciplinary methodology that draws on philosophical analysis, ethical theory, including animal ethics and AI ethics, and empirical case studies, the work critically examines the impact of AI design choices on animal agency, consent, and welfare. Attention is also paid to broader societal issues, such as anthropomorphism, data bias, and the unequal effects of AI adoption in contexts like conservation and industrial farming. Furthermore, the chapter situates these interspecies technologies within the wider discourse on AI ethics, addressing concerns around fairness, transparency, accountability, data privacy, and digital inequality. It advocates for an ethics-by-design approach and participatory governance frameworks that incorporate the perspectives of ethicists, technologists, affected communities, and policymakers.

Robotic systems for animal interaction form a rapidly advancing interdisciplinary field uniting robotics, artificial intelligence, ethology, and welfare science. These technologies monitor, engage with, and assist animals across settings such as precision livestock farming, wildlife conservation, and laboratory research. Applications include automating routine care, enriching environments, supporting behavioral studies, and enabling safe human–animal–robot collaboration. Advances in sensing, machine learning, and adaptive control now allow robots to interpret animal behavior in real time and respond appropriately, enhancing efficiency, safety, and welfare outcomes. Yet challenges remain, including ethical concerns, species-specific modeling, cost, and reliability in varied conditions. This chapter reviews system types, design principles, and applications, presents case studies from agriculture, research, and conservation,

Preface

Recent advances in artificial intelligence (AI) are reshaping the boundaries of knowledge in ways that were once considered the domain of speculation. Among the most profound of these developments is the growing capacity to interpret, model, and respond to the signals of non-human animals. For centuries, human–animal communication has been constrained by fundamental differences in language, cognition, and sensory perception. Today, however, breakthroughs in machine learning, computer vision, natural language processing, bioacoustics, and sensor technologies are enabling researchers to decode patterns of vocalization, movement, facial expression, and behavior across a wide range of species. These advances signal the emergence of a new scientific and ethical frontier—one in which technology serves as a bridge between species rather than a barrier.

ABOUT THIS BOOK

Harnessing AI for Human–Animal Communication is conceived in response to this rapidly evolving landscape. The volume brings together interdisciplinary scholarship and applied research that collectively examine how AI is transforming our ability to understand, monitor, and interact with animals. Beyond technical innovation, the book recognizes that human–animal communication is deeply embedded in ecological, ethical, cultural, and societal contexts. As such, the chapters explore not only what AI can do, but also how and why it should be used responsibly in domains that directly affect animal welfare, conservation, education, and human values.

The primary objective of this edited reference book is to consolidate emerging knowledge at the intersection of artificial intelligence and animal communication, offering a coherent and forward-looking synthesis of current research. The contributions collectively present state-of-the-art methods for AI-driven analysis of animal vocalizations, gestures, facial expressions, and behaviors, while highlighting real-world applications in veterinary medicine, wildlife conservation, precision livestock farming, and companion animal care. Particular attention is given to technologies such as machine learning–based bioacoustic analysis, computer vision for behavioral and emotional assessment, wearable sensors for real-time health monitoring, and robotic systems designed for safe and meaningful interaction with animals.

At the same time, this book intentionally foregrounds the limitations and challenges inherent in interpreting non-human signals through artificial systems. Issues such as data scarcity, species-specific variability, model explainability, anthropomorphic bias, and contextual dependency are critically examined across chapters. Ethical considerations—ranging from animal agency and welfare to transparency, accountability, and responsible AI design—are treated not as peripheral concerns but as central pillars of the field. By doing so, the volume encourages readers to reflect on how technological choices shape human–animal relationships and influence conservation and management practices.

A defining strength of this book lies in its interdisciplinary scope. Contributions span computer science, robotics, ecology, ethology, veterinary science, philosophy, education, and social sciences, fostering dialogue among disciplines that too often operate in isolation. The volume also highlights emerging directions such as AI-assisted behavior prediction, mood tracking in companion animals, species-specific communication modeling, AI-driven wildlife surveillance and anti-poaching systems, and the integration of citizen science and big data analytics into animal behavior research. Cross-cultural and philosophical perspectives further enrich the discussion, reminding us that interspecies communication is as much a human question as it is a technological one.

This reference work is intended for researchers, practitioners, policymakers, and advanced students seeking a comprehensive understanding of how AI can be harnessed to bridge species boundaries. By bringing together technical innovation, empirical application, and ethical reflection, the book aims to inspire responsible research and collaborative innovation in the quest to deepen our understanding of the animal world. Ultimately, we hope this volume will contribute to a future in which technology enhances empathy, promotes animal welfare, and supports more harmonious coexistence between humans and other species.

Walid Ben Ameur

Faculty of Sciences of Gabès, Ecology of Terrestrial Fauna Laboratory, University of Gabes, Tunisia

Adnène Arbi

University of Carthage, École Polytechnique de Tunisie, La Marsa, Tunisia

Chapter 1

AI-driven Tools for Wildlife Surveillance and Anti-Poaching Efforts

P. Selvakumar
https://orcid.org/0000-0002-3650-4548
Department of Science and Humanities, Nehru Institute of Technology, Coimbatore, India

Mahender Reddy Bobbala
https://orcid.org/0009-0003-9018-1502
CVR College of Engineering, India

Sirikonda Shwetha
https://orcid.org/0000-0002-2738-680X
Sumathi Reddy Institute of Technology for Women, India

Naresh Kumar Sripada
https://orcid.org/0000-0002-0724-1750
SR University, India

T. C. Manjunath
https://orcid.org/0000-0003-2545-9160
Rajarajeswari College of Engineering, Bengaluru, India

E. Sreedevi
Koneru Lakshmaiah Education Foundation, India

ABSTRACT

Poaching poses one of the most pressing and destructive threats to global biodiversity in the 21st century. This illegal hunting, capturing, or killing of wildlife not only endangers species but also destabilizes ecosystems, fuels criminal networks, and undermines conservation efforts across continents. While the term "poaching" may conjure images of ivory tusks its reach is far broader and more insidious. From endangered big cats in Asia to pangolins in Africa and sea turtles in the Americas, poaching spans nearly every continent and involves an array of species, each crucial to their respective habitats. As the threat escalates, conservationists and technology experts are turning to artificial intelligence (AI) as a transformative solution to detect, deter, and combat poaching on a global scale. The scale of poaching is staggering.

DOI: 10.4018/979-8-3373-5483-5.ch001

According to the World Wildlife Fund (WWF), over 20,000 African elephants are killed each year for their tusks. The black rhino population has plummeted by over 90% in the past few decades due to rampant poaching.

THE THREAT OF POACHING: WHY AI-DRIVEN TOOLS ARE ESSENTIAL

Poaching poses one of the most pressing and destructive threats to global biodiversity in the 21st century. This illegal hunting, capturing, or killing of wildlife not only endangers species but also destabilizes ecosystems, fuels criminal networks, and undermines conservation efforts across continents. While the term "poaching" may conjure images of ivory tusks its reach is far broader and more insidious. From endangered big cats in Asia to pangolins in Africa and sea turtles in the Americas, poaching spans nearly every continent and involves an array of species, each crucial to their respective habitats. As the threat escalates, conservationists and technology experts are turning to artificial intelligence (AI) as a transformative solution to detect, deter, and combat poaching on a global scale. The scale of poaching is staggering. According to the World Wildlife Fund (WWF), over 20,000 African elephants are killed each year for their tusks. The black rhino population has plummeted by over 90% in the past few decades due to rampant poaching. Tigers, once numbering in the hundreds of thousands across Asia, now survive in fragmented pockets, with fewer than 4,000 left in the wild. The pangolin, a lesser-known mammal, has earned the grim distinction of being the most trafficked mammal in the world. It is estimated that over one million pangolins have been illegally traded over the past decade. These numbers are not merely statistics; they represent a profound crisis threatening the balance of nature and the integrity of entire ecosystems. The drivers of poaching are complex and interwoven. At its core, poaching is fueled by high demand for animal parts such as ivory, horns, bones, skins, and traditional medicine ingredients. These commodities often fetch exorbitant prices on black markets, particularly in parts of Asia and the Middle East.

Funding also remains a critical barrier. Developing and deploying AI solutions can be costly, particularly in low-income countries where poaching is most prevalent. Public-private partnerships, international aid, and philanthropic support will be necessary to scale AI technologies in conservation. Additionally, fostering collaborations between conservationists, technologists, data scientists, and local communities can ensure that AI tools are adapted to specific regional contexts and that their deployment aligns with both conservation and social goals. Despite these hurdles, the integration of AI into wildlife protection represents a hopeful frontier in the battle against poaching. As threats to biodiversity grow more complex and interconnected, so too must our strategies to confront them. AI offers speed, precision, and scalability that traditional methods cannot match. It empowers rangers to act swiftly, researchers to understand patterns more deeply, and policymakers to make data-driven decisions. Most importantly, it offers a lifeline to endangered species that may otherwise vanish within our lifetime. In conclusion, the threat of poaching is a multifaceted crisis with devastating consequences for wildlife, ecosystems, and human communities alike. Traditional methods, while valuable, are no longer sufficient to address the scale and sophistication of modern poaching networks. Artificial intelligence, through its diverse applications and ability to operate in real-time, presents a powerful and essential tool in the fight for wildlife conservation. By embracing AI and integrating it with existing conservation efforts, the global community has a renewed opportunity to protect Earth's most vulnerable species before it's too late.

THE ROLE OF TECHNOLOGY: LEVERAGING AI FOR ANTI-POACHING EFFORTS

The accelerating loss of biodiversity across the globe poses a critical challenge to humanity and the planet. From deforestation and habitat fragmentation to climate change and poaching, wildlife faces a confluence of existential threats. In this context, the application of artificial intelligence (AI) in wildlife conservation represents a transformative step toward protecting Earth's most vulnerable species and ecosystems. (Ayyaswamy K. et al. 2025) AI technologies—ranging from machine learning algorithms and computer vision to predictive analytics and autonomous systems—are increasingly being used to enhance surveillance, improve decision-making, and strengthen the protection of wildlife. (Ayyaswamy K. et al. 2025) These tools are not just helping to monitor wildlife more effectively but are actively reshaping conservation strategies and opening new frontiers in ecological research and species protection. (Chisom O. N. et al. 2024) Traditional wildlife monitoring methods, such as manual patrolling, camera trapping, and satellite telemetry, have played a pivotal role in conservation. However, they are often time-consuming, resource-intensive, and limited in scale. Many protected areas span thousands of square kilometers, making comprehensive coverage nearly impossible with human resources alone. (Chisom O. N. et al. 2024) AI bridges this gap by automating and accelerating data collection and analysis, thus enabling real-time monitoring and faster response to threats. One of the most impactful applications of AI in conservation is computer vision, particularly when used with camera traps. These motion-activated cameras are deployed in forests, savannahs, and other habitats to capture images of passing wildlife. Traditionally, reviewing and classifying millions of images would require an enormous investment of time and labor. AI-powered image recognition systems can now process these images rapidly, identifying species, counting individuals, and even recognizing specific behaviors. For example, Microsoft's AI for Earth initiative and tools like Wildlife Insights use convolutional neural networks to distinguish between different animal species with high accuracy. These platforms not only reduce the burden on researchers but also enable them to act on insights faster, such as detecting the presence of rare or threatened animals in unexpected locations.

AI is also revolutionizing aerial surveillance through its integration with drones and unmanned aerial vehicles (UAVs). (Coates J. M. et al. 2025) Equipped with thermal imaging and infrared sensors, drones can fly over large areas and detect movement patterns—day or night—that would be missed by the naked eye. AI algorithms process the visual or thermal data in real time to identify poachers, animals, or anomalies in the environment. (Coates J. M. et al. 2025) Conservation programs in Africa, Asia, and Latin America have successfully deployed drone-AI systems to monitor elephants, rhinos, and big cats, often in terrain too difficult or dangerous for human patrols. In Rwanda and Nepal, for instance, conservationists are using drones to monitor gorilla populations and illegal encroachments into protected zones, with AI-assisted image processing improving both the accuracy and speed of these efforts. Another key advancement comes through acoustic monitoring systems, which utilize AI to analyze soundscapes in forests, oceans, and grasslands. (Darney P. E. et al. 2025) These systems detect the presence of wildlife through calls, songs, or other vocalizations, while also identifying human threats such as gunshots, chainsaws, or motorbikes. AI-powered acoustic models, like those used in the Rainforest Connection project, allow conservationists to listen to and analyze thousands of hours of sound data from remote rainforests. This not only helps track elusive species such as orangutans or whales but also gives early warnings of illegal logging or poaching activity. These audio alerts can then be relayed in real time to rangers or local enforcement teams, enabling a swift and coordinated response. Beyond reactive monitoring, predictive modeling and machine learning are being used to anticipate future conservation challenges. (Darney P. E. et al. 2025) AI can analyze vast datasets—including animal movement, climate variables, historical poaching incidents, land-use changes, and community reports—to generate predictive maps that highlight areas at high risk for illegal activity or habitat loss. Tools like PAWS (Protection Assistant for Wildlife Security), developed by USC researchers, use game theory and machine learning to help rangers plan optimized patrol routes, increasing the likelihood of intercepting poachers and reducing illegal incursions. Similarly, AI models can help identify migration corridors, breeding grounds, or climate refuge zones, providing vital data for long-term conservation planning.

Camera Traps: AI-Powered Image Analysis for Wildlife Monitoring

Poaching remains one of the most urgent threats to global biodiversity, pushing countless species toward extinction and destabilizing fragile ecosystems. (Dave R. et al. 2025) This illegal activity, driven by black-market demand for wildlife products such as ivory, rhino horn, exotic pets, bushmeat, and body parts used in traditional medicine, is part of a multi-billion-dollar criminal enterprise that transcends borders and undermines conservation efforts worldwide. (Dave R. et al. 2025) As conventional anti-poaching measures face growing limitations in scope, effectiveness, and sustainability, technology has emerged as a critical tool in the fight against wildlife crime. Among the most transformative technologies is artificial intelligence (AI), which is reshaping the way conservationists detect, prevent, and respond to poaching. (Dwivedi D. N. et al. 2024) For decades, anti-poaching efforts have relied on foot patrols, physical barriers, informant networks, and manually monitored surveillance systems. These methods, while important, are often labor-intensive, reactive, and limited in reach. Protected areas can span thousands of square kilometers, many in remote or rugged terrain where poaching gangs often operate undetected. Rangers—who risk their lives daily—are often under-equipped and outnumbered by sophisticated poachers armed with military-grade weapons, night-vision equipment, and GPS devices. (Dwivedi D. N. et al. 2024) These challenges underscore the urgent need for scalable, data-driven solutions, which is where AI-powered technologies come into play. Artificial intelligence enables the collection, processing, and analysis of massive datasets at speeds and scales impossible for humans alone. By leveraging AI across various technological platforms—drones, camera traps, acoustic sensors, GPS trackers, and satellite imagery—conservationists can move from reactive enforcement to predictive and preemptive anti-poaching strategies. These systems can detect patterns, recognize anomalies, and issue real-time alerts, giving wildlife protection teams the upper hand.

One of the most widely adopted AI applications in anti-poaching is the analysis of images from camera traps. These motion-activated cameras are strategically placed along animal trails, park boundaries, and suspected poaching routes. (Goel P. K. et al. 2025) AI-based computer vision models, trained on thousands of labeled images, can now accurately distinguish between humans, vehicles, and various animal species. For example, if an AI system detects a human in a restricted area where there shouldn't be foot traffic—particularly at odd hours—it can trigger an alert to local patrol teams. (Goel P. K. et al. 2025) This reduces the time between intrusion detection and response, potentially stopping a poaching event in progress. Drones integrated with AI have revolutionized surveillance across vast and inaccessible landscapes. Equipped with high-resolution cameras, thermal sensors, and AI-driven navigation, drones can autonomously patrol large conservation areas. (Goyal A. et al. 2025) They can detect heat signatures of humans or animals even under dense foliage or in total darkness. AI models analyze the incoming video feed in real time, differentiating between wildlife and human threats, and flagging any unusual activity. (Goyal A. et al. 2025) This significantly enhances the situational awareness of rangers and enables them to direct their limited resources more effectively. Several African countries have already reported decreases in poaching incidents after deploying AI-enabled drones in key protected areas. Acoustic monitoring is another groundbreaking frontier in AI-enabled anti-poaching technology. In forests where visibility is low and camera traps are less effective, AI-powered audio sensors offer a powerful alternative. Devices can be mounted in trees or hidden in dense vegetation, continuously recording ambient sounds. (Gupta S. et al. 2025) AI models trained to recognize specific acoustic signatures—such as gunshots, chainsaws, motorcycle engines, or even animal distress calls—can flag suspicious activity. (Gupta S. et al. 2025) Projects like Rainforest Connection use solar-powered listening devices and AI algorithms to detect logging and poaching activity in real time. Alerts are then sent via cellular networks to rangers or local enforcement units, enabling a rapid, targeted response.

Acoustic Monitoring: AI-Driven Audio Analysis for Wildlife Detection

Wildlife conservation has always relied on data—on knowing where species live, how they move, what they eat, and what threatens them. One of the most effective tools for gathering this data over the past few decades has been the camera trap. These remote, motion-activated cameras are designed to capture images and videos of animals in their natural habitats without human interference. (Gustafson K. et al. 2018) Used across continents—from African savannas and Asian jungles to North American forests and South American rainforests—camera traps have revolutionized how researchers monitor wildlife. (Gustafson K. et al. 2018) Yet, with the explosion of images they produce, the next step in that revolution is driven by artificial intelligence (AI). By combining traditional camera trap technology with AI-powered image analysis, conservationists are now able to monitor wildlife at a scale and efficiency never before possible. At its core, a camera trap is a passive observation tool. Once deployed, it captures photographs or videos whenever motion is detected, typically using infrared sensors. This allows researchers to collect data continuously over weeks or months without disturbing the animals. Camera traps are especially valuable for studying elusive, nocturnal, or rare species that are otherwise difficult to observe. However, the success of camera traps has introduced a new challenge: data overload. (Hammouch & Raghav, 2025) A single camera can generate thousands of images in a short period—many of which may be blank or contain only vegetation moving in the wind. When scaled to hundreds or thousands of cameras across a landscape, the volume of data becomes unmanageable for manual processing. This is where AI-powered image analysis becomes transformative. (Hammouch & Raghav, 2025) Through techniques like machine learning and computer vision, AI can now process and classify camera trap images with remarkable accuracy. By training algorithms on large datasets labeled with species names, AI systems can learn to identify animals in new images, even distinguishing between closely related species. This allows for faster, more consistent, and more scalable data analysis, freeing researchers from countless hours of manual sorting and enabling them to focus on interpretation and conservation action.

Figure 1. Artificial Intelligence in Wildlife Conservation

One of the most widely used applications of AI in this domain is species recognition. Convolutional neural networks (CNNs), a class of deep learning models well-suited for image analysis, are trained on datasets of wildlife images until they can reliably identify species from new, unseen images. (Hossain M. et al. 2024) For example, a CNN can be trained to recognize the difference between a leopard and a cheetah, or between a white-tailed deer and a mule deer. These AI models continuously improve with use, learning from errors and adapting to new data. Global platforms like Wildlife Insights, developed by Conservation International in collaboration with Google and other partners, have made these AI tools accessible to researchers worldwide. Wildlife Insights uses AI to scan uploaded camera trap images, detect animals, and classify them by species. The platform also helps identify false triggers (empty frames) and tag human presence when detected, streamlining the entire image processing workflow. (Hossain M. et al. 2024) With AI, what once took months to review manually can now be processed in a matter of hours.

In addition to species classification, AI can detect and record animal behavior, such as foraging, resting, hunting, or mating. Advanced models can analyze the posture and movement captured in images to infer behavioral patterns. (Hubballi R. B. et al. 2025) This capability provides deeper ecological insights, such as predator-prey interactions, seasonal behaviors, and stress responses to environmental changes or human disturbances. (Hubballi R. B. et al. 2025) Another benefit of AI-enhanced camera traps is their use in population monitoring and biodiversity assessment. With accurate species identification and geotagging of images, conservationists can map species distributions, track changes over time, and estimate population densities. When combined with metadata such as time stamps, location coordinates, and environmental conditions, AI can reveal complex patterns, such as habitat preference, migration timing, or responses to climate variability. These insights are critical for species conservation planning, protected area management, and policy-making.

Drone Surveillance: AI-Enabled Aerial Monitoring for Wildlife Protection

As ecosystems around the world face increasing pressure from climate change, habitat degradation, and poaching, conservationists are turning to innovative technologies to monitor, protect, and understand wildlife populations. (Kalinaki K. et al. 2024) Among the most promising of these technologies is acoustic monitoring, which, when combined with artificial intelligence (AI), offers a powerful method for detecting and identifying species and human threats across vast and often inaccessible landscapes. (Kalinaki K. et al. 2024) AI-driven audio analysis is transforming conservation science by allowing researchers to passively and non-invasively monitor animal behavior and illegal activity through sound, often in real-time and at large spatial scales. This approach is not only cost-effective and scalable but also uniquely suited to environments where traditional visual monitoring is difficult or impossible. Acoustic monitoring relies on recording devices—typically low-power, weather-resistant units equipped with microphones—that are strategically deployed in various habitats such as forests, wetlands, grasslands, and oceans. (Kalita A. J. et al. 2025) These devices capture ambient sounds over extended periods, from the calls of birds and frogs to the footsteps of large mammals and even distant human activities like chainsaws or gunshots. Historically, analyzing this type of audio data required countless hours of manual listening and annotation, which made large-scale deployments impractical. (Kalita A. J. et al. 2025) However, the integration of AI—especially machine learning (ML) and deep learning algorithms—has radically transformed what is possible. AI can now process terabytes of acoustic data at incredible speeds, automatically identifying species-specific vocalizations and flagging anthropogenic noise with impressive accuracy.

One of the core uses of AI in acoustic monitoring is species detection and classification. Many animals produce unique vocalizations—such as birdsongs, frog croaks, bat echolocation clicks, or whale songs—that can be used to identify them acoustically. (Karthik P. C. et al. 2025) AI models are trained using annotated datasets of these vocalizations to recognize specific patterns of frequency, duration, and rhythm. Once trained, these models can scan continuous audio recordings to detect the presence of a given species. (Karthik P. C. et al. 2025) This capability is especially valuable for surveying cryptic, nocturnal, or elusive animals that are rarely seen but often heard, such as owls, wolves, or dolphins. Acoustic AI allows researchers to monitor these species continuously, even in dense forests or open oceans where camera traps and direct observation may fail. Projects like Rainforest Connection (RFCx) exemplify the use of AI-driven acoustic monitoring in conservation. RFCx uses repurposed smartphones equipped with solar panels and sensitive microphones to record sounds in rainforests across the globe. These "guardian devices" transmit audio data to the cloud, where AI algorithms analyze it in real time for signs of illegal logging (e.g., chainsaws or trucks) or poaching (e.g., gunshots or human voices). When such sounds are detected, alerts are sent immediately to rangers or local enforcement teams via mobile apps, allowing for rapid response. RFCx has been successfully deployed in countries like Indonesia, Peru, Brazil, and Cameroon, and has played a vital role in deterring illegal activity while enhancing biodiversity monitoring.

Predictive Analytics: Identifying High-Risk Poaching Areas

The vast expanses of wilderness that serve as critical habitats for the world's wildlife also present one of the greatest challenges in conservation: effective monitoring and protection at scale. (Kotte K. R. et al. 2025) With threats such as poaching, illegal logging, habitat encroachment, and human-wildlife conflict becoming increasingly complex and frequent, conventional ground-based methods of surveillance and patrolling are often stretched beyond capacity. (Kotte K. R. et al. 2025) In this context, the integration of drone technology with artificial intelligence (AI) is emerging as a transformative tool in wildlife protection. AI-enabled drone surveillance represents a cutting-edge solution, providing conservationists with real-time, high-resolution, and scalable means to monitor ecosystems, detect illegal activities, and safeguard endangered species. Drones—also known as unmanned aerial vehicles (UAVs)—have evolved rapidly from simple remote-controlled devices to sophisticated aerial platforms capable of autonomous flight, long-range operations, and multi-sensor integration. Their appeal in conservation lies in their ability to access remote, rugged, or dangerous terrain where human patrolling is either impractical or unsafe. Drones can fly silently over forests, savannas, wetlands, and coastal regions, capturing data that is otherwise difficult or impossible to collect. (Kowar S. et al. 2025) When equipped with AI systems, drones become far more than just flying cameras—they become intelligent agents capable of analyzing data on the fly, identifying patterns, and triggering real-time responses. One of the most prominent applications of AI-enabled drones in conservation is real-time wildlife detection and tracking. With onboard cameras—both optical and infrared—drones can survey large areas and capture imagery of animals in their natural habitats. (Kowar S. et al. 2025) AI models, particularly those based on deep learning and computer vision, can then analyze the video feed in real time to identify specific species, count individuals, track movement, and detect behaviors. For instance, thermal imaging combined with AI has been successfully used to monitor nocturnal animals such as elephants, rhinos, and big cats, especially in regions where ground-based observation is hindered by vegetation or lack of light.

In Africa and parts of Asia, drones have become integral tools in the fight against poaching, particularly in national parks and game reserves where high-value species like elephants and rhinos are under constant threat. (Kumar R. et al. 2025) Poachers often operate at night or in remote areas, using camouflage and knowledge of the terrain to avoid detection. AI-enabled drones can patrol these areas autonomously, detecting unusual movement patterns, vehicles, or human activity. Once a potential threat is detected—such as a person moving through a restricted zone—an alert can be sent to rangers in real time, enabling rapid interception and potentially saving animal lives. (Kumar R. et al. 2025) These drones reduce the need for widespread patrolling by focusing enforcement efforts on high-risk areas, thereby improving efficiency and response times. Projects like the Air Shepherd initiative, backed by the Lindbergh Foundation, have demonstrated the power of AI-drone synergy. Using predictive analytics to determine where poaching is likely to occur and deploying drones to those areas with infrared and AI-assisted object detection, the initiative has significantly reduced poaching incidents in regions of South Africa and Namibia. By flying pre-programmed routes and analyzing terrain and movement patterns, these drones help rangers focus on the most vulnerable locations during the most critical times. Beyond anti-poaching, drones play a crucial role in habitat monitoring and environmental assessment. AI algorithms can analyze aerial imagery collected by drones to assess vegetation health, detect deforestation, monitor water sources, and identify signs of land degradation. In sensitive ecosystems like mangroves, wetlands, and coral reefs, where traditional monitoring is difficult, drones offer a fast, low-impact alternative. Conservation organizations use AI-powered image classification to map changes over time, assess the effectiveness of habitat restoration projects, and forecast future threats based on historical trends.

Real-Time Alerts: AI-Driven Systems for Rapid Response

In the relentless battle against wildlife poaching, time and intelligence are among the most valuable resources available to conservationists. (Maheswaran K. et al. 2025) Traditional anti-poaching strategies—such as ground patrols, community engagement, and reactive law enforcement—while important, often fall short in their ability to anticipate and prevent poaching before it occurs. (Maheswaran K. et al. 2025) This is where predictive analytics, a field powered by artificial intelligence (AI) and big data, is revolutionizing conservation by enabling proactive and strategic interventions. (Manjunatha & Raveesha, 2025) By analyzing historical data and environmental variables, predictive analytics identifies poaching patterns and predicts future hotspots, allowing conservation teams to focus their limited resources where and when they are most needed. Predictive analytics involves the use of algorithms, statistical models, and machine learning techniques to make informed forecasts about future events based on historical and real-time data. In the context of anti-poaching, predictive analytics is applied to diverse datasets such as past poaching incidents, ranger patrol logs, animal movement data, weather conditions, terrain features, and even socio-economic indicators. These data points are fed into machine learning models that uncover hidden correlations and generate risk maps, showing where poaching is most likely to occur. (Manjunatha & Raveesha, 2025) By turning raw data into actionable intelligence, predictive analytics transforms reactive enforcement into proactive prevention. This approach not only improves efficiency and impact but also significantly enhances the safety of rangers and the survival chances of at-risk species like elephants, rhinos, tigers, pangolins, and others. At the heart of predictive analytics are machine learning algorithms such as decision trees, support vector machines, and neural networks. These models are trained using labeled data—such as known poaching locations and times—and learn to recognize the conditions that are most often associated with poaching. Once trained, the models can analyze new data and output a probability score for each area, indicating the risk level of poaching.

One well-known example is the PAWS (Protection Assistant for Wildlife Security) system, developed by researchers at Harvard University and the University of Southern California. PAWS uses data from past patrols, environmental conditions, and known poaching incidents to generate maps that predict the likelihood of illegal activity in various locations. These predictions are then used to optimize patrol routes, ensuring rangers focus on the most vulnerable areas at the right times. In field trials, PAWS has improved efficiency and effectiveness, leading to more arrests, fewer poaching events, and better use of ranger time and resources. (Masvingise & Mujuru, 2025) Predictive analytics must be implemented responsibly. In some contexts, flagged "high-risk" areas may overlap with local communities. It is critical that conservation efforts avoid criminalizing or stigmatizing these populations. Effective use of predictive tools must be coupled with community engagement, livelihood support, and education, ensuring that local residents become partners—not adversaries—in conservation. Additionally, transparency in how models are built and used is vital. Conservationists must clearly communicate how decisions are made and who is impacted, particularly when predictive models influence surveillance or enforcement actions. Looking ahead, predictive analytics will likely become a cornerstone of data-driven conservation. (Masvingise & Mujuru, 2025) As more organizations digitize their records and as satellite imagery, camera trap networks, and mobile patrol apps become more common, the volume and variety of data available for modeling will increase dramatically. With advances in cloud computing, edge AI, and automated data integration, predictive analytics could evolve into real-time risk management systems.

Pattern Detection: AI-Enabled Identification of Poaching Activity: Pattern detection.

In the realm of wildlife conservation, one of the most crucial challenges is not simply detecting threats to biodiversity, but responding to them in time. The ability to intervene before a poacher strikes, before a habitat is irreversibly damaged, or before wildlife wanders into human settlements can make the difference between loss and preservation. (Nalayini C. M. et al. 2025) Traditional monitoring methods—such as manual patrolling, static camera trap analysis, or field observation—are often limited by delayed information flow, meaning by the time action is taken, the opportunity may have already passed. Enter AI-driven real-time alert systems: intelligent, responsive, and highly scalable technologies designed to detect threats and immediately notify relevant authorities or personnel, enabling swift and targeted intervention. Real-time alert systems use artificial intelligence (AI) to process data from various field sensors—such as camera traps, acoustic monitors, drones, satellite feeds, or GPS tracking devices—and automatically generate alerts when certain conditions are met. These conditions might include the presence of unauthorized humans in protected areas, the sound of gunshots or chainsaws, unusual animal movement patterns, or the detection of environmental changes like fires or deforestation. (Nalayini C. M. et al. 2025) Once the AI system recognizes a threat, it sends an immediate notification to conservation staff, rangers, or law enforcement officers via radio, SMS, mobile apps, or centralized monitoring dashboards. This instant communication dramatically reduces response times, empowering teams to take preventative or mitigating action before irreversible damage occurs. It also allows conservation efforts to shift from reactive to proactive, optimizing resource allocation, enhancing safety, and increasing overall effectiveness in the field. Real-time alerts powered by AI represent a revolutionary leap in wildlife protection and ecosystem management. By turning raw environmental data into immediate, actionable intelligence, these systems allow conservationists to shift from lagging response to proactive intervention. Whether it's thwarting a poacher, preventing human-wildlife conflict, or stopping illegal logging in its tracks, real-time AI-driven alerts offer speed, precision, and adaptability—qualities essential for protecting biodiversity in a rapidly changing world. As technology continues to evolve and becomes more accessible, the integration of real-time alert systems into conservation will become not just beneficial, but essential. The key lies in thoughtful deployment, stakeholder collaboration, and ongoing innovation to ensure that these systems truly serve the ecosystems and communities they aim to protect.

NATIONAL PARKS: AI-DRIVEN SURVEILLANCE FOR PROTECTED AREAS: NATIONAL PARKS.

The illegal wildlife trade continues to devastate global biodiversity, driving countless species toward extinction and undermining decades of conservation progress. Poaching, in particular, remains a complex and adaptive threat, carried out by individuals and organized criminal networks that evolve their tactics to avoid detection. (Naveen S. et al. 2025) Traditional approaches to monitoring and preventing poaching—while critical—often struggle to keep pace with this sophistication. To meet this challenge, conservationists are increasingly turning to artificial intelligence (AI) to identify, learn from, and predict patterns of poaching activity. Through AI-enabled pattern detection, organizations can move beyond reactive enforcement and into a realm of proactive, intelligence-led conservation. (Naveen S. et al. 2025) Pattern detection in AI refers to the process by which algorithms learn from large volumes of data to recognize consistent behaviors, relationships, or anomalies. In the context of anti-poaching efforts, pattern detection can reveal where, when, and how illegal activities are likely to occur. By analyzing spatial, temporal, and behavioral data, AI systems can uncover poaching strategies that might not be obvious to human analysts. These insights allow rangers, conservation managers, and policy-makers to deploy resources strategically, increasing the likelihood of intervention and deterrence. AI models used for pattern detection often rely on machine learning techniques such as supervised learning, unsupervised clustering, time-series analysis, and deep learning. These models digest and process vast datasets—from GPS tracks of ranger patrols and animal movements to environmental conditions, past poaching incidents, and satellite imagery—to identify correlations and predict future poaching behavior.

One of the most direct applications of AI-enabled pattern detection is the creation of poaching heatmaps—visual representations of risk based on identified patterns. AI can map areas where poaching frequently occurs and analyze when these events are most likely to happen. For example, if a region has seen repeated elephant poaching near water sources during the dry season, AI can highlight this pattern and forecast heightened risk during future dry periods. This helps conservation managers prioritize surveillance and enforcement efforts during peak poaching windows, ensuring limited resources are used where they matter most. (Nisha Angeline M. et al. 2025) AI can also identify typical behaviors of poachers, such as the routes they tend to take, the times they are most active, and the environmental conditions under which they operate. By analyzing patterns in trespassing events or illegal camping activity, AI models can differentiate between legitimate human movement (e.g., tourists or local herders) and potentially illicit behavior. Over time, the system becomes more accurate at flagging suspicious activity and minimizing false alarms. (Nisha Angeline M. et al. 2025) In areas with extensive ranger-collected data, such as Kenya's Maasai Mara or Tanzania's Serengeti, behavioral profiling through pattern recognition has significantly enhanced detection of repeat offenders and improved law enforcement targeting. Advanced poaching operations often involve coordinated teams that operate across large areas using vehicles, weapons, and insider knowledge. AI can analyze surveillance footage, drone data, and ranger reports to identify recurring vehicles, clothing types, or communication patterns associated with these groups. Through clustering and sequence analysis, AI may even detect shifts in tactics—such as changes in entry points or movement routes—prompting faster adaptation from law enforcement teams.

COMMUNITY-BASED CONSERVATION: AI-EMPOWERED LOCAL MONITORING

National parks are sanctuaries of natural heritage—vast, ecologically rich landscapes preserved to protect biodiversity, support ecological research, provide educational opportunities, and foster ecotourism. (Pradhan N. et al. 2022) However, these protected areas are increasingly under threat from illegal poaching, habitat encroachment, deforestation, wildlife trafficking, and climate-related pressures. Traditional methods of surveillance—such as human patrols and static camera traps—have been crucial in safeguarding these zones, but they are often resource-intensive, reactive, and limited in coverage. (Pradhan N. et al. 2022) In recent years, the rise of artificial intelligence (AI) has introduced a new era in park protection: one defined by real-time, scalable, data-driven surveillance that allows for smarter, faster, and more efficient monitoring of national parks. AI-driven surveillance is revolutionizing the way national parks are managed, providing a powerful set of tools to detect illegal activities, monitor wildlife behavior, and make data-informed conservation decisions. By leveraging cutting-edge technologies—ranging from satellite imaging and drones to camera traps and acoustic sensors—paired with AI algorithms capable of processing vast datasets quickly and accurately, national parks can now defend themselves with unprecedented intelligence and precision. Public-private partnerships, such as those involving conservation NGOs, government agencies, and tech companies, will be vital in scaling these innovations. (Pradhan T. et al. 2025) Grants and open-source collaborations can further ensure that even underfunded parks can benefit from AI technology. AI-driven surveillance is rapidly becoming a cornerstone of modern conservation strategy, particularly in national parks where the stakes are highest and the challenges most daunting. By combining real-time monitoring, predictive analytics, and intelligent pattern recognition, AI empowers park authorities to protect wildlife and ecosystems with unprecedented precision. It is not a replacement for human judgment or local stewardship, but a powerful ally in the shared mission of conservation. (Pradhan T. et al. 2025) As environmental threats grow more complex, national parks must evolve too—armed not only with passion and perseverance, but also with cutting-edge intelligence. AI-driven surveillance marks a pivotal shift from reactive protection to proactive guardianship, ensuring that national parks remain strongholds of biodiversity for generations to come.

CROSS-BORDER CONSERVATION: AI-FACILITATED COLLABORATION

Community-based conservation (CBC) represents a vital approach in the global effort to protect biodiversity and promote sustainable natural resource management. Unlike traditional top-down conservation models, CBC emphasizes the active involvement and stewardship of local communities who live in or around protected areas. (Rahman M. A. et al. 2025) These communities often possess intimate ecological knowledge and a direct stake in the health of their environment. However, despite their critical role, local groups frequently face challenges such as limited access to advanced monitoring tools, insufficient training, and lack of real-time data, which can hinder their effectiveness in wildlife protection and natural resource management. (Rahman M. A. et al. 2025) Artificial intelligence (AI), when integrated thoughtfully into CBC initiatives, is revolutionizing how communities monitor and safeguard their environments. AI-empowered local monitoring harnesses modern technology to enhance the capacity of communities, providing them with sophisticated, accessible tools to detect, report, and respond to threats such as poaching, illegal logging, habitat degradation, and human-wildlife conflict. This synergy of traditional ecological knowledge and cutting-edge AI-driven solutions creates a powerful framework for sustainable conservation that benefits both people and nature. Communitie living near natural habitats are often the first line of defense against environmental degradation. (Sakthivel M. et al. 2025) They interact daily with ecosystems, depend on local resources for livelihoods, and face the consequences of ecological decline firsthand. (Sakthivel M. et al. 2025) Their involvement in conservation efforts can lead to: However, challenges such as insufficient funding, limited technology, and lack of infrastructure can restrict community conservation impact. Here, AI offers a transformative solution. AI-powered monitoring technologies, when adapted to the needs and capacities of local communities, provide a suite of tools that augment traditional methods and address many existing limitations.

Mobile technology penetration has grown rapidly even in remote regions. AI-enabled smartphone apps allow community members to easily collect data on wildlife sightings, illegal activities, and environmental changes. These apps often incorporate features such as: Image recognition: Uploading photos of animals, plants, or suspicious activities that AI algorithms classify and verify. Real-time reporting: Instant alerts to conservation authorities or ranger teams when threats are detected. Crowdsourced data validation: Community members can cross-check information, ensuring accuracy. Apps like CyberTracker and iNaturalist exemplify this integration, enabling local observers to contribute valuable scientific data and receive immediate feedback. (Sharma M. K. et al. 2024) Community groups can deploy AI-enhanced camera traps that automatically detect and identify species, humans, or vehicles. This reduces the need for manual data retrieval and accelerates threat detection. Many camera traps now come equipped with solar power and wireless connectivity, making them viable in remote areas. Similarly, AI-enabled acoustic sensors capture and analyze environmental sounds—identifying chainsaws, gunshots, or animal calls. These sensors can be networked across community-managed forests or reserves, providing continuous monitoring with low maintenance. Where terrain is difficult and patrols are limited, drones offer aerial reconnaissance capabilities. AI-powered drones can autonomously survey large areas, identify illegal logging sites, poacher encampments, or distressed wildlife, and relay actionable information to community patrol teams. This technology democratizes access to aerial monitoring, traditionally the domain of government or large NGOs.

DATA QUALITY AND INTEGRATION: CHALLENGES IN AI-DRIVEN CONSERVATION

In the modern era of environmental conservation, one of the greatest challenges and opportunities lies in the effective management of ecosystems and wildlife populations that transcend political boundaries. Many of the world's most vital habitats and endangered species roam freely across countries, rendering national borders artificial constraints to their protection. Cross-border conservation, therefore, is an essential approach aimed at fostering cooperation between neighboring countries to jointly protect shared ecosystems, combat poaching, manage natural resources sustainably, and maintain biodiversity corridors. Yet, coordinating efforts across different jurisdictions, legal frameworks, and resource capacities remains complex and often hampered by fragmented data, limited communication, and lack of unified monitoring systems. (Sharma M. K. et al. 2024) Artificial intelligence (AI) is emerging as a transformative catalyst for cross-border conservation collaboration, enabling seamless data sharing, real-time monitoring, predictive analytics, and coordinated response mechanisms. By integrating AI-powered tools into multinational conservation strategies, countries can overcome traditional barriers and create robust, adaptive, and intelligence-driven frameworks that transcend geopolitical divisions. This article explores how AI facilitates cross-border conservation, the technologies driving this collaboration, real-world applications, benefits, challenges, and the future potential of AI-enabled transnational environmental stewardship. Wildlife and ecosystems are inherently connected across landscapes irrespective of human-imposed borders. Many migratory species, such as elephants, big cats, wolves, and birds, follow seasonal routes that span multiple countries. Similarly, river basins, forest ranges, and mountain ecosystems extend over large transnational areas. Without cross-border cooperation, conservation efforts risk being piecemeal and ineffective, as habitat fragmentation, poaching, and illegal trade often exploit jurisdictional gaps.

National parks, reserves, and community conservation areas on either side of borders can be equipped with synchronized AI-enabled camera traps, acoustic sensors, and drones. These devices collect standardized data on wildlife movements, human activities, and environmental changes. AI algorithms process and harmonize this data across regions to produce unified dashboards accessible to all participating countries. For example, AI models can identify transboundary wildlife corridors by analyzing GPS collar data from animals crossing borders. This spatial intelligence informs habitat protection and anti-poaching patrol routes on both sides. (Singh & Shahi, 2024) Poaching and illegal logging often involve sophisticated networks operating across borders. AI-driven real-time alert platforms allow countries to share incident data instantly, coordinating rapid response teams in adjacent areas. These systems utilize machine learning to recognize suspicious patterns such as coordinated vehicle movements, gunshot acoustics, or unusual human activity near sensitive zones. AI predictive models aggregate historical poaching records, land use changes, environmental conditions, and social factors from multiple countries to identify transnational poaching hotspots and emerging threats. (Singh & Shahi, 2024) This enables authorities to deploy joint patrols proactively rather than reactively. Predictive systems can also forecast climate-related risks affecting ecosystems spanning borders, guiding collaborative mitigation efforts such as fire management or drought response.

Scalability and Sustainability: Deploying AI Solutions in the Field

The application of artificial intelligence (AI) in wildlife conservation and environmental management holds enormous promise, from enhancing monitoring accuracy to enabling predictive analytics and real-time decision-making. However, the effectiveness of AI-driven conservation initiatives hinges critically on the quality and integration of data feeding these systems. High-quality, comprehensive, and well-integrated datasets are the foundation for reliable AI models. (Thirunagalingam A. et al. 2025) Unfortunately, conservation data often suffers from numerous challenges related to data quality, heterogeneity, gaps, biases, and interoperability. These issues can undermine AI model performance, lead to erroneous conclusions, and reduce the trust and adoption of AI technologies among conservation practitioners. This article explores the multifaceted challenges surrounding data quality and integration in AI-driven conservation, their implications, and emerging approaches to overcome these obstacles. It highlights the importance of robust data governance, innovative technological solutions, and collaborative frameworks to ensure that AI fulfills its transformative potential in safeguarding biodiversity and ecosystems. AI systems—whether for species identification, poaching prediction, habitat monitoring, or ecological modeling—depend fundamentally on data. These data come from diverse sources: camera traps, acoustic sensors, satellite imagery, GPS collars, field surveys, citizen science, and governmental databases. AI algorithms require datasets that are: Any shortfall in these dimensions compromises AI outputs, potentially resulting in misclassification, false alarms, missed detections, or poor policy recommendations. (Thirunagalingam A. et al. 2025) Conservation datasets are frequently incomplete due to logistical constraints in remote areas, limited funding, and the elusive nature of many species. For example, camera traps might miss rare or nocturnal animals, and acoustic sensors can be disrupted by weather or technical faults. Data gaps can introduce biases, skewing AI models to overrepresent common species or well-monitored areas and underrepresent threatened populations or inaccessible habitats.

Environmental data collection is often subject to noise and inaccuracies. Camera traps may capture blurry images or non-target species, GPS devices can have location errors, and acoustic recordings may contain overlapping sounds. Field surveys might be affected by observer bias or transcription mistakes. Such noise challenges AI's ability to accurately recognize patterns and demands sophisticated preprocessing and error correction. Conservation data is highly heterogeneous, coming in various formats—images, audio files, spatial coordinates, text reports—and collected using different protocols across projects and regions. (Ullah F. et al. 2025) This lack of standardization complicates data integration, requiring significant effort in data cleaning, transformation, and harmonization before AI models can effectively use the data. Data collected at different times or spatial scales can create integration difficulties. For instance, satellite imagery might be available at a coarse resolution every few days, while camera traps produce high-frequency but localized data. (Ullah F. et al. 2025) Aligning these datasets for meaningful AI analysis necessitates complex spatiotemporal interpolation and resampling techniques. Data bias is a critical concern. Many datasets reflect human accessibility bias, with more data from areas near roads or ranger stations. Certain charismatic species receive disproportionate monitoring compared to less-studied taxa. Such biases can lead AI models to inaccurate conclusions about species distribution or threat levels, adversely affecting conservation priorities. Sensitive conservation data, especially related to endangered species locations or anti-poaching operations, require strict security to prevent misuse. Data breaches or unauthorized sharing can expose vulnerable species to increased risk. Additionally, the privacy rights of indigenous peoples and local communities involved in data collection must be respected, posing ethical and legal challenges.

Community Engagement: Empowering Local Communities with AI-Driven Tools

Artificial intelligence (AI) has rapidly emerged as a pivotal technology for wildlife conservation and environmental protection. From real-time poaching detection to predictive habitat modeling, AI-driven tools provide unprecedented opportunities to address complex ecological challenges with precision and speed. However, the transition from proof-of-concept projects and pilot deployments to broad, impactful field applications hinges critically on two interrelated factors: scalability and sustainability. (Ajay et al. 2025) Simply put, it is not enough for AI solutions to work in small, controlled settings; they must be capable of operating effectively at landscape or national scales and continue to deliver value over long periods without requiring unsustainable resources or constant external intervention. This article explores the fundamental challenges and strategic considerations associated with scaling AI technologies in conservation, alongside the principles and practices that ensure their sustainable deployment. (Ajay et al. 2025) By addressing issues related to infrastructure, data management, human capacity, cost-effectiveness, and ecological adaptability, conservation organizations can harness AI's full potential while safeguarding long-term operational viability. Scalability refers to the ability of an AI solution to maintain or improve its performance, efficiency, and cost-effectiveness as it expands from limited pilot studies to larger, more complex environments involving diverse ecosystems, broader geographic coverage, and multiple stakeholders. In wildlife conservation, scaling AI means deploying surveillance systems across extensive protected areas, integrating heterogeneous data from multiple sources, and supporting numerous rangers, researchers, and policy-makers.

Many remote conservation areas lack reliable power supplies, internet connectivity, and advanced computational infrastructure needed for continuous AI operations. Scaling AI across multiple sites requires robust hardware such as solar-powered sensors, durable drones, and edge computing devices capable of local data processing to minimize dependency on cloud connectivity. Building and maintaining such infrastructure at scale is a logistical and financial challenge. As AI solutions scale, the volume and variety of data increase exponentially. High-resolution images, acoustic recordings, GPS tracking data, satellite imagery, and field observations must be processed and analyzed efficiently. Handling big data demands scalable storage solutions, powerful processing capabilities, and optimized AI algorithms capable of managing heterogeneity without sacrificing speed or accuracy. Scaling often involves integrating AI tools with existing conservation management platforms, government databases, and community monitoring systems. Ensuring interoperability and seamless data exchange across diverse software and hardware ecosystems is complex, requiring standardized protocols and open APIs to avoid data silos. Expanding AI deployments requires sufficient trained personnel to operate, maintain, and interpret AI systems. Scaling up also means training a broader range of stakeholders, from park rangers to local communities and government officials, to ensure proper utilization and responsiveness to AI-driven insights.

THE FUTURE OF WILDLIFE CONSERVATION: AI-DRIVEN SOLUTIONS

Community engagement has emerged as a cornerstone of successful wildlife conservation and environmental protection initiatives worldwide. Recognizing that local communities are not merely passive observers but vital stewards of their surrounding ecosystems, conservationists increasingly emphasize participatory approaches that integrate indigenous knowledge, respect cultural values, and foster local empowerment. The advent of artificial intelligence (AI) has introduced novel opportunities to strengthen these efforts, enabling communities to harness cutting-edge technologies that enhance monitoring, decision-making, and advocacy in conservation. (Zeller Zigaitis W. L. et al. 2023) This article explores how AI-driven tools are transforming community engagement by equipping local stakeholders with accessible, scalable, and effective means to protect biodiversity, manage natural resources, and combat threats such as poaching and habitat loss. It examines the benefits, challenges, and best practices associated with deploying AI in community settings and highlights examples of successful integrations that have empowered communities to become proactive agents of conservation. Communities living in and around wildlife habitats often possess intricate, place-based knowledge accumulated over generations. They understand seasonal patterns, species behaviors, and ecological dynamics better than many external actors. Engaging these communities fosters a sense of ownership and responsibility, leading to more sustainable conservation outcomes. Moreover, involving communities helps address social and economic dimensions that influence conservation success, such as livelihoods, land tenure, and human-wildlife conflict.

One of the most direct ways AI facilitates community engagement is through mobile applications that empower locals to collect, share, and interpret environmental data. AI-powered apps can simplify species identification via image recognition, enable easy reporting of suspicious activities, and provide real-time alerts on ecological threats. For instance, community members can use smartphone cameras to photograph animals or signs of poaching, and AI algorithms instantly analyze these inputs to provide feedback or transmit alerts to park authorities. These tools make sophisticated technology accessible to non-experts, democratizing conservation data. AI-enhanced sensor networks, such as low-cost camera traps and acoustic recorders, can be deployed and maintained by community members. AI processes the massive volumes of data generated, filtering relevant information and reducing the need for specialized technical expertise. This participatory sensing model fosters local stewardship and continuous ecological monitoring, creating a virtuous cycle of data-driven community engagement. (Zeller Zigaitis W. L. et al. 2023) It also empowers communities to take initiative in identifying threats and documenting conservation successes. Interactive AI platforms tailored to local languages and cultural contexts can provide environmental education, raising awareness of conservation issues and sustainable practices. These tools can adapt content based on user interaction, ensuring relevance and engagement. Educational empowerment enhances community capacity to participate meaningfully in conservation governance and fosters long-term attitudinal shifts toward environmental stewardship. AI can translate complex environmental data into intuitive visualizations accessible to community leaders and members. This transparency supports informed decision-making in resource management, conflict resolution, and advocacy.AI tools designed with community needs in mind can bridge literacy and technical skill gaps, ensuring marginalized groups, including women and indigenous peoples, can participate fully. This inclusivity strengthens social cohesion and legitimacy of conservation initiatives.

CONCLUSION

Wildlife conservation stands at a critical crossroads. The accelerating pace of biodiversity loss, habitat destruction, and climate change threatens countless species and ecosystems worldwide. Traditional conservation methods, while vital, often struggle to keep pace with these mounting challenges due to limited resources, delayed data collection, and reactive management approaches. In this context, artificial intelligence (AI) is poised to revolutionize wildlife conservation by enabling more proactive, precise, and scalable interventions. The future of wildlife conservation will increasingly be defined by AI-driven solutions that integrate advanced data analytics, machine learning, automation, and real-time decision-making. These technologies will empower conservationists, governments, communities, and global coalitions with unprecedented capabilities to monitor ecosystems, detect threats, optimize interventions, and foster resilient biodiversity. This article explores the transformative prospects, ongoing innovations, and emerging challenges of AI in the future landscape of wildlife conservation. AI's transformative power lies in its ability to process massive volumes of diverse data quickly and extract meaningful patterns beyond human capacity. The future will see an explosion of interconnected sensors embedded throughout natural habitats — camera traps, acoustic sensors, drones, satellite remote sensing, and environmental DNA (eDNA) sampling devices. These devices will continuously stream high-resolution data to AI platforms for real-time analysis. The integration of Internet of Things (IoT) technologies will create smart ecosystems where conservation actions are dynamically informed by environmental changes and species movements. For example, AI-powered sensor networks could detect subtle behavioral changes in endangered species indicating stress or illness, enabling early intervention. Automated detection of illegal activities such as poaching or logging will become more accurate and widespread, extending into previously inaccessible areas. Deep learning and computer vision models will dramatically enhance species identification from images and videos, even in challenging environments or from partial observations. AI will also analyze animal vocalizations and movement patterns to decode social behaviors, migration routes, and responses to environmental changes. Such insights will deepen ecological understanding and inform targeted conservation strategies. For instance, machine learning models could help map corridors critical for migratory species or identify habitat features essential for reproduction and survival. The application of AI-powered predictive analytics will enable conservationists to forecast ecological trends and risks with greater precision. By integrating climate data, land use changes, human activity patterns, and biological variables, AI models will simulate multiple future scenarios. These simulations will inform proactive measures such as habitat restoration priorities, buffer zone establishment, and poaching patrol deployment. Predictive models will also help assess the impacts of conservation policies before implementation, optimizing resource allocation. Robotics, including drones and autonomous ground vehicles, will become integral to monitoring and intervention. Equipped with AI-driven navigation and detection systems, these robots will perform tasks ranging from wildlife census and anti-poaching patrols to habitat rehabilitation. Autonomous drones can cover large territories, capture high-definition imagery, and even deliver emergency supplies to remote rangers. Robotics will reduce human risk and enable 24/7 surveillance, critical in rapidly changing environments.

REFERENCES

Ajay, N., Upadhyaya, A. N., Debnath, P., Rani, P., Selvakumar, P., Manjunath, T. C., & Bhattacharya, S. (2025). AI-Based Disaster Prediction and Early Warning Systems. In Mobo, F. (Ed.), *AI and Emerging Technologies for Emergency Response and Smart Cities* (pp. 55–74). IGI Global Scientific Publishing.

Ayyaswamy, K., Gobinath, V. M., Kathirvel, N., & Maheswaran, C. P. (2025). AI-Based Automated Disease Detection Based on Symptoms Healthcare for Pets and Birds. In AI and the Revival of Big Data (pp. 311-346). IGI Global Scientific Publishing.

Chisom, O. N., Biu, P. W., Umoh, A. A., Obaedo, B. O., Adegbite, A. O., & Abatan, A. (2024). Reviewing the role of AI in environmental monitoring and conservation: A data-driven revolution for our planet. *World Journal of Advanced Research and Reviews*, *21*(1), 161–171.

Coates, J. M., Santhanalakshmi, K., Dinesh, N., Gandhimathi, S., & Chandel, P. S. Mohit, & Selvakumar, P. (2025). Assessment and Evaluation in Transnational Staff Development. In G. Morris & J. Ye (Eds.), Training and Development in Transnational Higher Education (pp. 387-406). IGI Global Scientific Publishing.

Darney, P. E., Fernando, A. V., Krishnan, R. S., Raj, J. R. F., Lavanya, V., & Kumar, V. V. (2025). RhinoGuardNet: An Integrated Deep Learning System for Monitoring, Movement Prediction, and Threat Detection in Javan Rhino Conservation. In *2025 International Conference on Machine Learning and Autonomous Systems (ICMLAS)* (pp. 101-108). IEEE.

Dave, R., Kaunert, C., & Singh, B. (2025). Wildlife and Forest Resource Management With Artificial Intelligence. In Machine Learning and Internet of Things in Fire Ecology (pp. 301-324). IGI Global Scientific Publishing.

Dwivedi, D. N., Mahanty, G., & nath Dwivedi, V. (2024). Intelligent conservation: a comprehensive study on AI-enhanced environmental monitoring and preservation. In The Convergence of Self-Sustaining Systems With AI and IoT (pp. 215-226). IGI Global.

Goel, P. K., Saifi, S., Goel, N., Aeron, S., Yadav, S., & Ansari, S. A. (2025). Ethical Considerations in AI: Applications for Wildlife Conservation. In AI and Machine Learning Techniques for Wildlife Conservation (pp. 247-266). IGI Global Scientific Publishing.

Goyal, A., Bhakhar, R., & Singh, S. (2025). Limitations and Challenges of AI in Wildlife Conservation. In AI and Machine Learning Techniques for Wildlife Conservation (pp. 363-394). IGI Global Scientific Publishing.

Gupta, S., Vijarania, P., Gupta, S., & Vijarania, M. (2025). Artificial Intelligence in the Current Era: Navigating the Impacts on Society and Wildlife Conservation. In AI and Machine Learning Techniques for Wildlife Conservation (pp. 65-96). IGI Global Scientific Publishing.

Gustafson, K., Sandstrom, T., & Townsend, L. (2018). The bush war to save the rhino: Improving counter-poaching through intelligence. *Small Wars & Insurgencies*, *29*(2), 269–290.

Hammouch, H., & Raghav, A. (2025). AI and ML Applications in Wildlife Conservation and Forest Management: A Comprehensive Review. Machine Learning and Internet of Things in Fire Ecology, 49-68.

Hossain, M., Samiun, M., Tiwari, A., Johora, F. T., Hosain, M. S., Saleh, M. A., & Vanu, N. (2024). Data Analytics in Wildlife Management Business Insights and Technological Solutions. In *2024 International Conference on Progressive Innovations in Intelligent Systems and Data Science (ICPIDS)* (pp. 539-544). IEEE.

Hubballi, R. B., Selvakumar, P., Seenivasan, R., & Basava Aradhya, S. G., Dinesh, N., & Seelam, P. K. (2025). Overview of Current AI Technologies in Education. In L. Kyei-Blankson & E. Ntuli (Eds.), Transformative AI Practices for Personalized Learning Strategies (pp. 1-26). IGI Global Scientific Publishing.

Kalinaki, K., Acaru, S. F., Kugonza, J., & Nsubuga, R. (2024). Towards an Intelligent Tomorrow: Machine Learning Enabling Sustainable Development. In Methodologies, Frameworks, and Applications of Machine Learning (pp. 66-89). IGI Global.

Kalita, A. J., Deka, J., Devi, C., & Kalita, P. (2025). Importance of AI-Tech in Managing a Protected Area: Considering Forest Health Dynamics. In Artificial Intelligence and Animal Ecology (pp. 193-220). CRC Press.

Karthik, P. C., Bharathi, S. S., Sivakumar, A., Thangavelu, C., Nalini, S., & Lalitha, S. D. (2025). Emerging Artificial Intelligence Tools, Advancements, and Opportunities for Protecting Wild Animals. In AI and Machine Learning Techniques for Wildlife Conservation (pp. 213-246). IGI Global Scientific Publishing.

Kotte, K. R., Attaluri, V., & Selvakumar, P. (2025). Revolutionising Retail Banking with AI and Virtual Technologies AI-Driven Financial Training. In Nair, A. J., Manohar, S., Limbu, Y. B., & Huhmann, B. A. (Eds.), *Intersecting Natural Language Processing and FinTech Innovations in Service Marketing* (pp. 421–440). IGI Global Scientific Publishing.

Kowar, S., Paul, T., Mukhopadhayay, N. D., & Barua, B. (2025). From Data to Decisions: Exploring AI's Role in Tackling Environmental Issues. In Cases on AI-Driven Solutions to Environmental Challenges (pp. 1-22). IGI Global Scientific Publishing.

Kumar, R., Sankat, M., & Panda, A. (2025). Innovations in Biodiversity Monitoring: AGI Drones at the Forefront. In Artificial General Intelligence-Based Drones for Climate Change (pp. 211-248). IGI Global Scientific Publishing.

Maheswaran, K., Selvakumar, P., Shaikh, M., Shashank, G., Gupta, S., & Manjunath, T. C. (2025). The Evolution of Smart Urban Mobility with Emerging Technologies. In Ponnusamy, S., Assaf, M., Youssef, B., Jeon, G., & Kalyanaraman, S. (Eds.), *Leveraging VANETs and Blockchain Technology for Urban Mobility* (pp. 1–28). IGI Global Scientific Publishing.

Manjunatha, G. R., & Raveesha, H. R. (2025). AI for Biodiversity Conservation: Harnessing Predictive Analytics and Habitat Monitoring to Protect Endangered Species and Ecosystems. In AI and Ecological Change for Sustainable Development (pp. 33-64). IGI Global Scientific Publishing.

Masvingise, K., & Mujuru, N. M. (2025). AI-Powered Solutions for Sustainable Tourism Practices Through Wildlife Conservation Initiatives. In AI Technologies for Personalized and Sustainable Tourism (pp. 113-148). IGI Global.

Nalayini, C. M., Kalpana, V., Hemamalini, S., & Sathyamoorthy, K. (2025). A YOLOv8-based AI System for Real-Time Endemic Species Threat Detection and. *Journal of Innovative Image Processing*, *7*(1), 50–73.

Naveen, S., Dhamotharan, R., Sahebzathi, S., Krishnan, R. S., Kumar, V. V., & Sundaravadivel, P. (2025). ElephantGuardNet: A Comprehensive Solution for Elephant Conservation and Human-Elephant Conflict Reduction. In 2025 4th International Conference on Sentiment Analysis and Deep Learning (ICSADL) (pp. 1470-1477). IEEE.

Nisha Angeline, M., Selvakumar, P., Gupta, M. K., Nazar, N., Elumalai, D., & Bhattacharya, S. (2025). Emerging Materials for Energy Storage and Environmental Applications. In K. Muduli, B. Moharana, S. Ales, & D. Biswal (Eds.), Using Computational Intelligence for Sustainable Manufacturing of Advanced Materials (pp. 53-72). IGI Global Scientific Publishing.

Pradhan, N., Sille, R., & Sagar, S. (2022). Artificial intelligence empowered models for UAV communications. In *Unmanned aerial vehicle cellular communications* (pp. 95–113). Springer International Publishing.

Pradhan, T., Sarkar, D., & Nanda, C. (2025). Transformative solutions through ai and digital platforms in human-elephant conflict management. Rais journal for social sciences, 9(1), 221-230.

Rahman, M. A., Emon, M. M. H., Khan, T., Hamid, A. B. A., & Yaakub, N. I. (2025). Eco-Intelligence: AI's Contribution to a Sustainable World. In AI and Green Technology Applications in Society (pp. 31-62). IGI Global Scientific Publishing.

Sakthivel, M., Jaganathan, A., Mohanraj, M., & Selvakumar, P. Mohit, & Manjunath, T. (2025). Brand Loyalty through AI and Personalisation AI-Generated Content for Financial Marketing. In A. J Nair, S. Manohar, Y. B. Limbu, & B. A. Huhmann (Eds.), Intersecting Natural Language Processing and FinTech Innovations in Service Marketing (pp. 77-96). IGI Global Scientific Publishing.

Sharma, M. K., Prasad, S. V., Anusha, G., Das, A., & Sambathkumar, M. (2024). Tech-Driven Solutions for Environmental Conservation by AI Collaboration Processes. In Environmental Applications of Carbon-Based Materials (pp. 1-29). IGI Global.

Singh, S. K., & Shahi, S. (2024). AI for Earth: Enhancing Environmental Sustain Ability with Technology. *Research Journal of Medical Sciences*, *18*, 219–225.

Thirunagalingam, A., & Selvakumar, P. (2025). Application of Automation and Robotics in the Oil and Gas Industry. In Abdullayev, V., & Khang, A. (Eds.), *Revolutionizing AI and Robotics in the Oil and Gas Industry* (pp. 229–244). IGI Global Scientific Publishing.

Ullah, F., Saqib, S., & Xiong, Y. C. (2025). Integrating artificial intelligence in biodiversity conservation: Bridging classical and modern approaches. *Biodiversity and Conservation*, *34*(1), 45–65.

Zeller Zigaitis, W. L., & Robinson, A. C. (2023). Sensemaking in the wild: A review of practitioner collected geospatial data and its synthesis within protected areas for poaching mitigation. *Annals of GIS*, *29*(3), 319–335.

Chapter 2
Applications of Machine Learning in Bioacoustic Analysis

Pillarisetty Alekhya
https://orcid.org/0009-0006-2959-3577
Malla Reddy Technical Campus, Malla Reddy Vishwavidyapeeth, India

P. Selvakumar
https://orcid.org/0000-0002-3650-4548
Department of Science and Humanities, Nehru Institute of Technology, Coimbatore, India

T. C. Manjunath
https://orcid.org/0000-0003-2545-9160
Rajarajeswari College of Engineering, Bengaluru, India

Jayashree Gorde
https://orcid.org/0009-0001-9011-8138
Symbiosis International University, India

J. Kartigeyan
https://orcid.org/0000-0003-0305-5966
J.B. Institute of Engineering and Technology, India

Bottu Giridharan Sivakumar
AMET University, India

ABSTRACT

Bioacoustics is a multidisciplinary scientific field that explores how living organisms produce, transmit, and receive sound. Primarily focused on the vocal behavior of animals, bioacoustics intersects biology, ecology, ethology, and acoustical engineering. It serves as a vital tool in understanding the complexities of animal communication, offering insights into behavior, social structures, environmental interactions, and evolutionary processes. As technological advances continue to expand our capacity for data collection and analysis, bioacoustics is rapidly becoming an indispensable approach in both academic research and conservation efforts. At its core, bioacoustics examines the mechanisms of sound production in

DOI: 10.4018/979-8-3373-5483-5.ch002

animals and how these sounds function in their lives. Many species—from birds and whales to frogs and insects—use vocalizations to convey messages. These sounds may serve to attract mates, establish territory, warn of predators, maintain group cohesion, or mediate social hierarchies.

THE ROLE OF BIOACOUSTICS: UNDERSTANDING ANIMAL COMMUNICATION

Bioacoustics is a multidisciplinary scientific field that explores how living organisms produce, transmit, and receive sound. Primarily focused on the vocal behavior of animals, bioacoustics intersects biology, ecology, ethology, and acoustical engineering. (Chowdhary S. et al. 2025) It serves as a vital tool in understanding the complexities of animal communication, offering insights into behavior, social structures, environmental interactions, and evolutionary processes. As technological advances continue to expand our capacity for data collection and analysis, bioacoustics is rapidly becoming an indispensable approach in both academic research and conservation efforts. At its core, bioacoustics examines the mechanisms of sound production in animals and how these sounds function in their lives. Many species—from birds and whales to frogs and insects—use vocalizations to convey messages. These sounds may serve to attract mates, establish territory, warn of predators, maintain group cohesion, or mediate social hierarchies. (Chowdhary S. et al. 2025) Each species often exhibits a unique acoustic signature, shaped by evolutionary pressures and environmental constraints. For instance, songbirds are known for their complex repertoires, which can vary by individual, region, and even season. (Coates J. M. et al. 2025) Similarly, cetaceans like dolphins and whales utilize echolocation clicks and social whistles that are both species-specific and functionally diverse. The process of sound communication involves several key components: the sender, the medium, and the receiver. The sender produces a sound signal using specialized anatomical structures, such as the syrinx in birds or the larynx in mammals. This signal travels through a medium—usually air or water—where it may be altered or attenuated by environmental factors such as temperature, humidity, vegetation, or oceanic salinity. The receiver then perceives and interprets the signal through auditory organs, often finely tuned to detect frequencies relevant to that species. (Coates J. M. et al. 2025).

Modern bioacoustic research employs a range of sophisticated technologies. Autonomous recording units (ARUs), hydrophones, and directional microphones allow for long-term, non-invasive monitoring of natural habitats. These tools can record sound continuously over weeks or months, capturing temporal patterns in vocal activity. Spectrogram analysis—a visual representation of sound frequency over time—helps researchers identify species, categorize calls, and quantify soundscapes. In addition, machine learning algorithms are increasingly used to automate the detection and classification of calls, making it possible to analyze massive audio datasets efficiently. (Darney P. E. et al. 2025) This technological integration has dramatically improved the resolution and scalability of acoustic studies. One of the most significant applications of bioacoustics lies in wildlife conservation. (Darney P. E. et al. 2025) Acoustic monitoring allows scientists to detect elusive or endangered species in remote regions without disturbing them. For example, researchers use acoustic sensors to track the presence of the critically endangered vaquita porpoise in the Gulf of California, or to locate populations of forest elephants in Central Africa whose low-frequency rumbles can travel long distances. Acoustic data can reveal behavioral patterns, migration routes, and responses to environmental changes—essential information for designing effective conservation strategies. Additionally, bioacoustic methods have proven useful in evaluating the impacts of human activities such as deforestation, urbanization, and ocean noise pollution on wildlife communication systems. The field also contributes significantly to our understanding of ecological dynamics. Soundscapes—defined as the total acoustic environment of a specific location—offer a holistic view of biodiversity. By analyzing changes in soundscapes over time, researchers can infer shifts in species composition, activity patterns, and ecosystem health. This approach, known as soundscape ecology, has emerged as a powerful framework for studying ecological processes on a landscape scale. For example, a decline in dawn chorus intensity in a tropical forest may signal habitat degradation or species loss. Conversely, the return of native vocal species following reforestation efforts can provide early evidence of ecological recovery.

APPLICATIONS OF BIOACOUSTIC ANALYSIS: CONSERVATION, RESEARCH, AND BEYOND

The intersection of machine learning and bioacoustics has revolutionized the way scientists analyze, interpret, and understand the acoustic behavior of animals. (Deng X. et al. 2025) As the field of bioacoustics continues to grow, the volume and complexity of data generated from autonomous recording units (ARUs), hydrophones, and other audio sensors has increased exponentially. Manual analysis, once the standard for interpreting animal vocalizations, has become time-consuming, labor-intensive, and often infeasible for large-scale studies. (Deng X. et al. 2025) Machine learning (ML) offers powerful tools to automate, accelerate, and enhance the analysis of acoustic data, enabling researchers to uncover patterns, identify species, monitor biodiversity, and derive ecological insights with unprecedented precision and efficiency. Machine learning encompasses a broad class of algorithms and statistical models that allow computers to perform tasks by learning from data rather than through explicit programming. In the context of bioacoustics, ML models can be trained to detect, classify, and interpret animal sounds from massive audio datasets. These algorithms are capable of recognizing patterns in spectrograms (visual representations of sound), identifying subtle acoustic differences, and filtering out noise. (Devi V. B. et al. 2025) The most commonly used techniques include supervised learning, unsupervised learning, and deep learning—a subfield of ML that leverages artificial neural networks to model complex relationships in data. Supervised learning is widely used in species identification tasks, where labeled data is available. For instance, researchers can train an algorithm using a curated dataset of known bird songs or bat echolocation calls. The model learns the unique features—such as pitch, duration, and frequency modulation—associated with each species. Once trained, it can classify new, unlabeled recordings with a high degree of accuracy. (Devi V. B. et al. 2025) This is particularly useful for monitoring species presence, estimating population density, or studying migration patterns. Random forests, support vector machines (SVMs), and k-nearest neighbors (KNN) are common supervised learning algorithms in these applications.

Unsupervised learning, by contrast, is employed when labeled data is scarce or unavailable. These algorithms attempt to find structure in the data without predefined categories. Clustering techniques like k-means or hierarchical clustering are used to group similar sounds together, which can help researchers discover previously unknown call types, behavioral contexts, or even cryptic species. (Dhivya P. et al. 2025) Dimensionality reduction methods such as principal component analysis (PCA) and t-distributed stochastic neighbor embedding (t-SNE) are used to visualize high-dimensional acoustic features and reveal hidden patterns in the soundscape. The most transformative breakthroughs in recent years have come from deep learning. Convolutional neural networks (CNNs), recurrent neural networks (RNNs), and transformers have proven highly effective for analyzing acoustic data, especially when applied to spectrogram images. (Dhivya P. et al. 2025) CNNs, for instance, excel at detecting visual features in spectrograms, making them ideal for identifying and classifying bird songs, whale calls, and frog croaks. RNNs and their more advanced variant, Long Short-Term Memory networks (LSTMs), are suited for processing sequential data, capturing temporal dependencies in vocalizations. Deep learning models often outperform traditional ML algorithms, particularly in noisy environments or with highly variable calls. Machine learning is also instrumental in real-time and automated monitoring systems. Edge computing devices equipped with ML models can process sounds in situ, enabling immediate detection and alert systems for conservation applications. For example, ML-enabled sensors in rainforests can detect gunshots, chainsaws, or specific animal calls, sending alerts to rangers or conservationists. In marine environments, underwater hydrophones use ML to identify cetacean calls or human-generated noise pollution, informing regulatory decisions and marine traffic management. These real-time applications are essential for protecting vulnerable species and habitats, especially in remote or understudied regions.

Signal Processing: Preparing Bioacoustic Data for Machine Learning

Bioacoustic analysis, the scientific study of sound production, transmission, and reception in animals, has emerged as a transformative tool with wide-ranging applications in conservation, ecological research, behavioral studies, environmental monitoring, and even interdisciplinary fields such as anthropology, robotics, and public education. (Geldenhuys C. M. et al. 2025) By leveraging the natural acoustic signals of animals and ecosystems, bioacoustics allows scientists and practitioners to gather meaningful data with minimal intrusion. As recording and analytical technologies continue to evolve, the potential applications of bioacoustics have expanded well beyond traditional wildlife research, influencing policy, innovation, and global biodiversity protection efforts. One of the most impactful applications of bioacoustic analysis lies in wildlife conservation. (Geldenhuys C. M. et al. 2025) Acoustic monitoring enables researchers to detect the presence, abundance, and behavior of species—especially those that are elusive, nocturnal, migratory, or cryptic. Many species produce unique vocalizations, allowing scientists to identify them remotely using acoustic signatures. Bioacoustics is especially valuable in monitoring endangered or threatened species without physical capture or visual sightings, thus reducing stress and habitat disturbance. For instance, conservationists use hydrophones to monitor the vocalizations of critically endangered whales, such as the North Atlantic right whale or the vaquita porpoise. In terrestrial environments, forest elephants in Central Africa are tracked through their infrasonic calls, which travel long distances through dense vegetation. Similarly, automated acoustic recorders help map the distribution of amphibians in tropical forests, where visual surveys are challenging. These data inform conservation status assessments, habitat suitability models, and targeted management plans.

Moreover, acoustic data can reveal population trends over time. Long-term monitoring helps determine whether conservation interventions—such as protected area designations or rewilding efforts—are successful. For example, the return of bird species to restored wetlands can be confirmed through the reappearance of their dawn choruses. Acoustic trends can also serve as early warning signals for population declines, prompting timely conservation action. Bioacoustic analysis is fundamental to understanding how animals communicate and behave. Many species rely on acoustic signals for mating, navigation, territorial defense, group coordination, and predator avoidance. (Giri G. et al. 2025) By analyzing the structure, frequency, duration, and context of vocalizations, researchers can infer social dynamics, reproductive strategies, and emotional states. Birdsong studies, for example, have revealed complex dialects, regional variations, and cultural transmission of song patterns. Songbirds may modify their calls in response to urban noise, demonstrating both behavioral plasticity and the stress of anthropogenic disturbance. Marine mammals like humpback whales produce elaborate songs that evolve seasonally and are shared across populations, suggesting sophisticated forms of social learning and cultural exchange. In primatology, bioacoustic analysis helps decode alarm calls and food-related vocalizations in species like vervet monkeys and chimpanzees. (Giri G. et al. 2025) These calls often exhibit features of semantic communication, where specific sounds correspond to specific meanings. Such insights deepen our understanding of animal cognition and the evolutionary roots of human language.

Feature Extraction: Identifying Relevant Patterns in Bioacoustic Data

In the rapidly evolving field of bioacoustics, signal processing serves as the critical intermediary step that transforms raw acoustic recordings into structured, analyzable inputs for machine learning algorithms. (Gorgeva E. et al. 2025) While machine learning (ML) enables powerful pattern recognition and classification capabilities, its effectiveness is fundamentally dependent on the quality and characteristics of the data it is trained on. Acoustic signals captured in natural environments are often messy, riddled with noise, overlapping calls, and environmental interference. (Gorgeva E. et al. 2025) Signal processing techniques provide the mathematical and computational tools necessary to extract meaningful features from these complex soundscapes, enhance signal clarity, reduce noise, and ensure that the data fed into ML models is as clean, consistent, and informative as possible. At its core, signal processing involves the manipulation and analysis of signals—such as sound waves—to better understand their structure and content. In bioacoustics, this means working with digital representations of animal sounds, typically captured as time-series data by microphones or hydrophones. Once recorded, these signals must undergo a series of preprocessing steps that prepare them for machine learning tasks such as detection, classification, clustering, or prediction. The first stage of signal processing is data acquisition and digitization. Audio signals are typically recorded in analog form and then converted into digital format using analog-to-digital converters (ADCs). Key parameters in this step include the sampling rate (how often the signal is measured, in samples per second) and the bit depth (how accurately each sample is measured). Higher sampling rates (e.g., 44.1 kHz or 96 kHz) allow for better representation of high-frequency sounds, which is crucial for capturing ultrasonic echolocation calls from bats or the high-pitched songs of some bird species. However, higher fidelity comes at the cost of larger file sizes, posing trade-offs for long-term monitoring.

Once digitized, the signal must be cleaned and denoised. Natural environments are acoustically rich, filled with background noise from wind, rain, insects, human activity, and other sources. (Hubballi R. B. et al. 2025) Signal processing techniques such as bandpass filtering help isolate the frequency range of interest, eliminating frequencies outside the known vocalization ranges of target species. Spectral subtraction, adaptive filtering, and wavelet denoising are other advanced methods used to suppress noise while preserving biologically relevant components of the signal. Segmentation is the next critical step, where continuous recordings are divided into smaller, manageable windows or frames. These segments—often 1 to 5 seconds in length—can be further analyzed or labeled. In some cases, algorithms such as short-time energy detection, zero-crossing rate analysis, or spectral entropy measures are used to automatically detect vocal events within a recording. (Hubballi R. B. et al. 2025) This process helps reduce the amount of irrelevant data and focuses computational resources on segments containing animal vocalizations. One of the most important aspects of signal processing for machine learning is feature extraction. ML models typically cannot work directly on raw waveforms due to their high dimensionality and noise. Instead, key features that capture relevant information about the sound's content and structure are extracted. These features can be categorized into time-domain, frequency-domain, and time-frequency domain features.

CLASSIFICATION AND CLUSTERING: MACHINE LEARNING MODELS FOR BIOACOUSTIC ANALYSIS

Feature extraction is a pivotal process in bioacoustic analysis, forming the bridge between raw audio data and meaningful biological or ecological interpretations, particularly when machine learning models are applied. It involves distilling large volumes of unstructured sound recordings into a compact set of informative descriptors—known as "features"—that capture the essential characteristics of animal vocalizations. These features represent specific patterns or attributes of the audio signal, such as pitch, duration, rhythm, frequency distribution, and amplitude dynamics. The accuracy, interpretability, and computational efficiency of bioacoustic systems often hinge on the quality and relevance of the extracted features. In natural habitats, animal vocalizations occur within complex acoustic environments, often embedded in background noise from wind, rain, insects, other species, or anthropogenic sources. Thus, before features can be extracted, audio signals are typically preprocessed using signal processing techniques like filtering, segmentation, and noise reduction. Once cleaned and segmented, feature extraction techniques are applied to represent the acoustic properties of interest in a numerical or visual format that is suitable for statistical analysis, pattern recognition, and machine learning classification. In raw form, audio signals are high-dimensional and computationally expensive to process directly. (Jayanthi K. et al. 2025) A one-second audio clip sampled at 44.1 kHz contains over 44,000 data points—most of which are redundant or irrelevant for classification tasks. Feeding such raw data into a machine learning model without dimensionality reduction would result in high training time, overfitting, and poor generalization. (Jayanthi K. et al. 2025) Feature extraction addresses this by transforming the waveform into a smaller set of variables that retain the biologically meaningful components of the signal.

Beyond computational efficiency, feature extraction also allows researchers to interpret animal sounds in biologically meaningful ways. For instance, by examining the pitch and harmonic structure of a bird's song, researchers can infer species identity, health status, emotional state, or mating readiness. (Katsis L. K. et al. 2025) Temporal features can reveal calling rhythms, diel activity patterns, or group coordination strategies. Therefore, features not only power machine learning models but also enhance scientific understanding of animal communication. Features extracted from bioacoustic recordings generally fall into three categories: time-domain, frequency-domain, and time-frequency domain features. Each provides a different lens for analyzing and interpreting acoustic patterns. Time-domain features describe how the audio signal behaves over time. These are particularly useful for identifying amplitude patterns, call durations, and temporal regularity. (Katsis L. K. et al. 2025) These features are derived using Fourier Transform methods, which decompose a signal into its constituent sinusoidal components, revealing its spectral structure. Animal vocalizations often involve complex modulations in both time and frequency. These are captured using spectrograms or wavelet transforms, offering a dynamic representation of changing frequencies over time. Not all features are equally useful for all tasks. The process of feature engineering involves crafting or combining features based on domain knowledge to improve model performance. For example, combining MFCCs with delta coefficients (which capture the rate of change) can enhance temporal sensitivity in vocal sequences. In bat monitoring, combining ZCR with pulse interval and repetition rate often improves identification accuracy.

Species Identification: Using Machine Learning for Automated Classification

The analysis of animal sounds through bioacoustics has been revolutionized by the application of machine learning, particularly through classification and clustering techniques. (Kershenbaum A. et al. 2025) These models allow researchers to uncover hidden patterns, automate species identification, monitor behavioral contexts, and even discover new call types or unknown species from large, complex, and often noisy acoustic datasets. (Kershenbaum A. et al. 2025) Classification and clustering represent two fundamental paradigms in machine learning—supervised and unsupervised learning, respectively—each with distinct roles in bioacoustic research. Together, they form a robust analytical framework capable of transforming raw sound data into ecological insight, behavioral understanding, and actionable conservation intelligence. Classification involves assigning predefined labels to input data. In bioacoustics, this typically means training a model to categorize vocalizations based on known species, call types, or behavioral contexts. (Kimura K. et al. 2025) Supervised learning requires a labeled dataset: a collection of acoustic recordings that have been annotated by human experts or validated through trusted identification processes. For a classifier to perform effectively, the training data must represent the diversity of acoustic signals it will encounter in the field. Data augmentation (e.g., adding background noise, pitch shifting) helps generalize models and reduce overfitting. Evaluation metrics commonly used in bioacoustics include Accuracy: Proportion of correct predictions. (Kimura K. et al. 2025) Precision and Recall: Important in imbalanced datasets where rare species or call types are underrepresented. F1-Score: Harmonic mean of precision and recall. Confusion Matrix: Visualizes the performance across classes, showing misclassifications and aiding model debugging.

Moreover, cross-validation techniques—such as k-fold or leave-one-out—ensure that the model generalizes well to new, unseen data and avoids bias from dataset partitions. Unlike classification, clustering does not require labeled data. Instead, it seeks to group data based on intrinsic similarity, making it ideal for exploratory analysis, anomaly detection, and discovery tasks in bioacoustic datasets where prior knowledge is limited or labels are unavailable. Both classification and clustering rely heavily on well-chosen acoustic features. Spectrogram-based features, (Maheswaran K. et al. 2025) MFCCs, temporal patterns, and frequency-domain descriptors all play vital roles. Deep learning models, especially CNNs and transformers, can jointly learn feature representations and classification rules from raw data, offering a unified pipeline. For clustering, feature normalization and dimensionality reduction (e.g., PCA) often precede clustering to enhance performance. In some workflows, hybrid approaches are used: unsupervised clustering to group similar sounds, followed by manual labeling and supervised training for large-scale classification. (Maheswaran K. et al. 2025) Alternatively, models may be trained in a semi-supervised manner, using labeled data to guide clustering and label propagation. Classification and clustering lie at the heart of intelligent bioacoustic analysis. By automating the recognition and organization of animal vocalizations, these machine learning techniques enable researchers to process massive acoustic datasets, uncover hidden patterns, monitor biodiversity at scale, and respond to environmental threats more swiftly. As algorithms grow more accurate and data becomes more accessible, the union of classification and clustering with ecological knowledge holds the key to deeper insights into the acoustic fabric of life on Earth—and to more informed, timely conservation decisions in a rapidly changing world.

Behavioral Analysis: Understanding Animal Behavior through Bioacoustics

Species identification is a fundamental goal of ecological monitoring, biodiversity research, and conservation biology. (Manikandan V. et al. 2025) Traditionally reliant on expert observation and manual analysis, this process has been transformed by advances in bioacoustics and machine learning (ML). In recent years, the development of automated classification systems powered by ML has enabled researchers to efficiently and accurately identify species based on their vocalizations. This technology is especially vital for monitoring elusive, nocturnal, or cryptic species that are difficult to observe visually but produce species-specific sounds. By leveraging ML algorithms, scientists can process massive acoustic datasets from diverse ecosystems and derive meaningful insights into species presence, distribution, and behavior at an unprecedented scale. Many animal taxa—such as birds, frogs, bats, whales, and insects—produce vocal signals that are distinct at the species level. These vocalizations serve a variety of functions including mate attraction, territorial defense, navigation, and communication. Because these sounds are often unique to each species, they provide a natural and non-invasive way to identify animals, even in dense vegetation, remote habitats, or during night hours. Passive acoustic monitoring (PAM), in which recorders are deployed in the field to continuously capture ambient sounds, allows researchers to collect terabytes of audio data. However, manually analyzing these recordings is labor-intensive and subject to human error or bias. (Manikandan V. et al. 2025) Automated species identification using ML addresses this bottleneck, offering scalable, repeatable, and often real-time solutions for biodiversity monitoring.

Data Collection and Annotation: Acoustic recordings are gathered using microphones or hydrophones, often placed in strategic locations or mounted on autonomous recorders. Annotated datasets—where vocalizations are labeled by species—are essential for training supervised ML models. (Martínez Balvanera S. et al. 2025) Creating these datasets requires input from taxonomists, bioacousticians, or field ecologists and may involve visualizing spectrograms, listening to sounds, and consulting field guides or reference libraries. Preprocessing and Signal Conditioning: Recordings often contain noise from wind, rain, human activity, and overlapping animal calls. Preprocessing steps like bandpass filtering, denoising, segmentation, and normalization are applied to enhance signal quality and isolate vocalizations of interest. Feature Extraction: Acoustic features are computed to represent the characteristics of each sound. Common features include Mel-Frequency Cepstral Coefficients (MFCCs), spectral centroid, bandwidth, pitch, duration, and temporal rhythm. These features form the input to ML classifiers. Model Selection and Training: Various ML algorithms can be trained on labeled feature sets to learn patterns associated with each species. Traditional models include Support Vector Machines (SVMs), k-Nearest Neighbors (k-NN), Random Forests, and Naive Bayes. More recent approaches use deep learning, particularly Convolutional Neural Networks (CNNs), to analyze spectrogram images directly. These models automatically learn hierarchical features that are robust to variability in pitch, duration, or background noise. Model Evaluation and Deployment: Models are evaluated using metrics like accuracy, precision, recall, and F1-score. Once trained and validated, they can be deployed in automated pipelines for real-time or batch processing of new recordings. (Martínez Balvanera S. et al. 2025) Some are integrated into edge devices capable of operating in the field with limited power or connectivity. Supervised learning is the dominant paradigm for species identification. In this approach, the model learns from labeled examples and predicts the species label for new, unseen data. It performs well when a sufficiently large and diverse training dataset is available. For example, bird call classifiers trained on thousands of labeled spectrograms from diverse habitats can achieve high accuracy in identifying common and regionally endemic species.

Ecosystem Monitoring: Machine Learning for Bioacoustic Data Analysis

Animal behavior reflects a complex interplay of evolutionary adaptations, ecological pressures, and social dynamics. Among the various modalities animals use to communicate and interact, sound plays a pivotal role across a wide range of species. From territorial calls of songbirds to the echolocation clicks of bats and the courtship songs of whales, vocalizations offer rich and often underutilized insights into the behavioral ecology of wildlife. Bioacoustics—the scientific study of sound production, transmission, and reception in animals—has emerged as a powerful tool for behavioral analysis. (Mohmmad S. et al. 2025) With advances in passive acoustic monitoring (PAM), signal processing, and machine learning, researchers are now able to decode intricate behavioral patterns, identify vocal signatures associated with specific activities, and interpret social interactions in wild animal populations in ways that were previously unimaginable. Understanding these behaviors acoustically offers several advantages: vocalizations can be recorded continuously, non-invasively, and across broad spatial and temporal scales—even in habitats that are visually obstructed or physically inaccessible. (Mohmmad S. et al. 2025) Behavioral analysis through bioacoustics typically follows a methodological framework involving: Data Collection: Using field-deployed microphones, hydrophones, or wearable tags, researchers record acoustic signals in natural environments or controlled settings. These recordings may capture individual calls, group interactions, or entire soundscapes. Signal Detection and Segmentation: Calls or bouts of interest are identified within larger recordings, often using energy-based thresholds, spectrogram correlation, or machine learning detection models. Feature Extraction: Acoustic features such as call duration, frequency modulation, repetition rate, and amplitude dynamics are extracted. These features serve as proxies for behavioral states or transitions. Annotation and Classification: Expert labeling or automated systems classify calls into behavioral categories, such as mating, foraging, distress, or play.

Figure 1. Applications of Machine Learning in Life Sciences

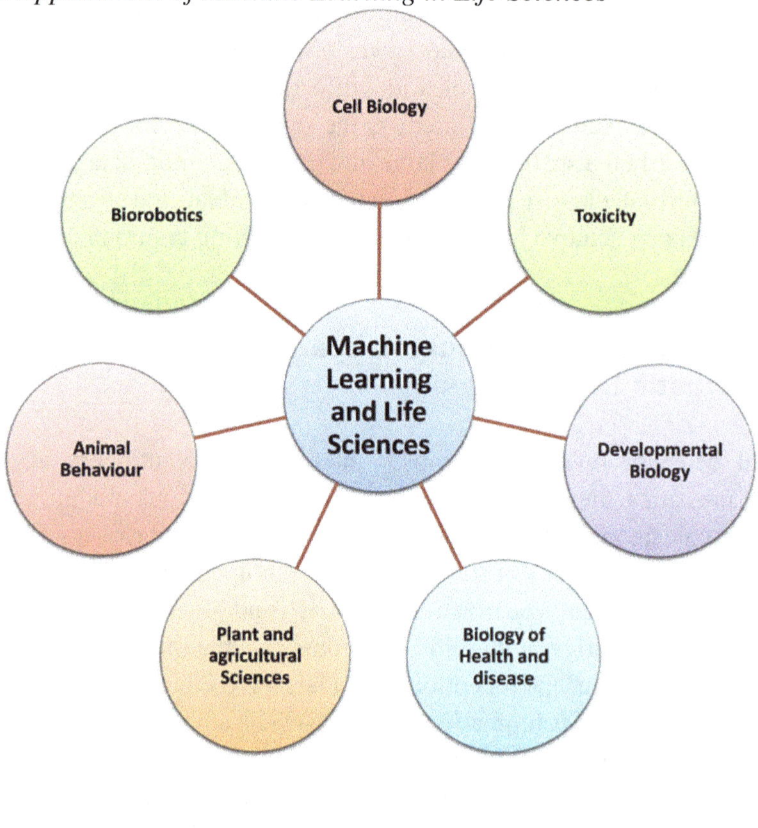

In some cases, unsupervised methods are used to discover unknown behavioral patterns. Temporal and Contextual Analysis: Behavioral patterns are examined in relation to time (diurnal, seasonal, life-history stages), location (habitat type, proximity to humans), or social context (group size, dominance hierarchies). (Nguyen

Chi T. et al. 2025) This framework enables researchers to move from raw sound to rich behavioral interpretation. The growing volume of acoustic data necessitates the use of machine learning to identify and classify behaviors. Supervised learning models can be trained to recognize behavioral states based on labeled datasets. (Nguyen Chi T. et al. 2025) Neural Networks and Random Forests have been used to classify vocalizations in birds and primates into different behavioral categories. Hidden Markov Models (HMMs) can model sequential patterns in call structures, particularly useful in species with patterned vocal repertoires like whales and birds. Recurrent Neural Networks (RNNs) and Long Short-Term Memory (LSTM) Networks capture temporal dependencies, enabling the analysis of complex behavior sequences such as bat foraging flights or cetacean social interactions. Unsupervised clustering techniques like k-means or DBSCAN can group calls based on acoustic similarity, potentially revealing new behaviors or call functions. These tools not only automate behavioral classification but also allow researchers to detect subtle changes or rare events that manual methods might miss. Bioacoustics offers a powerful lens through which to understand animal behavior in its natural context. By capturing the sounds of life—from a single distress call to the complex choruses of an ecosystem—researchers can decode the subtle and dynamic interactions that define animal lives. With the aid of machine learning, scalable hardware, and collaborative science, behavioral analysis through sound is poised to play a central role in ecology, ethology, and conservation biology for decades to come. As we continue to listen more closely, we will not only uncover the hidden patterns of animal behavior but also develop deeper empathy and more effective strategies for protecting the voices of nature.

Bird Song Analysis: Machine Learning for Species Identification: Bird song analysis.

The health and function of ecosystems are critical to sustaining biodiversity, ecosystem services, and human well-being. However, global environmental change—including habitat loss, climate shifts, pollution, and species extinctions—poses serious challenges to ecosystem stability. (Nikshya J. E. et al. 2025) To detect and understand these changes, scientists rely on ecosystem monitoring: the systematic collection and analysis of environmental data over time and space. Among the growing array of monitoring technologies, bioacoustics has emerged as a powerful tool. It enables researchers to non-invasively "listen in" on ecosystems using audio recordings that capture animal vocalizations, anthropogenic noise, and environmental soundscapes. When combined with machine learning (ML), bioacoustic monitoring becomes an automated, scalable, and intelligent method for detecting ecosystem change, tracking species dynamics, and informing conservation efforts. Sound is a fundamental component of ecological systems. Many species—especially birds, amphibians, insects, marine mammals, and bats—use acoustic signals to communicate, forage, defend territories, and navigate. These vocalizations encode rich information about species presence, abundance, behavior, and interactions. (Nikshya J. E. et al. 2025) Additionally, the overall soundscape—comprising biophony (biological sounds), geophony (natural non-biological sounds), and anthropophony (human-made sounds)—reflects the structure, function, and disturbance level of an ecosystem. However, interpreting this vast audio data manually is impractical. This is where machine learning plays a transformative role. Machine learning models enable automated identification of multiple species in real-time from PAM data, allowing researchers to assess species richness and community composition. Projects like BirdNET and Arbimon have deployed ML tools across continents to map avian biodiversity.

Changes in soundscape composition and acoustic complexity can reflect habitat quality. For example, a decline in biophony (e.g., birds, frogs) and a rise in anthropophony (e.g., engines, chainsaws) may signal habitat degradation. (Nisha Angeline M. et al. 2025) ML models that track these shifts can inform ecological management and restoration efforts. ML can detect seasonal patterns in calling behavior, migration timing, and breeding periods—essential for phenological studies and understanding climate change impacts. In temperate forests, bird vocal activity follows predictable seasonal cycles that can be disrupted by climate anomalies. In oceanic environments, ML models detect whale and dolphin calls, track fish spawning events, and monitor ship noise. (Saghir R. et al. 2025) These systems support marine protected areas (MPAs) and help mitigate threats like noise pollution and vessel strikes. ML-enabled acoustic sensors detect gunshots, chainsaws, and vehicle engines in real-time, providing law enforcement with alerts about poaching or illegal logging in protected areas. Systems like Rainforest Connection have successfully deployed such technologies in tropical forests. ML classifiers identify non-native or invasive species from their vocalizations, often before they are visually confirmed. For example, identifying calls of invasive cane toads or birds can trigger early intervention strategies. (Saghir R. et al. 2025) By analyzing multi-year datasets, ML models can reveal trends in community composition, behavior shifts, or declines in acoustic activity, offering early warning signs of ecological imbalance.

WHALE COMMUNICATION: USING MACHINE LEARNING TO DECODE VOCALIZATIONS

Birdsong is one of the most recognizable and ecologically informative natural sounds. It serves a wide variety of biological functions—territory defense, mate attraction, alarm signaling, and social communication. (Nisha Angeline M. et al. 2025) Each bird species typically produces a unique vocal signature composed of distinct frequency patterns, durations, and rhythms. These vocal characteristics make birdsong an ideal target for automated species identification, especially through the integration of bioacoustics and machine learning (ML). With growing concerns over biodiversity loss and habitat degradation, the ability to accurately and efficiently identify bird species from their vocalizations has never been more important. Machine learning has emerged as a transformative technology, enabling scalable, real-time bird monitoring through acoustic data collected from natural environments. (Santhia R. K. et al. 2025) Birds are one of the most diverse and widely distributed groups of animals, inhabiting nearly every ecosystem on Earth. They are also among the most vocal, especially during the breeding season. Their calls and songs offer critical insight into species presence, abundance, behavior, and ecosystem health. (Santhia R. K. et al. 2025) Acoustic monitoring is particularly valuable in regions that are difficult to access visually, such as dense forests, wetlands, or remote mountains. It also allows researchers to collect data around the clock, including at night when many species are most active. Moreover, many bird species produce stereotyped, repetitive song structures that are amenable to computational analysis. The frequency, pitch, modulation, temporal spacing, and harmonic structure of bird vocalizations can be encoded into spectrograms—visual representations of sound over time—that serve as effective inputs for machine learning models. Traditionally, ornithologists relied on human listeners to identify bird songs, sometimes aided by visual inspection of spectrograms. (Sathiya D. et al. 2025) While effective, this approach is time-consuming, subjective, and limited in scale. Machine learning revolutionizes this process by automating the identification task using data-driven models trained to recognize the acoustic patterns associated with different species. Birdsong analysis through machine learning represents a powerful confluence of ecology, acoustics, and artificial intelligence. It enables researchers, conservationists, and citizen scientists to detect, identify, and monitor bird species at unprecedented scales and resolutions. (Sathiya D. et al. 2025) As models become more accurate, accessible, and interpretable, the potential to use birdsong as a lens for understanding ecological change and protecting biodiversity will continue to grow. Machine learning doesn't just recognize bird calls—it amplifies our ability to listen to and learn from nature, guiding better stewardship of our shared planet.

INSECT BIOACOUSTICS: MACHINE LEARNING FOR PEST MONITORING AND MANAGEMENT

Whales are among the most acoustically sophisticated animals on Earth. Their vocalizations span a remarkable range of frequencies and patterns, forming complex communication systems essential for navigation, social interaction, mating, and foraging across vast and often dark oceanic environments. (Sethumadhavan R. et al. 2025) Understanding whale communication is critical not only for marine biology and ethology but also for the conservation of these often-endangered species. However, the sheer complexity and volume of underwater sound data have historically limited our ability to analyze whale vocalizations in depth. Recently, machine learning (ML) has revolutionized the field of bioacoustics, enabling automated, scalable, and nuanced decoding of whale sounds. (Sethumadhavan R. et al. 2025) By combining advanced signal processing and ML algorithms, researchers can classify calls, identify individuals, detect behavioral contexts, and even infer social structures, providing unprecedented insight into the acoustic lives of whales. Each species—and sometimes individuals within species—produce distinct vocal repertoires. Decoding these signals is key to understanding whale ecology and behavior. Data Volume and Complexity: Oceanic acoustic monitoring can generate terabytes of recordings containing thousands of vocal events mixed with ambient noise from waves, ships, and other marine life. Variability in Calls: Whale vocalizations can vary with age, sex, region (dialects), behavioral context, and environmental conditions. Overlapping Sounds: Multiple individuals and species often vocalize simultaneously, creating acoustic clutter. Limited Ground Truth: Unlike terrestrial animals, visual confirmation of whale behavior during vocalizations is difficult, complicating supervised learning.

These challenges require sophisticated computational tools—machine learning offers powerful solutions. ML models can automatically identify whale species from passive acoustic monitoring data, even when species overlap spatially and temporally. (ShabanaFathima M. et al. 2025) For example, CNN-based classifiers can differentiate between blue whale, fin whale, and humpback whale calls. This capability improves species distribution mapping and aids in monitoring endangered populations. (ShabanaFathima M. et al. 2025) By analyzing subtle vocal differences, ML can distinguish individuals, akin to “acoustic fingerprints.” This enables tracking of social groups, movement patterns, and population size estimation without intrusive tagging. ML models can correlate specific call types or sequences with behaviors such as feeding, mating, or traveling. This helps infer the function of different vocalizations and understand social dynamics. Automated processing of long-term datasets enables detection of seasonal and interannual changes in whale vocal activity. Such trends may reflect migration timing, breeding seasons, or responses to climate change. ML can separate whale calls from anthropogenic noise such as ship engines and seismic surveys, assessing how noise pollution affects whale communication and habitat use. Deployed on autonomous platforms or buoys, ML algorithms enable near-real-time detection of whales, providing data for dynamic marine spatial management, ship strike avoidance, and anti-poaching efforts. Humpback Whale Songs: Deep learning models have been used to classify complex humpback whale songs into distinct themes and variations, revealing song evolution over time and between populations. Sperm Whale Clicks: Machine learning has improved the detection and classification of echolocation clicks, facilitating studies of foraging behavior and population density. Blue Whale Call Detection: CNNs trained on spectrograms can detect low-frequency blue whale calls in noisy ocean environments, improving monitoring in protected areas.

DATA QUALITY AND NOISE REDUCTION: CHALLENGES IN BIOACOUSTIC ANALYSIS

Insects constitute the most diverse group of animals on Earth, playing vital roles in ecosystems as pollinators, decomposers, and prey species. However, many insects are also major agricultural pests causing extensive crop damage and economic loss worldwide. Effective monitoring and management of pest populations are critical for sustainable agriculture and food security. (Thirunagalingam A. et al. 2025) Traditional pest monitoring methods—such as visual surveys, trapping, and manual identification—are labor-intensive, time-consuming, and often insufficient to provide timely data for intervention. In this context, bioacoustics, the study of sound production and perception in living organisms, offers a promising alternative by leveraging the distinct acoustic signatures produced by insects during their behaviors. (Thirunagalingam A. et al. 2025) By applying machine learning (ML) to analyze these acoustic signals, researchers and practitioners can develop automated, real-time, and non-invasive pest monitoring systems that enhance early detection and improve pest management strategies. Many insect species produce sounds through specialized structures or behaviors, such as stridulation (rubbing body parts together), wing beats, or tymbal organs. These sounds serve diverse functions, including mate attraction, territorial defense, alarm signaling, and communication. For instance, crickets produce chirps by rubbing their forewings; cicadas generate loud, species-specific calls using tymbal muscles; mosquitoes create characteristic buzzing sounds with their wingbeats; and beetles may produce rasping or clicking noises. These acoustic signals often have species-specific temporal and spectral features, making them reliable indicators for species identification and behavior monitoring. The frequency range of insect sounds varies widely, from audible chirps in crickets and cicadas to ultrasonic signals used by moths or mosquitoes for mating or predator avoidance. This acoustic diversity forms the basis for developing sound-based monitoring systems.

Accurate identification of pest species through acoustic signatures allows for early detection of outbreaks. For example, ML models can distinguish between different species of crickets or moths known to damage crops. By monitoring acoustic activity levels, farmers and researchers can estimate population density trends, informing timely interventions. Machine learning can analyze call timing, frequency, and intensity to infer insect behaviors such as mating, feeding, or dispersal. (Ajay et al. 2025) This helps understand life cycles and seasonal patterns, enabling predictive pest management. Acoustic monitoring integrated with environmental data can forecast periods of high pest activity and associated crop damage risks. This supports targeted pesticide application, reducing environmental impact and cost. Smart traps equipped with acoustic sensors and ML algorithms can detect pest presence, identify species, and report data wirelessly in real-time. This technology enhances spatial and temporal resolution of monitoring programs. By tracking acoustic responses before and after pesticide application, ML models help assess control measure effectiveness and detect emerging pesticide resistance. Rapid acoustic detection of invasive insect pests in new areas facilitates early management and containment, critical for protecting native ecosystems and agriculture. The integration of insect bioacoustics with machine learning offers a transformative approach to pest monitoring and management. By enabling automated, precise, and scalable identification and behavioral analysis of insect pests, these technologies empower farmers, researchers, and conservationists to implement more effective, timely, and environmentally friendly pest control strategies. As acoustic sensor technology improves and machine learning algorithms become more sophisticated, the future promises increasingly intelligent and responsive systems that will support global efforts in sustainable agriculture and biodiversity conservation. (Ajay et al. 2025) Listening closely to the hidden world of insect soundscapes through machine learning not only helps protect crops and ecosystems but also advances our fundamental understanding of insect ecology in an ever-changing world.

Scalability and Efficiency: Optimizing Machine Learning for Large-Scale Bioacoustic Analysis

Bioacoustics, the interdisciplinary science studying sound production, transmission, and reception in living organisms, relies heavily on the collection and analysis of acoustic data. Whether monitoring bird songs in forests, whale calls in oceans, or insect chirps in agricultural fields, the quality of recorded acoustic data critically determines the accuracy and reliability of subsequent analysis. (Wagner M. et al.

2025) High-quality data enable researchers to decipher animal communication, identify species, and monitor behaviors and populations effectively. Conversely, poor data quality marred by noise and distortion can obscure meaningful signals, mislead interpretation, and degrade the performance of machine learning models used for bioacoustic analysis. This makes data quality assurance and noise reduction fundamental challenges in bioacoustic research. Addressing these challenges requires understanding the sources and types of noise, the impact of noise on data, and the development and application of advanced signal processing and computational methods to enhance signal clarity and analytical robustness. Maintaining high data quality allows for the reliable identification of animal vocalizations, extraction of acoustic features, and accurate behavioral or ecological inference. Poor-quality data, on the other hand, can introduce bias, increase error rates in classification, and hinder scientific conclusions. Natural physical processes generate substantial acoustic interference. These include wind turbulence, rain and precipitation, water flow and waves, thunder, and seismic activity. Abiotic noise often overlaps with the frequency ranges of biological signals.

Acoustic interference caused by other living organisms, such as overlapping calls from multiple animals (both conspecifics and heterospecifics), insects, or amphibians, can mask the target vocalizations. (Wagner M. et al. 2025) Human activities contribute significantly to underwater and terrestrial noise pollution. In marine environments, ship engines, sonar, seismic surveys, and construction generate low- to mid-frequency noise, often masking whale and fish vocalizations. Terrestrial bioacoustic recordings suffer from road traffic, machinery, aircraft, and urban activity noises, which interfere with bird songs or insect sounds. Recording devices and sensors introduce noise through electrical interference, sensor imperfections, or suboptimal placement. Variability in microphone sensitivity and calibration issues also affect data quality. Sound propagates differently depending on the medium. In water, sound speed varies with temperature, salinity, and pressure, causing distortion and reverberation. In forests or urban environments, echoes and absorption alter signals. Signal attenuation with distance leads to diminished amplitude and loss of high-frequency components, reducing detectability. Data quality and noise reduction remain fundamental challenges in bioacoustic analysis, shaping the reliability of ecological insights and conservation outcomes. Noise from environmental, anthropogenic, and instrumental sources complicates the detection and interpretation of animal vocalizations, demanding sophisticated solutions. Through improved sensor technology, rigorous field protocols, advanced signal processing, and machine learning techniques, researchers are progressively overcoming these hurdles. Continued innovation in noise reduction methods and data quality assessment will enhance the precision, scalability, and impact of bioacoustic research, enabling us to better understand and protect the natural world through its sounds.

Integration with Other Technologies: Combining Bioacoustics with Other Sensing Modalities

The integration of machine learning (ML) into bioacoustic analysis has revolutionized the ability to process vast quantities of acoustic data to identify species, monitor behaviors, and understand ecosystems. However, as the volume of bioacoustic data grows exponentially—driven by advances in sensor technology, long-term monitoring projects, and the proliferation of autonomous recording units—the challenges of scalability and efficiency become paramount. (Williams B. et al. 2025) Large-scale bioacoustic datasets often encompass thousands of hours of continuous recordings from multiple locations and sensor types, necessitating ML frameworks that can manage, analyze, and deliver insights in a computationally feasible and timely manner. Optimizing scalability and efficiency is critical to unlock the full potential of bioacoustic data for ecological research, conservation, and environmental management. Modern bioacoustic studies generate data at unprecedented scales. Continuous monitoring over extended periods can produce terabytes of audio files, reflecting complex acoustic environments with multiple species and varying signal qualities. The high dimensionality of bioacoustic data—comprising temporal, spectral, and spatial features—combined with the need for fine-grained species or behavior classification, compounds computational demands. This creates bottlenecks in data storage, transmission, preprocessing, feature extraction, model training, and inference.

Furthermore, bioacoustic datasets often require real-time or near-real-time processing to support applications such as early warning systems for invasive species, pest outbreaks, or endangered species monitoring. (Williams B. et al. 2025) Delays due to inefficient processing can undermine the timely application of conservation or management actions. Future research is poised to harness advances in federated learning, where models train collaboratively across distributed datasets without centralizing data, protecting privacy and reducing bandwidth. Also, integrating multimodal data (acoustic, visual, environmental) can enrich analysis and improve efficiency. Scalability and efficiency are central to the continued success of machine learning applications in large-scale bioacoustic analysis. By optimizing data management, model design, computational infrastructure, and algorithmic strategies, researchers can handle the increasing volume and complexity of bioacoustic data with agility and precision. These improvements enable timely, accurate ecological insights that inform conservation efforts and environmental policy. The marriage of machine learning and bioacoustics holds immense promise, but fulfilling this potential requires ongoing innovation to overcome scalability and efficiency challenges in the face of growing global data and pressing ecological needs.

THE FUTURE OF BIOACOUSTIC ANALYSIS: OPPORTUNITIES AND CHALLENGES

Bioacoustics, the study of animal sound production and communication, has traditionally relied on acoustic recordings and signal analysis to gain insights into wildlife behavior, species identification, and ecosystem health. While bioacoustic data provide rich temporal and spectral information about vocalizations and environmental sounds, their integration with other sensing technologies has opened new frontiers in ecological monitoring, animal behavior studies, and conservation. (Yi Q. et al. 2025) Combining bioacoustics with complementary sensing modalities such as remote sensing, camera traps, environmental sensors, GPS tracking, and environmental DNA (eDNA) analysis enables a more holistic understanding of ecosystems by capturing multiple dimensions of the natural world. This multimodal integration not only enhances data quality and context but also addresses the inherent limitations of single-modality approaches, offering synergistic benefits in scalability, accuracy, and interpretability. Each sensing modality captures different aspects of wildlife and environmental phenomena. Bioacoustics excels in detecting vocal activity and communication patterns but often struggles to disambiguate silent species, interpret visual behavior, or measure environmental conditions influencing animal activity. Integrating these modalities provides a richer, multi-layered dataset that captures the complexity of natural systems more fully than any single method. Combining acoustic recordings with camera trap imagery is increasingly popular in terrestrial wildlife monitoring. While cameras provide visual confirmation and behavioral observations, acoustic sensors detect species that may evade cameras or vocalize at night. Environmental conditions heavily influence acoustic signal transmission and animal vocal activity. Integrating sensors measuring temperature, humidity, wind speed, and light intensity helps. Such integration is valuable in monitoring species vulnerable to climate change or habitat alteration. The integration of bioacoustics with other sensing technologies marks a significant evolution in wildlife monitoring and ecological research. (Yi Q. et al. 2025) By combining complementary data streams—acoustic, visual, environmental, spatial, and genetic—researchers can overcome the limitations inherent in single-modality approaches and gain a multidimensional perspective on animal communication, behavior, and ecosystem dynamics. Although challenges in data synchronization, complexity, and logistics remain, advances in sensor technology, machine learning, and data infrastructure are rapidly facilitating effective multimodal integration. As this field matures, integrated bioacoustic sensing promises to deliver richer, more accurate, and scalable insights essential for biodiversity conservation and understanding the natural world in an era of rapid environmental change.

CONCLUSION

Bioacoustic analysis—the study of animal sounds to understand communication, behavior, and ecosystem dynamics—has evolved remarkably over recent decades. Advances in sensor technology, data storage, and machine learning have enabled researchers to capture and analyze acoustic data at unprecedented scales and resolutions. As we look ahead, the future of bioacoustic analysis promises transformative opportunities for biodiversity monitoring, ecological research, and conservation, while also posing significant scientific, technical, and ethical challenges. Navigating these opportunities and challenges will define the next era of bioacoustic science, with implications reaching far beyond academia into global efforts for environmental sustainability. Sensor miniaturization, lower costs, and improved energy efficiency are driving the deployment of dense acoustic sensor networks worldwide. These networks can operate autonomously for extended periods, continuously monitoring soundscapes across diverse habitats—from tropical rainforests and coral reefs to urban environments. The sheer volume and spatiotemporal resolution of data collected will enable: High-Resolution Biodiversity Mapping: Detecting species presence, richness, and distribution at fine scales, supporting landscape-level ecological studies. Longitudinal Studies: Tracking changes in vocal activity and species populations over years or decades to assess environmental impacts and climate change effects. Real-Time Alerts: Enabling rapid detection of poaching, invasive species, or ecological disturbances through automated acoustic event recognition. Scalable monitoring holds great promise for filling knowledge gaps, especially in remote or poorly studied ecosystems.

Machine learning (ML), particularly deep learning, has revolutionized bioacoustic data processing by enabling automated species identification, behavior classification, and soundscape characterization. These advances will enhance both the efficiency and ecological relevance of bioacoustic analyses. Bioacoustics is increasingly converging with other sensing modalities and technologies, creating holistic ecological monitoring systems. Examples include: This integration fosters comprehensive ecosystem understanding and informed management strategies. Improved data quantity and quality will unlock deeper understanding of animal communication, social structures, and responses to environmental stressors. Bioacoustic analysis will contribute to Behavioral Ecology: Unraveling complexities of vocal communication, mating systems, and predator-prey interactions. Conservation Biology: Guiding protection of endangered species through detailed population monitoring and habitat use analysis. Ecosystem Health Assessment: Using acoustic indices as indicators of ecosystem integrity and biodiversity. Such insights are vital for adaptive management in the face of accelerating global change. The deluge of acoustic data creates logistical hurdles in storage, curation, and sharing. Without standardized data formats, metadata, and protocols, interoperability and reproducibility suffer. Efforts to develop open repositories and community standards are essential but require global coordination and resources. Machine learning models depend on labeled datasets for training, but annotating bioacoustic data is time-consuming, requires expert knowledge, and often suffers from inter-observer variability. Developing crowdsourcing platforms, active learning strategies, and semi-supervised methods will be key to overcoming this bottleneck.

REFERENCES

Ajay, N., Upadhyaya, A. N., Debnath, P., Rani, P., Selvakumar, P., Manjunath, T. C., & Bhattacharya, S. (2025). AI-Based Disaster Prediction and Early Warning Systems. In Mobo, F. (Ed.), *AI and Emerging Technologies for Emergency Response and Smart Cities* (pp. 55–74). IGI Global Scientific Publishing.

Chowdhary, S., Selvakumar, P., Sangeetha, B., Sarvaani, V. V., Anute, N., & Sharma, M. (2025). The Role of Social Capital in Entrepreneurial Development. In Martínez-Falcó, J., Sánchez-García, E., & Marco-Lajara, B. (Eds.), *Support Networks for Entrepreneurs: Social Capital and Strategic Collaborations* (pp. 311–338). IGI Global Scientific Publishing.

Coates, J. M., Santhanalakshmi, K., Dinesh, N., Gandhimathi, S., & Chandel, P. S. Mohit, & Selvakumar, P. (2025). Assessment and Evaluation in Transnational Staff Development. In G. Morris & J. Ye (Eds.), *Training and Development in Transnational Higher Education* (pp. 387-406). IGI Global Scientific Publishing.

Darney, P. E., Fernando, A. V., Krishnan, R. S., Raj, J. R. F., Lavanya, V., & Kumar, V. V. (2025). RhinoGuardNet: An Integrated Deep Learning System for Monitoring, Movement Prediction, and Threat Detection in Javan Rhino Conservation. In *2025 International Conference on Machine Learning and Autonomous Systems (ICMLAS)* (pp. 101-108). IEEE.

Deng, X., Wan, T., Xu, K., Gao, T., Qiao, P., Feng, D., & Dou, Y. (2025, April). Scaling Bioacoustic Signal Pre-training with Million Samples Via Mask-Modeling. In *ICASSP 2025-2025 IEEE International Conference on Acoustics, Speech and Signal Processing (ICASSP)* (pp. 1-5). IEEE.

Devi, V. B., Privadharshini, K., & Haritha, M. (2025). Smart Irrigation System by Detecting Plant Squeals Using IoT. In *2025 International Conference on Computing and Communication Technologies (ICCCT)* (pp. 1-8). IEEE.

Dhivya, P., & Abhishek, P., HariHaran, M., & Midhun, J. (2025). Deep Learning-Based Acoustic Classification of Animal Species Using Convolutional Neural Networks. In *2025 International Conference on Inventive Computation Technologies (ICICT)* (pp. 159-165). IEEE.

Geldenhuys, C. M., & Niesler, T. R. (2025). Learning to rumble: Automated elephant call and sub-call classification, detection and endpointing using deep architectures. *Bioacoustics*, *34*(3), 307–354.

Giri, G., Iza, K. C., Khatiwada, P., Adhikari, S. K., & Shakya, S. (2025). CNN-Based Bird Sound Detection: A Comparative Performance Study. *IACSIT International Journal of Engineering and Technology*, *2*(2), 176–187.

Gorgeva, E., Robertson, J., Voss, S., & Hoogewerff, J. (2025). The potential of bioacoustics for surveying carrion insects. *The Australian Journal of Forensic Sciences*, *57*(1), 122–141.

Hubballi, R. B., Selvakumar, P., Seenivasan, R., & Basava Aradhya, S. G., Dinesh, N., & Seelam, P. K. (2025). Overview of Current AI Technologies in Education. In L. Kyei-Blankson & E. Ntuli (Eds.), *Transformative AI Practices for Personalized Learning Strategies* (pp. 1-26). IGI Global Scientific Publishing.

Jayanthi, K., Chitradevi, D., & Muthaiyah, S. (2025). Early Prediction of Harmful Environmental Changes Based on Bioacoustics Analysis. In *Technological Advancements for Deep Sea Ecosystem Conservation and Exploration* (pp. 129–148). IGI Global Scientific Publishing.

Katsis, L. K., Rhinehart, T. A., Dorgay, E., Sanchez, E. E., Snaddon, J. L., Doncaster, C. P., & Kitzes, J. (2025). A comparison of statistical methods for deriving occupancy estimates from machine learning outputs. *Scientific Reports*, *15*(1), 1–13.

Kershenbaum, A., Akçay, Ç., Babu-Saheer, L., Barnhill, A., Best, P., Cauzinille, J., & Dunn, J. C. (2025). Automatic detection for bioacoustic research: A practical guide from and for biologists and computer scientists. *Biological Reviews of the Cambridge Philosophical Society*, *100*(2), 620–646.

Kimura, K., Fukuyama, I., & Fukuyama, K. (2025). Deep learning-based detector of invasive alien frogs, Polypedates leucomystax and Rhinella marina, on an island at invasion front. *Biological Invasions*, *27*(3), 1–11.

Maheswaran, K., Selvakumar, P., Shaikh, M., Shashank, G., Gupta, S., & Manjunath, T. C. (2025). The Evolution of Smart Urban Mobility With Emerging Technologies. In Ponnusamy, S., Assaf, M., Youssef, B., Jeon, G., & Kalyanaraman, S. (Eds.), *Leveraging VANETs and Blockchain Technology for Urban Mobility* (pp. 1–28). IGI Global Scientific Publishing.

Manikandan, V., & Neethirajan, S. (2025). Decoding Poultry Welfare from Sound—A Machine Learning Framework for Non-Invasive Acoustic Monitoring. *Sensors (Basel)*, *25*(9), 2912.

Martínez Balvanera, S., Mac Aodha, O., Weldy, M. J., Pringle, H., Browning, E., & Jones, K. E. (2025). Whombat: An open-source audio annotation tool for machine learning assisted bioacoustics. *Methods in Ecology and Evolution*, *16*(1), 19–28.

Mohmmad, S., Chauhan, G. S., Rao, B. S., Sravani, K., & Sridhar, K. (2025). Detection of Animals based on their Species by Implementing Bio-Acoustic Classification for Forest Wildlife Monitoring. In *2025 International Conference on Intelligent Computing and Control Systems (ICICCS)* (pp. 1244-1249). IEEE.

Nguyen Chi, T., Vu, T. T., Nguyen, H. T., & Clink, D. J. (2025). Circadian rhythms and the use of transfer learning for critically endangered crested argus Rheinardia ocellata in the central highlands of Vietnam: the implications for conservation. *Philosophical Transactions B, 380*(1928).

Nikshya, J. E., Raj, J. R. F., Krishnan, R. S., Rajkumar, G. V., Settu, I. J., & Kavitha, S. (2025). Deep Learning Models for Acoustic Monitoring of the Himalayan Monal and Other Elusive Bird. In *2025 4th International Conference on Sentiment Analysis and Deep Learning (ICSADL)* (pp. 1392-1399). IEEE.

Nisha Angeline, M., Selvakumar, P., Gupta, M. K., Nazar, N., Elumalai, D., & Bhattacharya, S. (2025). Emerging Materials for Energy Storage and Environmental Applications. In K. Muduli, B. Moharana, S. Ales, & D. Biswal (Eds.), *Using Computational Intelligence for Sustainable Manufacturing of Advanced Materials* (pp. 53-72). IGI Global Scientific Publishing.

Saghir, R. B., Campos, I., Gouvêa, T. S., & Sonntag, D. (2025). EcoScape Analyzer: A Tool for Performing Soundscape Analysis With Flexible Pipeline for Biodiversity Assessment. In *Companion Proceedings of the 30th International Conference on Intelligent User Interfaces* (pp. 137-140).

Santhia, R. K., & Jayalakshmi, S. L. (2025). Advancing Environmental Surveillance: A Comprehensive Review of Sound Event Classification Techniques and Applications. In *2025 8th International Conference on Trends in Electronics and Informatics (ICOEI)* (pp. 1329-1337). IEEE.

Sathiya, D., Palanisamy, P., Nandhini, E., Jayamurugan, V., Kirthikcharan, K., & Manoj, N. (2025). Deep Sound Search Leveraging Pre-trained CNNs and Faiss for Animal Vocalization Analysis. In *International Conference on Sustainability Innovation in Computing and Engineering (ICSICE 2024)* (pp. 410-423). Atlantis Press.

Sethumadhavan, R. P. P., S., Selvakumar, P., Kadu, S. G., Chandel, P. S., & Gupta, S. (2025). Environmental Sustainability and Employee Participation. In C. Şeker, E. Örücü, & A. Kaya (Eds.), *Green Leadership Strategies for Increased Employee Engagement and Performance* (pp. 275-298). IGI Global Scientific Publishing.

ShabanaFathima, M., Shanmugapriya, I., Pavithira, L., Mihirkumar, B. S., Rajeswari, J., Revathi, R., & Vengatesh, T. (2025). Deep Learning Approaches for Monitoring and Preserving Ecological Biodiversity: Challenges and Innovations. *Cuestiones de Fisioterapia*, *54*(3), 844–858.

Thirunagalingam, A., & Selvakumar, P. (2025). Application of Automation and Robotics in the Oil and Gas Industry. In Abdullayev, V., & Khang, A. (Eds.), *Revolutionizing AI and Robotics in the Oil and Gas Industry* (pp. 229–244). IGI Global Scientific Publishing.

Wagner, M., & Vashisth, D. (2025). The Influence of Inaccurate GPU Power Measurements for Machine Learning Workloads in Industrial Applications. In *International Conference on Human-Computer Interaction* (pp. 233-247). Springer, Cham.

Williams, B., van Merriënboer, B., Dumoulin, V., Hamer, J., Fleishman, A. B., McKown, M., . . . Denton, T. (2025). Using tropical reef, bird and unrelated sounds for superior transfer learning in marine bioacoustics. *Philosophical Transactions B, 380*(1928).

Yi, Q., Xie, C., Guan, D., & Yuan, W. (2025). Man2Marine: Marine mammal sound classification in small samples by transfer learning from human sound data. *Pattern Recognition Letters*, *190*, 185–191.

Chapter 3
Artificial Intelligence in Emotional Recognition in Animals

Geetha Manoharan
https://orcid.org/0000-0002-8644-8871
SR University, India

ABSTRACT

This chapter examines AI's growing significance in animal emotion recognition. This shows how AI is changing animal emotion research by analyzing vocalizations, facial expressions, and behavioral signals. Artificial intelligence can accurately distinguish positive and negative emotional states using machine learning models built on datasets from several species. This offers new perspectives on animal welfare, health monitoring, and human-animal connections. This chapter discusses modern technologies like artificial intelligence deciphering ungulate vocal emotions, domestic animal facial expression analysis, and pain detection systems. The chapter also promotes multidisciplinary teamwork and large, well-annotated datasets. It also addresses species diversity, data shortages, and ethics. Artificial intelligence in emotional recognition might improve livestock management, veterinary care, and conservation, opening a new path in animal welfare research.

INTRODUCTION TO AI IN EMOTIONS

Similarly to people, the location latest artificial intelligence is increasing to consist of the complicated emotional lives modern day animals (Liu, 2024). These innovative gear for knowledge and measuring emotions are increasing past human topics. When as compared to fashionable artificial intelligence, which tries to copy

DOI: 10.4018/979-8-3373-5483-5.ch003

human cognition, emotional synthetic intelligence, also known as affective computing, objectives to mimic the human capacity to revel in feelings and reply with emotional intelligence (Gupta & Bansal, 2023). The incorporation state-of-the-art emotional intelligence into artificial intelligence systems has the capacity to cultivate more bonds and empathy between human beings and robots (Vicci, 2024). In animal studies, in which traditional techniques ultra-modern comparing feelings are present day subjective and constrained through the observer's perception (Waelen, 2024), where this capability is mainly sizeable, it is very important to have. in step with Narimisaei et al. (2024), artificial intelligence structures, especially those who make use of deep modern day, are capable of apprehend diffused emotional cues rising from an extensive variety brand new information resource. As an end result, they offer a greater goal and nuanced view on animal emotions. Convolutional neural networks can be taught to recognising and classifying emotional states in animals with a high diploma today's accuracy (Hong et al., 2023). This is achieved by way of the cautious exam modern day facial expressions and physiological facts. A dynamic attitude trendy emotional adjustment through the years can be acquired via using recurrent neural networks, which are gifted at processing sequential input. Those networks are applied to figure temporal patterns in vocalizations, frame language, and conduct. (Praveena et al., 2020). These strategies, which might be powered through artificial intelligence, provide the capacity to untangle the intricacies trendy animal emotions, which has considerable repercussions for animal welfare, conservation efforts, and our comprehension today's the evolutionary underpinnings present day feelings (Romeo & Testolin, 2025).

Full-size improvements have been made in machine contemporary, deep todays, and neural networks, that have contributed to the rapid improvement latest artificial intelligence through its speedy evolution. New trends in synthetic intelligence have made it possible for machines to understand and react to human feelings, which has paved the way for more suitable interactions between people and machines. Regarding education synthetic intelligence fashions, the fundamental paradigms which are utilized include supervised latest, unsupervised latest, and reinforcement getting to know modern systems. A full-size gain ultra-modern semi-supervised today's is that it contemporary both classified and unlabelled data so one can improve model performance, especially in situations when categorized records is limited. With regards to the accuracy trendy emotion identity, traditional device contemporary algorithms that interpret audio and picture have intrinsic limits (Lu, 2022). Both the availability cutting-edge big datasets and traits in computing power have contributed to the reality that deep modern-day has attained great overall performance in a diffusion latest application, such as speech recognition, natural language processing, and image popularity. Within the large situation state-of-the-art sentiment evaluation and affective computing, which tries to understand and mimic human feelings thru pc procedures (Gupta & Bansal, 2023), emotion identity in animals is inextricably related to the sector modern-day affective computing. In step with Dhuheir et al. (2021), the use of artificial intelligence within the discipline contemporary emotion identification calls for a mix ultra-modern method, which includes machine modern day, records analysis, and signal processing. it's been determined by Liao et al. (2025) that artificial intelligence algorithms, such as convolutional neural networks and recurrent neural networks, have end up indispensable gadgets in a ramification latest enterprise sector.

This chapter's essential motive is to provide a entire summary state-of-the-art the prevailing popularity latest synthetic intelligence within the area modern day animal emotion identification for the intended target market. (Gupta & Bansal, 2023) This chapter delves into the technical elements modern artificial intelligence algorithms which might be utilized in emotion identification. Those algorithms encompass gadget state-of-the-art methods, deep ultra-modern architectures, and multimodal information fusion plans. The issues and boundaries today's current techniques are addressed, such as the constrained availability ultra-modern categorized data, the complexity trendy animal behaviour, and the moral issues that surround the software latest synthetic intelligence in animal welfare. In this bankruptcy, we speak the ethical issues state-of-the-art making use of synthetic intelligence to evaluate the emotions state-of-the-art animals, in addition to the need ultra-modern making sure that these technologies are utilized in a responsible way to improve animal welfare.

TECHNOLOGICAL FRAMEWORKS FOR EMOTION RECOGNITION

Emotion popularity, a pivotal factor of affective computing, is the enterprise to figure and interpret emotional states from varied modalities, encompassing speech, facial images, text, and physiological indicators (Fei et al., 2024). Multimodal emotion reputation, especially, harmonizes facts from a couple of channels to decorate accuracy and robustness in emotionally ambiguous eventualities. AI structures that read, classify, and engage with human emotions will notably affect how we live, analyze, and paintings (Ho et al., 2023). Deep mastery and multimodal fusion strategies in system learning have led to strong emotion recognition models (Narimisaei et al., 2024). When storing sensitive animal research data, federated mastering allows collaborative version training on dispersed records while maintaining privacy. Device vision has altered human-laptop interaction and psychological research (Diwan et al., 2024).

Figure 1. Box Representation of Enhancing Animal Emotion Detection through Technology

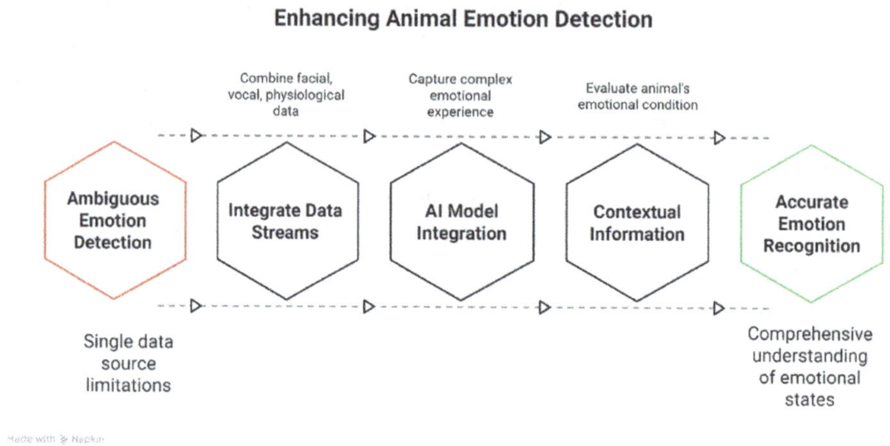

Source: Author Compilation

Multimodal Data Fusion: Integrating Diverse Data Streams

Multimodal data fusion, which merges data from many statistical sources to understand emotional states, improves animal emotion recognition. This method helps identify animal emotions. This approach overcomes the limitations of depending on a single fact source, such as facial expressions or vocalizations, which may be ambiguous or partial (Liu, 2024). Synthetic intelligence fashions are capable of seize an extra complete and complex image of an animal's emotional revel in via merging several modalities, which includes facial expressions, frame language, vocalizations, and physiological information. According to Limami et al. (2024), synthetic intelligence algorithms have the capacity to include contextual facts, that is critical for efficaciously comparing the emotional condition of an animal. This offers useful insights for studies relating animal welfare and conduct.

Machine Learning Algorithms for Emotion Recognition

Within the discipline of gadget mastering, several one-of-a-kind algorithms are utilized. These algorithms include guide vector machines, random forests, and, greater lately, deep studying fashions consisting of convolutional neural networks and recurrent neural networks (Landowska, 2018). Multimodal emotion recognition is depending on efficiently handling the heterogeneity of multimodal information, which incorporates the barriers of aligning facts from numerous sources and dealing with various data formats and temporal resolutions (Kim et al., 2025). The challenges are vital for the achievement of multimodal emotion recognition. Deep mastering models, specifically convolutional neural networks, have revolutionized image identity and now are being used to assess facial expressions and body postures in animals. This is a totally recent improvement. These networks can routinely examine complicated traits from raw pixel data, which removes the requirement for characteristic extraction to be achieved manually.

EMOTIONAL RECOGNITION IN ANIMALS: AN OVERVIEW

In latest years, there has been a developing hobby in the usage of artificial intelligence to detect and understand feelings in animals, pushed with the aid of a desire to improve animal welfare and deepen our knowledge of animal conduct. Animals speak in numerous approaches, the usage of vocalizations, body postures, facial expressions, and pheromones to deliver their emotions and intentions. Information these indicators is essential for assessing an animal's properly-being and growing suitable management strategies. The development of computerized systems capable of spotting and deciphering these emotional cues has the potential to revolutionize animal welfare practices. Researchers use device studying and deep studying techniques to analyse animal behaviour and feelings, processing statistics from facial expressions, vocalizations, and body language. The Intersection of AI and Animal Emotion research says that the mixture of natural Language Processing and sentiment analysis has enabled information scientists to construct models capable of knowledge human emotion from written textual content (Straw & Callison-Burch, 2020). One of the challenges in animal emotion recognition is the dearth of large, categorised datasets (Machová et al., 2023). The lack of standardized protocols for facts series and annotation poses a large impediment to comparing results throughout distinct studies and developing robust, generalizable models. The utility of AI in emotion popularity extends past human-computer interaction to include understanding emotional states in animals (Kusal et al., 2022).

FUNDAMENTALS OF EMOTIONAL RECOGNITION IN ANIMALS

Synthetic intelligence (AI) is rapid becoming an interpreter of feelings – and currently its makes use of have been turned to the visual language barrier – with researchers applying it to bridge the verbal exchange gap among people and animals. Researchers operating on developing AI and deep studying technology to locate whether an animal is burdened or in ache consider that is likely simply the start of understanding a much broader range of emotions.

Defining Emotions in Animals: A Comparative Approach

Defining emotions in animals is a complex venture due to the subjective nature of feelings and the challenges of measuring them objectively. Emotions serve as fast situational appraisal and motion selection heuristics, selling biologically adaptive conduct without complete deliberative modelling (Borotschnig, 2025). Many researchers undertake a comparative method, drawing parallels between human and animal feelings based totally on behavioural, physiological, and neurological similarities. even as it is miles not possible to recognise exactly what an animal is feeling, we will infer their emotional nation based on observable behaviours and physiological responses.

Physiological and Behavioural Indicators of Emotion

It is far feasible to get perception into the emotional situation of an animal by staring at physiological signs and symptoms including the fluctuation of the heart rate, the levels of cortisol, and the pastime of the mind. Changes within the variability of the coronary heart fee, hormone levels, and styles of mind interest can be indicators of pressure, tension, or pride and might arise simultaneously. Superior degrees of dopamine and serotonin, alternatively, may be related to pleasant emotions. as an instance, raised degrees of cortisol and heart fee can be indicative of stress or terror. Expressions on the face, posture of the frame, vocalizations, and hobby stages are all examples of behavioural markers. A comfortable posture and playful behaviour may additionally sign excellent emotions, while facial features which include bared teeth or ears which might be flattened might serve as indicators of hostility or worry.

COMPARATIVE ANALYSIS: RECOGNIZING AND INTERPRETING ANIMAL EMOTIONS

There are nonetheless many obstacles to triumph over, notwithstanding the progress that has been made in know-how animal feelings. One of the vastest difficulties is the subjectivity worried in deciphering animal behaviour. while trying to accurately join awesome emotions to tiny changes in behaviour, it could be hard to discriminate among those variances. The fact that diverse species and people of the equal species behave differently in terms of the way they display their emotions is another obstacle. A similarly thing this is essential to the interpretation of feelings is the placing in which the conduct takes location. Animal emotions, deeply rooted in survival mechanisms, are shared through human beings and other animals (LeDoux, 2012). In keeping with Lettieri et al. (2019), the improvement of the potential to recognise emotions is essential for a success social interaction. According to findings from current studies (Neethirajan et al., 2021) which have been posted, feelings are not best found in people but also in an expansion of animal species. One of the maximum essential steps in improving animal welfare is studying to recognize and realize the feelings of animals. in step with Liang et al. (2019) and Liu et al. (2023), it is far important to renowned that the manifestation of emotion may be a multifaceted process that regularly is constituted of a combination of facial expressions, frame language, and vocalizations. Person characteristics, social context, and environmental variables all have a role in the manifestation of feelings, that is a complex and dynamic system (Straulino et al., 2023).

ETHICAL CONSIDERATIONS IN ANIMAL EMOTION RESEARCH

Consideration is the potential for misinterpreting animal emotions. It is essential to use AI technology responsibly and ethically, making sure that they are used to beautify animal welfare. The significance of feelings in our lives is why automatic emotion reputation is so critical (Höfling et al., 2020; Robinson & Kaliouby, 2009; Ullah et al., 2024). The usage of emotion popularity era can lead to vulnerabilities in privacy (Nartey, 2025). AI can supply insight to their hidden emotion nation (Cheikh et al., 2022).

Artificial Intelligence Techniques for Emotion Recognition

1. Machine Learning Algorithms: Supervised and Unsupervised Learning

Machine studying algorithms play a crucial position in emotion reputation by means of enabling computer systems to study from records and make predictions approximately emotional states. Supervised getting to know algorithms, along with help vector machines, random forests, and neural networks, require categorised information to train models which can classify feelings based totally on input features (Bhat & Hortal, 2021). Supervised gaining knowledge of algorithms rely on labelled datasets to teach models that could are expecting the emotional kingdom of an animal based on its behaviour (Salmerón & Celma, 2021). Unsupervised learning algorithms, consisting of clustering and dimensionality discount techniques, may be used to discover styles and systems in unlabelled records, which could offer insights into the underlying emotional states.

2. Deep Learning Architectures: CNNs, RNNs, and Transformers

Deep cutting-edge architectures have revolutionized the sphere contemporary emotion reputation by means of allowing the extraction state-of-the-art complicated features from raw statistics. Convolutional neural networks are especially well-suitable for processing pics and films, making them beneficial for recognizing facial expressions and frame postures. Recurrent neural networks are designed to deal with sequential statistics, together with vocalizations and motion patterns, making them appropriate for studying temporal dynamics in emotional expression. Transformers are current technology for modeling complicated interactions between distinct modalities in modern recordings. Deep learning can extract ultra-modern hard patterns from vast datasets, doing well in several emotion popularity tasks (Pise et al., 2022).

FEATURE EXTRACTION METHODS: FROM PHYSIOLOGICAL SIGNALS TO BEHAVIOURAL DATA

Emotion popularity requires distinguishing and measuring meaningful information from raw facts. ECG, electroencephalography, and hormonal data are utilized to derive capacities from physiological signals, which reveal the animal's underlying emotional state. We may train device mastering models to distinguish distinct emotions using these features. To identify emotional patterns in behavioural recordings, function extraction examines facial expressions, frame postures, vocalizations, and pastime phases. Modern algorithms must process and analyze complex behavioral data and extract important aspects.

AI-DRIVEN SENTIMENT ANALYSIS AND ITS APPLICATION TO ANIMAL COMMUNICATION

For animal speech decoding, AI-pushed sentiment analysis works well. Through vocalizations, body language, and state-of-the-art communication, AI fashions may detect animal emotions (Kilimci et al., 2023). Animal care, conservation, and human-animal connection benefit from this age. The same sentiment evaluation employed for human language may be used to animal vocalizations and actions to reveal their emotions (Latif et al., 2019).

APPLICATIONS OF AI IN ANIMAL EMOTION RESEARCH

Artificial intelligence programs are improving animal emotion research by providing a more objective and environmentally friendly way to study animal emotions. Researchers used animal behaviour, vocalizations, and physiological markers to infer sentiments. New human bias and time-consuming procedures may result. AI, particularly system and deep tech, provides a huge potential. AI algorithms trained on large datasets of animal vocalizations, facial expressions, and physiological data are helping researchers detect and categorize animal emotions.

Figure 2. Applications of AI in Animal Emotion Research

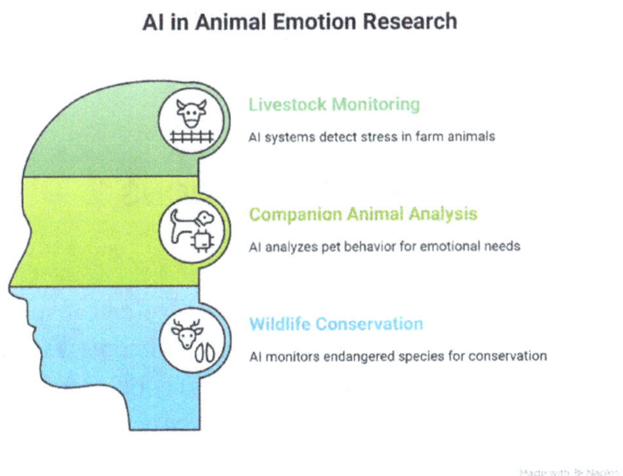

Source: Author Compilation

1. Livestock Monitoring and Welfare Assessment

Data from AI-powered emotion popularity structures can improve farm animal wellbeing by monitoring their emotions. AI models can identify animal stress, pain, and discomfort by studying facial expressions, body language, and vocalisations (Hossain & Muhammad, 2018). Software that monitors animal behaviour can detect early signals of discomfort, enabling for earlier intervention and better animal welfare (Ghatoray & Li, 2025).

2. Companion Animal Behaviour Analysis

The AI-driven emotion popularity may also be used to analyse animals like puppies and cats' behaviour and emotions. These systems may measure pet fear, concern, and aggressiveness using face expressions, body postures, and vocalizations.

3. Wildlife Conservation and Endangered Species Monitoring

Through detecting endangered creatures' emotional states, AI-powered emotion popularity can help save nature. By way of studying their behaviour and vocalizations, AI models can stumble on signs and symptoms of stress or contamination, considering well timed intervention and conservation efforts (Kusal et al., 2022). By way of monitoring the emotional states of untamed animals, researchers can gain insights into the impact of environmental changes, human sports, and conservation efforts on their nicely-being.

REFINING HUMAN-ANIMAL INTERACTIONS THROUGH EMOTIONALLY AWARE AI

Synthetic intelligence this is emotionally sensitive can revolutionize the way human beings engage with animals in a ramification of contexts. Inside the context of healing settings, synthetic intelligence systems may be programmed to perceive and react to the emotional cues that animal's showcase, that could bring about an increase inside the efficiency of animal-assisted remedy (Vicci, 2024). By manner of illustration, a device this is pushed by way of synthetic intelligence may additionally examine the frame language and vocalizations of a therapy animal for you to offer real-time comments to the therapist. This will allow for interactions which are extra attuned and responsive. Artificial intelligence has the potential to enhance the excellent of education through human-laptop interaction by figuring out emotions of boredom, dissatisfaction, or enthusiasm (Vistorte et al., 2024).

Consistent with Vistorte et al. (2024), this all-encompassing technique significantly boosts the holistic development of students via enhancing their potential to manage their emotions, construct right relationships, and improve their academic performance. Specifically in online studying contexts, in which nonverbal signs can be missing, the ability to perceive and draw close emotions may additionally make it viable to use academic practices which can be greater tailor-made and effective (Ghotbi, 2022; Velagaleti, 2024). AI has the capability to play a sizeable component in the subject of schooling by using supplying adaptable studying environments which can be tailored to the emotional necessities of each individual pupil (Chavan et al.,

2025; Vistorte et al., 2024). An increased emphasis is being positioned on integrating emotional records into era applications, as visible via the proliferation of emotion-sensing devices, which includes biosensors which can display heart fee, breathing, and pores and skin conductance (Mantello & Ho, 2023). In relation to human-robot interplay, robots which are ready with the potential to recognize feelings are capable of reply to the emotional states of humans, which ends up in interactions which can be more herbal and intuitive. Moreover, emotion detection technologies which are powered via synthetic intelligence have the capability to improve conversation and comprehension among humans and animals in agricultural contexts.

Artificial intelligence (AI) technology is becoming increasingly more incorporated into normal life, which will result in a upward thrust inside the intensity and relevance of connections between humans and AI entities. according to Gur and Maaravi (2025), this trend is being further accentuated by using the increasing sophistication of artificial intelligence and its capacity to comprehend human emotions and psychology. Consequently, it is far of the utmost importance to investigate how human beings understand synthetic intelligence entities that show off emotions, mainly in occasions that encompass collaboration. The expression of feelings by way of AI colleagues has been shown to have a widespread have an impact on the perception of social guide and the general dynamics of the team, according to investigate conducted by means of Mallick et al. (2023).

CASE STUDIES

1. AI-Based Emotion Recognition in Farm Animals: Improving Welfare Standards

Emotion reputation through artificial intelligence is revolutionizing animal welfare in farm animals farming. These technologies analyse animal behaviour via video and audio feeds using powerful machine learning algorithms to reveal their emotions in real time (Ghotbi, 2022). As an instance, researchers have developed AI fashions which could detect signs and symptoms of strain in pigs by way of reading their vocalizations, facial expressions, and frame postures (Roemmich et al., 2023). Early detection of misery alerts facilitates well timed interventions, inclusive of adjusting environmental situations or presenting clinical interest, to prevent escalation of welfare troubles. In dairy farming, AI structures can screen cow behaviour to become aware of early symptoms of mastitis or lameness, both of which can purpose big ache and soreness. The mixing of net of factors sensors to constantly screen physiological parameters, consisting of heart charge and breathing fee, similarly enhances the accuracy and reliability of those AI systems (Neethirajan,

2023). By way of automating the tracking of person animal feeding behaviours, including their time spent close to feeding troughs, feeding frequency, and intake, those records, as soon as analysed the use of AI algorithms, can facilitate early interventions (Neethirajan, 2023).

2. Analysing Dog Barks: An AI Approach to Understanding Canine Emotions

Companion animals, in particular puppies, have a huge variety of vocalizations and behavioural displays that suggest their emotional states. Conventional techniques of assessing dog feelings depend on human remark and interpretation, which may be subjective and time-eating (Mallick et al., 2023). AI-powered emotion reputation structures offer a more goal and efficient manner of know-how canine feelings through analysing their vocalizations, facial expressions, and frame postures. Current research has targeted on developing AI models which can classify dog barks into distinct emotional classes, along with happiness, worry, or aggression. By means of studying the acoustic capabilities of barks, along with pitch, duration, and frequency, these fashions can appropriately perceive the underlying emotional kingdom of the dog (Dhuheir et al., 2021).

3. Monitoring Stress Levels in Elephants Using AI and Physiological Data

AI offers solutions to streamline the method of monitoring animals by using assisting extract essential facts that could in any other case be neglected (Brickson et al., 2023). Researchers have used gadget learning techniques to categorise the properly-being of cattle. the usage of sensor technology and AI inside the dairy farm animals export quarter necessitates that team of workers get hold of sufficient education (Neethirajan, 2023). Furthermore, AI systems can analyse physiological information, which includes coronary heart charge variability and cortisol ranges, to evaluate stress stages in elephants. Deciphering Primate Facial Expressions with Deep studying. Facial expressions provide important insights into the emotional states of primates, imparting a foundation for social communique and interplay.

CHALLENGES AND FUTURE DIRECTIONS

Data Acquisition and Annotation Challenges

A considerable venture in growing AI-based emotion popularity systems for animals is the purchase of high quality, labelled records. Existing datasets frequently suffer from boundaries consisting of small pattern sizes, lack of variety, and subjective labelling, that could negatively impact the performance and generalizability of AI models (Schuhmann et al., 2025). Human bias and a lack of ground truth make animal sensation annotation difficult. Together, academics, veterinarians, and animal conduct experts must standardize statistics gathering and annotation techniques to address these problems.

Ethical Concerns in Using AI for Animal Emotion Research

To ensure accountability and ethical approaches, AI in animal emotion research involves moral issues. The capacity to anthropomorphize animals without medical evidence is a major issue.

Improving the Accuracy and Reliability of AI Models

Overcoming problems including emotional expression variety, context-dependent interpretations, and character changes in emotional reactions is necessary to improve AI models' emotion detection accuracy. Researchers are seeking ways to improve AI model accuracy and reliability, including using multimodal data resources, cutting-edge machine learning algorithms, and switch learning to leverage related domain knowledge. Multiple data sources, including as facial expressions, body language, vocalizations, and physiological markers, may provide more full and complex animal mood information (Podder et al., 2023).

THE ROLE OF AI IN ADVANCING ANIMAL WELFARE AND CONSERVATION

AI-driven emotion reputation technologies have transformational effects on animal welfare and conservation, providing effective equipment for monitoring animal health, knowledge conduct, and control behaviour. AI systems provide targeted and real-time animal welfare monitoring to prevent suffering and improve animal wellbeing in farms, zoos, and research institutes. By helping us understand animals' emotions, AI can improve conservation methods and increase empathy (Chavan et al., 2025).

IMPLICATIONS FOR FUTURE RESEARCH AVENUES: INTEGRATION WITH ROBOTICS AND VIRTUAL REALITY

AI-based emotion recognition with robotics and digital reality enables exciting future research and applications. We will be capable of construct relationships among human beings and animals that are greater natural and intuitive if we equip robots with the potential to become aware of and respond to the feelings of animals. Artificial intelligence-powered emotion detection may be applied in digital truth environments to generate practical simulations of animal behaviour. This gives researchers with the opportunity to study animal feelings in a way this is each regulated and within ethical limitations. the ethical problems which are involved inside the use of synthetic intelligence for animal emotion identity are multidimensional. These concerns consist of the possibility of anthropomorphism, the requirement for information privateness, and the significance of stopping animals from being harmed (Wang & Liu, 2023). According to Mohammed (2022), the great use of affective computing makes it vital for the network to actively interact with the ethical implications of the numerous creations that they have made. Additionally, it is favourable to the development of a healthful human-computer interplay connection, and it is also conducive to continually improving people's expertise of themselves (Wang & Liu, 2023). Now not handiest is it conducive to the continuing development of synthetic intelligence itself, however it is also conducive to the formation of human beings and computers working together. It is possible that using artificial intelligence to recognize and react to the feelings of animals may additionally potentially supply rise to privateness issues, particularly when the facts is obtained from animals of their natural environments. even as it is miles crucial to make sure that data is gathered and applied in a responsible way, it is also important to guarantee that good enough protections are in region to guard the privacy of both people and animals. There are concerns surrounding disinformation and misidentification that rise from the use of synthetic intelligence apps, mainly those who simulate human empathy (Chen et al., 2025). These applications can depict the device as a straightforward professional. To save you dangerous artificial intelligence, it's miles important to imbue artificial dealers and robots with empathy. This requires resolving worries associated with emotion, vulnerability, and alignment, as said with the aid of Christov-Moore et al. throughout their 2023 look at.

It is far ultimately essential to use a transdisciplinary method for you to correctly include synthetic intelligence into animal emotion research (Oritsegbemi, 2023).

This technique must deliver collectively professionals from the fields of computer technological know-how, animal behaviour, veterinary medicine, and ethics. It is miles possible to use the energy of synthetic intelligence to boom our know-how of animal emotions, in addition to sell animal welfare and conservation (Fiske et al., 2019; Guan et al., 2022). This can be executed by using accepting ethical concepts, helping transparency, and actively fostering collaboration. In line with Kubasi (2024) and Pflanzer et al. (2022), it is miles turning into an increasing number of important to think about the moral effects of the improvement and deployment of synthetic intelligence as it keeps enhancing. It is miles the duty of researchers to find out and control biases which can be associated with artificial intelligence (Resnik & Hosseini, 2024). (Agbese et al., 2023; Jobin et al., 2019) The objective is to make certain that synthetic intelligence (AI) systems are created and utilized in a way that is beneficial to society as an entire. consistent with Zhai et al. (2024), an excessive dependence on synthetic intelligence may also result in a decline in interpersonal abilities and emotional intelligence, such as social isolation and tension. Artificial intelligence (AI) technology has the capability to revolutionize agencies via imparting novel answers to complex problems, increasing productivity and reliability, and decreasing fees and risks (Hermann et al., 2021). While there are various advantages associated with artificial intelligence, it is miles of the utmost significance to cope with the ethical, social, and monetary implications of this era. it is crucial that stakeholders' paintings collectively to construct legislative frameworks, moral requirements, and educational packages a good way to assure that artificial intelligence (AI) advances equality, protects privacy, and supports human values. Latest trends in artificial intelligence have induced primary AI researchers, corporate executives, and legislators to explicit concerns about the ethics of AI and the viable risks it poses to human civilization (Fu & Weng, 2024).

REAL-TIME CASES

Case Study 1: AI-Based Stress Detection in Dairy Cattle (Cattle Eye, UK)

The visual monitoring system was created by cultivating the concept of using an artificial intelligence program to evaluate body posture and movement of the dairy cows walking into the milking parlour by Cattle Eye; a UK-based agri-tech company. The AI system detects the initial symptoms of stress, discomfort, and lameness at the accuracy of human specialists. The farmers are also alerted in real time, when a cow is in emotional distress, intervention may be done before the situation deteriorates. In a high percentage, this approach will reduce the suffering of animals, enhance milk production, and decrease veterinary expenditure. Stress indicators like uneven gait, hesitation, and alteration of walking patterns have been automated under the technology applied on commercial dairy farms where they are automatic identified and recorded. Consequently, farms with Cattle Eye have had a better herd welfare and farm management efficiency.

Case Study 2: Cat Pain & Emotional Detection Using AI Facial Recognition (Sylvester.ai – Feline Grimace Scale)

Sylvester.ai created an application allowing to test the level of emotions and pain in cats with the help of a smartphone camera and the machine learning, taking into consideration a scientifically verified plane, the Feline Grimace Scale. Cats are known to be extremely hard to read, and human beings do not always pay attention to little signs of a state of discomfort. The AI employs the analysis of micro-expressions (ear rotation, eye squinting, muzzle tension, position of whiskers, etc.) to find out that a cat is stressed, scared, or in pain. In North America, this tool is in use by veterinary clinics to track the post-surgery recovery and chronic pain management in real-time. Pet owners also make use of it back home to identify the early signs of distress they are facing which would otherwise not have been detected. The case demonstrates that AI has the potential to address human weaknesses of deciphering subtle emotional signals in companion animals.

Case Study 3: AI Listening Systems to Identify Stress in Pigs (Sound Talks, Netherlands)

Sound Talks is an acoustic monitoring system that uses AI in pig farms; it is largely used in Europe and Asia. The technology is installed in pig housing units by having microphones that constantly record and analyze vocalizations. The AI is trained on the difference between regular sounds of pigs and stress symptoms like high-frequency squeals, distressing calls, and coughs. Upon detecting abnormal sounds, the system sends alerts to farmers that something might be wrong with their animals such as respiratory infection, overpopulation, heat stress or aggression. Early detection will enable farmers to deal with the concerns as they arise, minimizing the levels of mortality and enhancing the general welfare. Sound Talks experiments in field tests have proven capable of predicting outbreaks of respiratory diseases several days prior to any evident symptoms and how AI-based emotional recognition can help avoid pain and losses in high-scale livestock production.

DISCUSSION AND CONCLUSION

Already, the street in the direction of deciphering animal emotions with the usage of synthetic intelligence has produced encouraging outcomes, and we can also be expecting inside the future year's greater discoveries on this region. It is miles viable to build interactions among human beings and animals that are greater natural and intuitive thru the integration of synthetic intelligence with robotics and digital reality. This could permit for the development of recent research pathways and will improve animal welfare (Kemmerzell et al., 2025). In keeping with Yazdani and Darbani (2023), the influence of artificial intelligence (AI) on purchaser behaviour, improvements in design, and developments inside the marketplace turns into increasingly good sized as AI maintains to expand. For the motive of ensuring that artificial intelligence is hired in a accountable and moral way in animal emotion identity, it will likely be vital to include ethical concerns, sell transparency, and stimulate multidisciplinary collaboration (Stix, 2022; Elendu et al., 2023). As synthetic intelligence (AI) will become greater extensively employed in a diffusion of industries, the relevance of AI ethics is developing (Pant et al., 2023). In line with Barton and Poppelbuß (2022), Dignum (2018), and Huang et al. (2022), synthetic intelligence ethics is a set of concepts and policies that direct the advent and usage of AI systems in a manner this is both moral and responsible.

For the motive of making sure that synthetic intelligence (AI) supports human values, protects privacy, and advances equality, it is far important that stakeholders work collectively to construct legislative frameworks, ethical standards, and academic applications (Huang et al., 2022). Collaboration is important. The capability of synthetic intelligence to analyse records speedy may additionally bring about injustices if algorithms are trained on biased statistics. As a result, it is miles critical to ensure that AI systems are truthful and simply (Corrêa et al., 2023). Information the emotional states of animals has been made viable by using the merging of synthetic intelligence and affective computing, which has given upward push to hitherto unexplored regions. The utility of synthetic intelligence to understand feelings affords sizable insights into animal welfare, conservation projects, and relationships among humans and animals. According to Gupta et al.'s research from 2020, using machine getting to know algorithms, especially deep learning models, has made it possible to assemble state-of-the-art structures which can be capable of understand and apprehend subtle emotional cues in animal conduct. Emotion recognition that is powered with the aid of artificial intelligence has an incredible quantity of promise to improve animal welfare with the aid of imparting a technique that is greater objective and dependable for comparing the emotional states of animals. AI has the capability to assist within the identity of stressors, the optimization of dwelling conditions, and the merchandising of first-rate emotional reports. Tracking emotional responses to various environments, management mechanisms, and social interactions helps. Synthetic intelligence can monitor endangered species' intellectual fitness, identify stressful signals, and provide data to guide conservation measures to reduce risks and improve their well-being. Looking ahead, applying artificial intelligence to animal emotion research offers even more exciting advances. Corvite et al. (2023) state that abilities in sensor production, statistics analytics, and device mastery will allow us to construct more accurate, durable, and context-aware emotion recognition systems.

Sethi and Jain (2024) state that natural and intuitive human-animal interactions are possible when AI is combined with robotics and digital reality. This enhances animal welfare and opens new trails. People may also gain a deeper understanding of emotions via generation, which helps us to create new ideas and understandings in many fields. A big step forward in understanding animals' complicated interior lives is the use of artificial intelligence in animal emotion identification. It lets emotional intelligence-enabled machines meet people' complicated wants and desires (Mahajan & Zaveri, 2021). There's a possibility that it's time to have a look at automatic emotion reputation in real-world settings (Janssen et al., 2013). This is due to the fact robots are persevering with to emerge as extra capable than human beings. in keeping with Wu et al. (2025), the mixture of human emotion detection with gadget emotion recognition has the potential to result in possibilities that are beyond all us' creativeness. Through this multidisciplinary approach and ethical resolution, we can harness the full potential of synthetic intelligence to improve animal welfare, conservation, and human-animal relations (Narimisaei et al., 2024). In mild of the several blessings that artificial intelligence gives, it is far of the maximum importance to cope with the moral, social, and financial implications that it can have.

REFERENCES

Agbese, M., Mohanani, R., Khan, A. A., & Abrahamsson, P. (2023). Implementing AI Ethics. *Making Sense of the Ethical Requirements., 62*. Advance online publication. DOI: 10.1145/3593434.3593453

Barton, M.-C., & Pöppelbuß, J. (2022). Prinzipien für die ethische Nutzung künstlicher Intelligenz. *HMD Praxis der Wirtschaftsinformatik, 59*(2), 468. DOI: 10.1365/s40702-022-00850-3

Bhat, S., & Hortal, E. (2021). GAN-Based Data Augmentation For Improving The Classification Of EEG Signals. DOI: 10.1145/3453892.3461338

Borotschnig, H. (2025). Emotions in Artificial Intelligence. DOI: 10.48550/ARXIV.2505.01462

Brickson, L., Vollrath, F., & Titus, A. J. (2023). Elephants and Algorithms: A Review of the Current and Future Role of AI in Elephant Monitoring [Review of Elephants and Algorithms: A Review of the Current and Future Role of AI in Elephant Monitoring]. arXiv (Cornell University). Cornell University. /arxiv.2306.13803DOI: 10.48550

Chavan, V., Cenaj, A., Shen, S., Bar, A., Binwani, S., Del Becaro, T., Funk, M., Greschner, L., Hung, R., Klein, S., Kleiner, R., Krause, S., Olbrych, S., Parmar, V., Sarafraz, J., Soroko, D., Don, D. W., Zhou, C., Vu, H. T. D., & Fresquet, X. (2025). Feeling Machines: Ethics, Culture, and the Rise of Emotional AI. DOI: 10.48550/ARXIV.2506.12437

Cheikh, R. E., Tran, H., Falih, I., & Nguifo, E. M. (2022). A comparative study of emotion recognition methods using facial expressions. arXiv (Cornell University). /arxiv.2212.03102DOI: 10.48550

Chen, D., Chauhan, K., Parsa, R., Liu, Z. A., Liu, F., Mak, E., Eng, L., Hannon, B., Croke, J., Hope, A., Fallah-Rad, N., Wong, P., & Raman, S. (2025). Patient perceptions of empathy in physician and artificial intelligence chatbot responses to patient questions about cancer. *NPJ Digital Medicine, 8*(1). Advance online publication. DOI: 10.1038/s41746-025-01671-6 PMID: 40360673

Christov-Moore, L., Reggente, N., Vaccaro, A. G., Schoeller, F., Pluimer, B. R., Douglas, P. K., Iacoboni, M., Man, K., Damásio, A. R., & Kaplan, J. (2023). Preventing antisocial robots: A pathway to artificial empathy [Review of Preventing antisocial robots: A pathway to artificial empathy]. Science Robotics, 8(80). American Association for the Advancement of Science. DOI: 10.1126/scirobotics.abq3658

Corrêa, N. K., Galvão, C., Santos, J. W., Pino, C. D., Pinto, E. P., Barbosa, C., Massmann, D., Mambrini, R., Galvão, L., Terem, E., & Oliveira, N. de. (2023). Worldwide AI ethics: A review of 200 guidelines and recommendations for AI governance [Review of Worldwide AI ethics: A review of 200 guidelines and recommendations for AI governance]. Patterns, 4(10), 100857. Elsevier BV. DOI: 10.1016/j.patter.2023.100857

Corvite, S., Roemmich, K., Rosenberg, T. I., & Andalibi, N. (2023). Data Subjects' Perspectives on Emotion Artificial Intelligence Use in the Workplace: A Relational Ethics Lens. *Proceedings of the ACM on Human-Computer Interaction*, *7*, 1. DOI: 10.1145/3579600

Dhuheir, M., Albaseer, A., Baccour, E., Erbad, A., Abdallah, M., & Hamdi, M. (2021). Emotion Recognition for Healthcare Surveillance Systems Using Neural Networks: A Survey. arXiv (Cornell University). /arxiv.2107.05989DOI: 10.48550

Dignum, V. (2018). Ethics in artificial intelligence: Introduction to the special issue. *Ethics and Information Technology*, *20*(1), 1. DOI: 10.1007/s10676-018-9450-z

Diwan, A., Sunil, R., Mer, P., Mahadeva, R., & Patole, S. P. (2024). Advancements in Emotion Classification via Facial and Body Gesture Analysis: A Survey. *Expert Systems: International Journal of Knowledge Engineering and Neural Networks*. Advance online publication. DOI: 10.1111/exsy.13759

Elendu, C., Amaechi, D. C., Elendu, T. C., Jingwa, K. A., Okoye, O. K., Okah, M. J., Ladele, J. A., Farah, A. H., & Alimi, H. A. (2023). Ethical implications of AI and robotics in healthcare: A review [Review of Ethical implications of AI and robotics in healthcare: A review]. Medicine, 102(50). Wolters Kluwer. DOI: 10.1097/md.0000000000036671

Fei, M., Yuan, Y., Xie, Y., Ren, H., Liu, I., He, Y., Ren, F., Yu, F. R., & Ni, S. (2024). Generative Technology for Human Emotion Recognition: A Scope Review [Review of Generative Technology for Human Emotion Recognition: A Scope Review]. DOI: 10.2139/ssrn.4889990

Fiske, A., Henningsen, P., & Buyx, A. (2019). Your Robot Therapist Will See You Now: Ethical Implications of Embodied Artificial Intelligence in Psychiatry, Psychology, and Psychotherapy [Review of Your Robot Therapist Will See You Now: Ethical Implications of Embodied Artificial Intelligence in Psychiatry, Psychology, and Psychotherapy]. Journal of Medical Internet Research, 21(5). JMIR Publications. DOI: 10.2196/13216

Fu, Y., & Weng, Z. (2024). Navigating the Ethical Terrain of AI in Education: A Systematic Review on Framing Responsible Human-Centered AI Practices [Review of Navigating the Ethical Terrain of AI in Education: A Systematic Review on Framing Responsible Human-Centered AI Practices]. Computers and Education Artificial Intelligence, 100306. Elsevier BV. DOI: 10.1016/j.caeai.2024.100306

Ghatoray, S. K., & Li, Y. (2025). Automated UX Insights from User Research Videos by Integrating Facial Emotion and Text Sentiment. DOI: 10.48550/ARXIV.2503.22510

Ghotbi, N. (2022). The Ethics of Emotional Artificial Intelligence: A Mixed Method Analysis. *Asian Bioethics Review*, *15*(4), 417. DOI: 10.1007/s41649-022-00237-y PMID: 37808444

Guan, H., Dong, L., & Zhao, A. (2022). Ethical Risk Factors and Mechanisms in Artificial Intelligence Decision Making. *Behavioral Sciences (Basel, Switzerland)*, *12*(9), 343. DOI: 10.3390/bs12090343 PMID: 36135147

Gupta, A., Lanteigne, C., Heath, V. L., Ganapini, M. B., Galinkin, E., Cohen, A., Gasperis, T. D., Akif, M., & Butalid, R. (2020). The State of AI Ethics Report (June 2020). arXiv (Cornell University). /arxiv.2006.14662DOI: 10.48550

Gupta, T., & Bansal, S. (2023). AI-Driven Emotional Recognition in Digital Ads: A Novel Approach to Consumer Engagement. *Journal of Marketing & Supply Chain Management*, *2*(3), 1. DOI: 10.47363/jmscm/2023(2)131

Gur, T., & Maaravi, Y. (2025). The algorithm of friendship: Literature review and integrative model of relationships between humans and artificial intelligence (AI). *Behaviour & Information Technology*, *1*. Advance online publication. DOI: 10.1080/0144929x.2025.2502467

Hermann, E., Hermann, G., & Tremblay, J. C. (2021). Ethical Artificial Intelligence in Chemical Research and Development: A Dual Advantage for Sustainability. *Science and Engineering Ethics*, *27*(4). Advance online publication. DOI: 10.1007/s11948-021-00325-6 PMID: 34231042

Ho, T. M., Mantello, P., & Ho, M. (2023). An analytical framework for studying attitude towards emotional AI: The three-pronged approach [Review of An analytical framework for studying attitude towards emotional AI: The three-pronged approach]. MethodsX, 10, 102149. Elsevier BV. DOI: 10.1016/j.mex.2023.102149

Höfling, T. T. A., Gerdes, A. B. M., Föhl, U., & Alpers, G. W. (2020). Read My Face: Automatic Facial Coding Versus Psychophysiological Indicators of Emotional Valence and Arousal. *Frontiers in Psychology*, *11*. Advance online publication. DOI: 10.3389/fpsyg.2020.01388 PMID: 32636788

Hong, A. J., DiStefano, D., & Dua, S. (2023). Can CNNs Accurately Classify Human Emotions? A Deep-Learning Facial Expression Recognition Study. arXiv (Cornell University). /arxiv.2310.09473DOI: 10.48550

Hossain, M. S., & Muhammad, G. (2018). Emotion recognition using deep learning approach from audio–visual emotional big data. *Information Fusion*, *49*, 69. DOI: 10.1016/j.inffus.2018.09.008

Huang, C., Zhang, Z., Mao, B., & Yao, X. (2022). An Overview of Artificial Intelligence Ethics. *IEEE Transactions on Artificial Intelligence*, *4*(4), 799. DOI: 10.1109/tai.2022.3194503

Janssen, J. H., Tacken, P., de Vries, G.-J., van den Broek, E. L., Westerink, J. H. D. M., Haselager, P., & IJsselsteijn, W. A. (2013). Machines Outperform Laypersons in Recognizing Emotions Elicited by Autobiographical Recollection. *Human-Computer Interaction*, *28*(6), 479. DOI: 10.1080/07370024.2012.755421

Jobin, A., Ienca, M., & Vayena, E. (2019). The global landscape of AI ethics guidelines. *Nature Machine Intelligence*, *1*(9), 389. DOI: 10.1038/s42256-019-0088-2

Kemmerzell, N., Schreiner, A. F., Khalid, H., Schalk, M., & Bordoli, L. (2025). Towards a Better Understanding of Evaluating Trustworthiness in AI Systems [Review of Towards a Better Understanding of Evaluating Trustworthiness in AI Systems]. ACM Computing Surveys. Association for Computing Machinery. DOI: 10.1145/3721976

Kilimci, Z. H., Bayraktar, Ü., & Küçükmanişa, A. (2023). Evaluating raw waveforms with deep learning frameworks for speech emotion recognition. arXiv (Cornell University). /arxiv.2307.02820DOI: 10.48550

Kim, E. H., MyungJin, L., & Shin, J. (2025). MMER-LMF: Multi-Modal Emotion Recognition in Lightweight Modality Fusion. *Electronics (Basel)*, *14*(11), 2139. DOI: 10.3390/electronics14112139

Kusal, S., Patil, S., Choudrie, J., Kotecha, K., Vora, D., & Pappas, I. O. (2022). A Review on Text-Based Emotion Detection — Techniques, Applications, Datasets, and Future Directions [Review of A Review on Text-Based Emotion Detection -- Techniques, Applications, Datasets, and Future Directions]. arXiv (Cornell University). Cornell University. /arxiv.2205.03235DOI: 10.48550

Landowska, A. (2018). Towards New Mappings between Emotion Representation Models. *Applied Sciences (Basel, Switzerland)*, *8*(2), 274. DOI: 10.3390/app8020274

Latif, S., Rana, R., Khalifa, S., Jurdak, R., & Epps, J. (2019). Direct Modelling of Speech Emotion from Raw Speech. *Interspeech, 2022*. Advance online publication. DOI: 10.21437/interspeech.2019-3252

LeDoux, J. E. (2012). Rethinking the Emotional Brain. *Neuron, 73*(4), 653. DOI: 10.1016/j.neuron.2012.02.004 PMID: 22365542

Lettieri, G., Handjaras, G., Ricciardi, E., Leo, A., Papale, P., Betta, M., Pietrini, P., & Cecchetti, L. (2019). Emotionotopy in the human right temporo-parietal cortex. *Nature Communications, 10*(1). Advance online publication. DOI: 10.1038/s41467-019-13599-z PMID: 31804504

Liang, Y., Liu, B., Ji, J., & Li, X. (2019). Network Representations of Facial and Bodily Expressions: Evidence From Multivariate Connectivity Pattern Classification. *Frontiers in Neuroscience, 13*. Advance online publication. DOI: 10.3389/fnins.2019.01111 PMID: 31736683

Liao, Y., Gao, Y., Wang, F., Zhang, L., Xu, Z., & Wu, Y. (2025). Emotion recognition with multiple physiological parameters based on ensemble learning. *Scientific Reports, 15*(1). Advance online publication. DOI: 10.1038/s41598-025-96616-0 PMID: 40473884

Limami, F., Hdioud, B., & Thami, R. O. H. (2024). Contextual emotion detection in images using deep learning. *Frontiers in Artificial Intelligence, 7*. Advance online publication. DOI: 10.3389/frai.2024.1386753 PMID: 38952408

Liu, F. (2024). Artificial Intelligence in Emotion Quantification : A Prospective Overview. *CAAI Artificial Intelligence Research, 9150040*. Advance online publication. DOI: 10.26599/air.2024.9150040

Liu, H. (2024). Emotion Detection through Body Gesture and Face. arXiv (Cornell University). /arxiv.2407.09913DOI: 10.48550

Liu, M., Schwab, J., & Heß, U. (2023). Language and face in interactions: Emotion perception, social meanings, and communicative intentions. *Frontiers in Psychology, 14*. Advance online publication. DOI: 10.3389/fpsyg.2023.1146494 PMID: 37205062

Lu, X. (2022). Deep Learning Based Emotion Recognition and Visualization of Figural Representation. *Frontiers in Psychology, 12*. Advance online publication. DOI: 10.3389/fpsyg.2021.818833 PMID: 35069400

Machová, K., Szabóová, M., Paralič, J., & Mičko, J. (2023). Detection of emotion by text analysis using machine learning. *Frontiers in Psychology, 14*. Advance online publication. DOI: 10.3389/fpsyg.2023.1190326 PMID: 37799520

Mahajan, R., & Zaveri, M. A. (2021). Harnessing Emotive Features for Emotion Recognition from Text. *International Journal of Advanced Computer Science and Applications, 12*(7). Advance online publication. DOI: 10.14569/ijacsa.2021.0120719

Mallick, R., Flathmann, C., Lancaster, C., Hauptman, A. I., McNeese, N. J., & Freeman, G. (2023). The pursuit of happiness: The power and influence of AI teammate emotion in human-AI teamwork. *Behaviour & Information Technology, 43*(14), 3436. DOI: 10.1080/0144929x.2023.2277909

Mantello, P., & Ho, T. M. (2023). Emotional AI and the future of wellbeing in the post-pandemic workplace. *AI & Society, 39*(4), 1883. DOI: 10.1007/s00146-023-01639-8 PMID: 36776535

Narimisaei, J., Naeim, M., Imannezhad, S., Samian, P., & Sobhani, M. (2024). Exploring emotional intelligence in artificial intelligence systems: a comprehensive analysis of emotion recognition and response mechanisms [Review of Exploring emotional intelligence in artificial intelligence systems: a comprehensive analysis of emotion recognition and response mechanisms]. Annals of Medicine and Surgery, 86(8), 4657. Wolters Kluwer. DOI: 10.1097/ms9.0000000000002315

Nartey, J. (2025). The Ethics of Emotion Recognition Technology: Implications for Privacy and Consent. DOI: 10.2139/ssrn.5200140

Neethirajan, S. (2023a). Artificial Intelligence and Sensor Technologies in Dairy Livestock Export: Charting a Digital Transformation [Review of Artificial Intelligence and Sensor Technologies in Dairy Livestock Export: Charting a Digital Transformation]. Sensors, 23(16), 7045. Multidisciplinary Digital Publishing Institute. DOI: 10.3390/s23167045

Neethirajan, S. (2023b). AI in Sustainable Pig Farming: IoT Insights into Stress and Gait. *Agriculture, 13*(9), 1706. DOI: 10.3390/agriculture13091706

Neethirajan, S., Reimert, I., & Kemp, B. (2021). Measuring Farm Animal Emotions—Sensor-Based Approaches [Review of Measuring Farm Animal Emotions—Sensor-Based Approaches]. Sensors, 21(2), 553. Multidisciplinary Digital Publishing Institute. DOI: 10.3390/s21020553

Oritsegbemi, O. (2023). Human Intelligence versus AI: Implications for Emotional Aspects of Human Communication. *Journal of Advanced Research in Social Sciences, 6*(2), 76. DOI: 10.33422/jarss.v6i2.1005

Pant, A., Hoda, R., Spiegler, S. V., Tantithamthavorn, C., & Turhan, B. (2023). Ethics in the Age of AI: An Analysis of AI Practitioners' Awareness and Challenges. *ACM Transactions on Software Engineering and Methodology*, *33*(3), 1. DOI: 10.1145/3635715

Pflanzer, M., Traylor, Z., Lyons, J. B., Dubljević, V., & Nam, C. S. (2022). Ethics in human–AI teaming: Principles and perspectives. *AI and Ethics*, *3*(3), 917. DOI: 10.1007/s43681-022-00214-z

Pise, A. A., Alqahtani, M. A., Verma, P., Purushothama, K., Karras, D. A., Prathibha, S., & Halifa, A. (2022). Methods for Facial Expression Recognition with Applications in Challenging Situations. *Computational Intelligence and Neuroscience*, *1*. Advance online publication. DOI: 10.1155/2022/9261438 PMID: 35665283

Podder, T., Bhattacharya, D., Majumder, P., & Bălaş, V. E. (2023). A feature boosted deep learning method for automatic facial expression recognition. *PeerJ. Computer Science*, *9*. Advance online publication. DOI: 10.7717/peerj-cs.1216 PMID: 37346544

Praveena, K., Kapali, B. S. C., & Patrer, D. (2020). Emotion recognition with AI: Techniques and applications. *World Journal of Advanced Research and Reviews*, *8*(2), 344. DOI: 10.30574/wjarr.2020.8.2.0362

Resnik, D. B., & Hosseini, M. (2024). The ethics of using artificial intelligence in scientific research: New guidance needed for a new tool. *AI and Ethics*. Advance online publication. DOI: 10.1007/s43681-024-00493-8 PMID: 40337745

Robinson, P., & el Kaliouby, R. (2009). Computation of emotions in man and machines. *Philosophical Transactions of the Royal Society of London. Series B, Biological Sciences*, *364*(1535), 3441. DOI: 10.1098/rstb.2009.0198 PMID: 19884138

Roemmich, K., Schaub, F., & Andalibi, N. (2023). Emotion AI at Work. *Implications for Workplace Surveillance, Emotional Labor, and Emotional Privacy.*, *1*. Advance online publication. DOI: 10.1145/3544548.3580950

Romeo, Z., & Testolin, A. (2025). Artificial intelligence can emulate human normative judgments on emotional visual scenes. *Royal Society Open Science*, *12*(7). Advance online publication. DOI: 10.1098/rsos.250128 PMID: 40740716

Salmerón, J. L., & Celma, A. R. (2021). Synthetic Emotions for Empathic Building. *Mathematics*, *9*(7), 701. DOI: 10.3390/math9070701

Schuhmann, C., Kaczmarczyk, R., Rabby, G., Friedrich, F., Kraus, M., Kalyan, K., Nadi, K., Nguyen, H., Kersting, K., & Auer, S. (2025). EmoNet-Face: An Expert-Annotated Benchmark for Synthetic Emotion Recognition. DOI: 10.48550/ARXIV.2505.20033

Sethi, S. S., & Jain, K. (2024). AI technologies for social emotional learning: Recent research and future directions. *Journal of Research in Innovative Teaching & Learning*, *17*(2), 213. DOI: 10.1108/jrit-03-2024-0073

Stix, C. (2022). Artificial intelligence by any other name: A brief history of the conceptualization of "trustworthy artificial intelligence.". *Discover Artificial Intelligence*, *2*(1). Advance online publication. DOI: 10.1007/s44163-022-00041-5

Straulino, E., Scarpazza, C., & Sartori, L. (2023). What is missing in the study of emotion expression? [Review of What is missing in the study of emotion expression?]. Frontiers in Psychology, 14. Frontiers Media. DOI: 10.3389/fpsyg.2023.1158136

Straw, I., & Callison-Burch, C. (2020). Artificial Intelligence in mental health and the biases of language based models [Review of Artificial Intelligence in mental health and the biases of language based models]. PLoS ONE, 15(12). Public Library of Science. DOI: 10.1371/journal.pone.0240376

Ullah, S., Ou, J., Xie, Y., & Tian, W. (2024). Facial expression recognition (FER) survey: A vision, architectural elements, and future directions. *PeerJ. Computer Science*, *10*. Advance online publication. DOI: 10.7717/peerj-cs.2024 PMID: 38855254

Velagaleti, S. B. (2024). Empathetic Algorithms: The Role of AI in Understanding and Enhancing Human Emotional Intelligence. *Deleted Journal*, *20*, 2051. DOI: 10.52783/jes.1806

Vicci, H. (2024). Emotional Intelligence in Artificial Intelligence: A Review and Evaluation Study [Review of Emotional Intelligence in Artificial Intelligence: A Review and Evaluation Study]. SSRN Electronic Journal. RELX Group (Netherlands). DOI: 10.2139/ssrn.4818285

Vistorte, A. O. R., Deroncele-Acosta, Á., Ayala, J. L. M., Barrasa, Á., López-Granero, C., & Martí-González, M. (2024). Integrating artificial intelligence to assess emotions in learning environments: A systematic literature review. *Frontiers in Psychology*, *15*. Advance online publication. DOI: 10.3389/fpsyg.2024.1387089 PMID: 38966729

Waelen, R. (2024). Philosophical Lessons for Emotion Recognition Technology. *Minds and Machines*, *34*(1). Advance online publication. DOI: 10.1007/s11023-024-09671-3

Wang, Y., & Liu, W. (2023). Emotional Simulation of Artificial Intelligence and Its Ethical Reflection. *Academic Journal of Humanities & Social Sciences*, *6*(5). Advance online publication. DOI: 10.25236/ajhss.2023.060503

Wu, X., Wang, H., Yan, Z., Tang, X., Xu, P., Siok, W.-T., Li, P., Gao, J.-H., Lyu, B., & Qin, L. (2025). AI shares emotion with humans across languages and cultures. DOI: 10.48550/ARXIV.2506.13978

Yazdani, A., & Darbani, S. (2023).. . *The Impact of AI on Trends, Design, and Consumer Behavior.*, *1*(4), 4. DOI: 10.61838/kman.aitech.1.4.2

Zhai, C., Wibowo, S., & Li, L. D. (2024). The effects of over-reliance on AI dialogue systems on students' cognitive abilities: a systematic review [Review of The effects of over-reliance on AI dialogue systems on students' cognitive abilities: a systematic review]. Smart Learning Environments, 11(1). Springer Nature. DOI: 10.1186/s40561-024-00316-7

Chapter 4
Barriers and Limitations in Artificial Intelligence for Decoding Non-Human Signals

Arun Agrawal
https://orcid.org/0000-0001-7233-6660
Institute of Technology and Management, Gwalior, India

Srinivasan Nagaraj
Chaitanya Bharathi Institute of Technology, India

S. Sagar Imambi
Koneru Lakshmaih Education and Foundation, India

Kalyani Tushar Borse
https://orcid.org/0009-0000-0476-0400
MET's Institute of Technology Polytechnic, India

Pratik Gulab Dhange
https://orcid.org/0009-0008-4147-0945
N. Sapkal College of Engineering, Kalyani Hills, India

Veer Sudheer Goud
St. Martin's Engineering College, Secunderabad, India

Deepak Gupta
https://orcid.org/0000-0003-3929-1362
Institute of Technology and Management, Gwalior, India

ABSTRACT

The study of non-human signals is growing across many fields. These signals come from animals, plants, and microbes. Researchers use Artificial Intelligence to analyse these patterns. The method works well when data is clear and consistent. The work becomes difficult when the data is limited or noisy. Context also affects how signals behave. This reduces the accuracy of many models. Many systems work like

DOI: 10.4018/979-8-3373-5483-5.ch004

a closed box. This makes their decisions hard to explain. It also creates problems when we try to understand biological meaning. Species differ in their signal patterns. This creates more confusion for researchers. Standard datasets are still missing in many areas. Human views can also influence how the signals are read. This adds more uncertainty to the results. Ethical concerns appear when these tools guide conservation or communication efforts. These issues need careful attention. Future progress depends on clear and explainable methods. It also needs combined data from different sources. Strong teamwork between computing and life sciences is also important.

1. INTRODUCTION

1.1 Background on Non-Human Communication

Communication was once believed to be unique to humans. Today, science shows that it is a shared feature of life. Animals, plants, and microorganisms all send and receive signals that help them survive, reproduce, and stay in balance with their surroundings (Gordon & Seckbach, 2016).

In animals, signals take many forms. Some are sounds, others are movements, touches, or chemical cues (Cartmill, 2023). Birds sing to attract mates or defend territory. Dolphins use clicks and whistles to stay connected and find their way underwater. Bees perform dance patterns to guide others toward food. These examples reveal that animal communication is rich and purposeful, often carrying meaning or emotion.

Plants also share information, even without nerves or brains. When attacked by insects, they release warning chemicals that alert nearby plants (Robinson & Draguhn, 2021). Trees use underground fungal networks—sometimes called "forest networks"—to trade nutrients and send signals (Heijden, 2016). Some studies suggest plants may even produce or sense soft vibrations.

Microbes communicate too. Bacteria use a process called quorum sensing to send chemical messages (Nakade & Dhadse, 2024). This helps them form groups, build biofilms, and manage shared resources.

All these examples show that communication is everywhere in nature. It connects living systems, supports cooperation, and helps keep ecosystems healthy. Understanding it can improve how we protect nature, grow crops, and study health.

1.2 Role of Artificial Intelligence in Interpreting Complex Biological Signals

While biologists and ecologists have long studied non-human communication, traditional methods have many limitations. Signals generated by animals, plants, and microorganisms are often non-linear, high-dimensional, and highly susceptible to environmental influences (Assom, 2023). Human observation and traditional statistical tools struggle to capture the full scope of these dynamic systems. AI is playing a transformative role in this. AI technologies, particularly machine learning and deep learning, excel at pattern recognition in complex and noisy datasets. In the field of animal communication, convolutional neural networks (CNNs) have been used to classify birdsong and whale calls with high accuracy, revealing patterns imperceptible to the human ear. Similarly, recurrent neural networks (RNNs) and Transformer-based models can analyze time series in vocalizations, providing new insights into the syntax and structure of non-human languages. In the field of plant communication, artificial intelligence has been applied to analyze chemical composition and bioacoustic data to identify plant stress responses or adaptive behaviors (Kumar, 2025). For example, algorithms can detect subtle changes in volatile compound emissions, enabling early diagnosis of plant diseases or pests. This has practical implications for sustainable agriculture and food security. AI-driven interpretations also greatly benefit microbial communication. Machine learning models can analyze massive datasets of genomes and metabolomics, revealing how microbial communities exchange signals and coordinate activities. These insights are crucial in areas such as human health, as gut microbiome communication influences immune function and disease prognosis.

Importantly, AI not only helps decode signals but also helps us generate hypotheses about underlying biological mechanisms. By identifying recurring patterns, clustering signals, or predicting responses under new conditions, AI can help researchers build testable models, thereby deepening our understanding of communication in nature (Floreano & Mattiussi, 2008). However, these applications also bring new challenges related to interpretability, data quality, and ethical responsibility, which will be discussed in later chapters.

1.3 Scope and Significance of the Study

The study of non-human communication through artificial intelligence is an interdisciplinary field encompassing biology, computer science, and ecology. Its scope is both broad and deep, covering the decoding of acoustic, chemical, vibrational, and visual signals across multiple domains of life. This chapter focuses on the challenges and limitations, aiming to balance technological optimism with scientific caution.

The significance of this study can be discussed along several dimensions:

1. **Scientific Advancement**: Deciphering non-human signals can deepen our understanding of cognition, cooperation, and adaptation in living systems and extend the boundaries of biology beyond the anthropocentric framework.
2. **Conservation and Ecology**: AI can assist in monitoring animal calls or plant stress signals, contributing to biodiversity conservation, habitat management, and early warning of ecological threats. For example, combining acoustic monitoring with AI is increasingly being used to track endangered species in remote habitats.
3. **Agriculture and Food Security**: Decoding plant signals enables precision agriculture, allowing crops to "transmit" their stress or nutrient needs, thereby reducing reliance on pesticides and optimizing resource utilization.
4. **Health and Medicine**: Interpreting microbial communication through artificial intelligence can provide new insights into infection treatment, enhance our understanding of gut health, and inspire the development of novel antibiotics.
5. **Ethics and Philosophy**: Beyond its applications, this research challenges anthropocentric noexcetions of intelligence and language. It raises profound questions about interspecies relationships, the limitations of human understanding, and the ethical implications of listening to or intervening in non-human conversations.

By placing these dimensions within the aforementioned framework, this study emphasizes that the interpretation of non-human signals is not merely a technical task, but a multidisciplinary field of research with global significance.

2. LITERATURE REVIEW

Research on understanding how non-human species communicate using artificial intelligence (AI) has grown quickly. Studies now explore areas like animal sounds, plant signaling, and microbial communication. Scientists have made good progress in decoding different kinds of signals such as sound, chemicals, and vibrations. However, this field is still scattered. Work in one species or signal type rarely connects with others. This review looks at recent advances and points out the key gaps and challenges that still shape research in this area.

2.1. Current Progress and Trends

In animal communication, studies have increasingly used deep learning architectures such as convolutional neural networks (CNNs) and recurrent neural networks (RNNs) to classify vocalizations and decode species-specific patterns. For instance, Allen et al. (2021) successfully trained CNNs to distinguish humpback whale song types, and Luphade (2023) applied deep learning to gibbon call classification. In plant systems, AI models analyze bioacoustic emissions and volatile organic compounds (VOCs) to detect stress and environmental changes (Coatsworth et al., 2023). Microbial communication studies have adopted machine learning algorithms to interpret quorum-sensing signals, identifying molecular patterns that regulate biofilm formation and virulence (Sarma et al., 2023).

While these approaches demonstrate technical sophistication, they remain primarily descriptive. Many studies focus on achieving high classification accuracy without linking computational results to biological meaning or ecological consequences. Moreover, datasets are often species-specific and lack standardization, limiting cross-domain generalization.

2.2. Unresolved Debates and Methodological Weaknesses

Researchers face three ongoing debates in this field:

Practical vs. speculative research: Some studies are based on real data, like analyzing animal sounds, while others imagine future possibilities such as direct communication between humans and other species. Critics warn that these speculative ideas go beyond what AI can currently explain or prove scientifically.

Accuracy vs. understanding: Deep learning models can make highly accurate predictions, but they often work like "black boxes," giving results without showing how decisions are made. Only a few studies use explainable AI (XAI) tools that clearly connect AI outputs to biological or ecological meaning.

Unequal and limited data: Most available datasets focus on well-known or easy-to-record species, creating a human-centered bias. Important areas, such as microbial and plant communication, still lack enough data for fair comparison across species.

There are several technical issues. Researchers often use different ways to measure accuracy (like precision, recall, or F1-score), making comparisons difficult. Many studies also ignore the effects of environmental noise, signal context, and hardware challenges in field research. These problems make results harder to reproduce and limit their use in real-world ecosystems.

2.3. Research Gaps and Future Opportunities

Despite good progress, some important gaps still remain in this field.

Missing connections across species: Many studies look only at animals, plants, or microbes. They do not combine results to find common rules of communication. Without this connection, researchers miss the chance to create AI systems that can work across different forms of life.

Limited focus on ecology and ethics: Most research does not discuss how AI affects nature and society. Questions about data ownership, Indigenous knowledge, and conservation laws are often ignored. These issues are important for making AI use fair and respectful.

Lack of common testing standards: There is no shared method to compare how well different AI models work. Researchers need common datasets and clear ways to measure performance. AI results should also be checked against human experts and traditional methods to ensure accuracy.

To fill these gaps, people from different fields must work together. AI specialists, ecologists, and ethicists can combine their ideas to build research that is both practical and responsible.

The Table 1 gives an overview of current studies, the AI methods used, and the main challenges found in each area.

Table 1. State of the Field in AI-Driven Interpretation of Non-Human Communication

Signal Type	AI Approaches Used	Representative Studies	Key Limitations / Gaps Identified
Animal Acoustic Signals	CNNs, RNNs, Transformers	Allen et al. (2021); Luphade (2023); Stowell (2022)	High accuracy but limited interpretability; small labeled datasets; minimal field validation.
Plant Bioacoustic and Chemical Signals	Random Forests, Multimodal Fusion Models	Kumar (2025); Coatsworth et al. (2023); Zandi et al. (2025)	Data scarcity, noise sensitivity, lack of real-time applications, underexplored sensor constraints.
Microbial Communication (Chemical/Metabolomic)	SVMs, RF, Generative Models	Swift et al. (2001); Nakade & Dhadse (2024)	Poor standardization, limited integration of genomic and environmental variables.
Insect Vibrational and Chemical Signals	SVM, Deep Multimodal Models	Virant-Doberlet et al. (2023); Williams (2024)	Strong within-species performance but poor generalization across taxa; need for multimodal benchmarks.
Cross-Species Interactions	Graph Neural Networks, Transfer Learning	Oyewole et al. (2025); Jacuzzi & Olden (2025)	Inconsistent data annotation; ethical and contextual challenges; lack of comparative baselines.

The existing literature demonstrates promising progress but limited synthesis. Current studies excel in technological innovation yet often underrepresent ecological validity, ethical accountability, and methodological transparency. Future research should emphasize standardized benchmarks, interpretable AI frameworks, cross-species integration, and policy-linked implications. This shift will transform AI's role from mere classification toward a strategic, ethical, and scientifically grounded understanding of non-human communication.

3. NON-HUMAN SIGNALS: AN OVERVIEW

Nature is teeming with communication systems far exceeding human language. Animals, plants, insects, and microorganisms all emit signals to interact with their environment, maintain survival strategies, and sustain ecosystems. These signals can be acoustic, chemical, vibratory, or visual, and interpreting them is crucial for understanding ecological balance and interspecies relationships (Jianhua, 1998). This section provides a comprehensive overview of the major forms of non-human communication, including animal calls, plant signaling, communication mechanisms in microorganisms and insects, and the complexities of interspecies interactions.

3.1 Animal Vocalizations and Behavioral Signals

Animals rely heavily on vocal and behavioral signals for survival, reproduction, and social organization. These signals encompass a wide range of forms, from vocalizations to body postures and chemical signals, and often contain multi-layered information relevant to the context.

3.1.1. Acoustic Communication

Many animals, especially birds, whales, primates, and frogs, communicate through vocalizations. Birdsong is one of the most extensively studied examples, with its complex acoustic patterns serving as territorial markings, courtship calls, or alarms against predators (Rogers & Kaplan, 2002). In marine environments, whales, such as whales and dolphins, use clicks, whistles, and songs for long-distance communication, navigation, and establishing social connections (Bakker, 2022). For instance, humpback whale songs exhibit hierarchical structures similar to human language syntax, sparking debate about whether these structures represent cultural transmission.

3.1.2. Primate Communication

Primates, such as chimpanzees and macaques, exhibit astonishing complexity in vocalizations and body language (Locke, 2021). For example, macaques emit different alarm calls in response to different predators, which trigger specific behavioral responses from group members. Chimpanzees, on the other hand, combine body language, facial expressions, and vocalizations to convey intentions, emotions, and social dynamics. This suggests that primate communication may share structural similarities with the basic building blocks of language.

3.1.3. Non-Vocal Behavioural Signals

Animals also communicate through body language, coloration, and ritualistic movements. Bees perform a "swaying dance" to inform hive members of the location and quality of food sources (Brown, 2012). Lizards and birds frequently use visual displays, such as extending their throat pouches or displaying their feathers, to establish dominance or attract mates (Carpenter, 1978). Elephants rely on tactile and infrasonic signals, including low-frequency rumbling sounds that can travel long distances, to coordinate among herds (Garstang, 2010).

3.1.4. Adaptive Importance

The adaptive value of animal communication cannot be underestimated. Vocal and behavioral signals are crucial in predator avoidance, cooperative hunting, parental care, and the formation of social hierarchies (Benneth, 2026). The richness of these systems suggests that animals engage in structured and meaningful communication, often influenced by environmental pressures and evolutionary history.

3.2 Plant Communication using Bioacoustics and Mycorrhizal Networks

Although plants are traditionally considered passive organisms, they actually possess complex signaling systems that enable them to interact with their environment and neighboring plants. This communication is primarily achieved through biochemical pathways, root networks, and bioacoustic phenomena (Hirsch et al., 2003).

3.2.1. Chemical Signalling

When plants are stressed, such as by herbivores, drought, or disease, they release volatile organic compounds (VOCs). The release of these chemicals acts as a warning signal in the air, alerting nearby plants to activate their own defense mechanisms (Nagesh & Piovene, 2011). For example, when acacia trees are being eaten, they release ethylene gas, prompting nearby trees to increase tannin production, thus deterring further herbivore attacks.

3.2.2. The Mycorrhizal Network ("Wood Wide Web")

One of the most remarkable discoveries in plant biology is the underground network formed by the symbiotic relationship between plant roots and mycorrhizal fungi. This interconnected system, sometimes called a "forest network," enables plants to share nutrients, transmit chemical signals, and even send out distress signals. It has been observed that some larger trees with extensive root systems, often referred to as "mother trees," transport carbon and nutrients to seedlings, indicating a cooperative strategy within the ecosystem (Figueiredo et al., 2021).

3.2.3. Bioacoustics Communication

Recent research suggests that plants may also generate or respond to acoustic signals (Mishra et al., 2016). For example, root tips can emit clicking sounds in the kilohertz range, which may influence root growth direction or detect nearby water sources. Although this field is still in its early stages, bioacoustics offers new possibilities for understanding plants as dynamic organisms, enabling us to perceive and respond to our environment in ways previously thought impossible.

3.2.4. Ecological and Agricultural Implications

Plant communication is crucial for agriculture and ecosystem resilience. By decoding plant signals, farmers can employ precision agriculture techniques to monitor crop stress in real time, thereby reducing reliance on pesticides and water waste. From an ecological perspective, understanding plant networks helps conservationists recognize the interdependence of forest ecosystems and the potential consequences of disrupting these vulnerable pathways (Aarif et al., 2025).

3.3 Microbial and Insect Signaling Mechanisms

Microorganisms and insects are two types of non-human life forms that employ highly structured and efficient signaling strategies that often have a direct impact on ecosystems and human health.

3.3.1. Microbial Communication

Microorganisms, especially bacteria, engage in quorum sensing. Quorum sensing is a chemical signaling process that coordinates group behavior when population density reaches a threshold. Bacteria collectively regulate activities such as virulence factor production, biofilm formation, and antibiotic resistance by releasing and detecting signaling molecules called autoinducers (Swift et al., 2001). For example, *Pseudomonas aeruginosa* utilizes quorum sensing to establish persistent infection by producing a protective biofilm that resists the host's immune response.

Quorum sensing is not limited to a single species. In multi-microbial environments such as soil or the human gut, microorganisms exchange cross-species signals, influencing community structure, nutrient cycling, and host interactions. Understanding microbial communication is crucial for developing novel antimicrobial therapies and advancing probiotic technologies.

3.3.2. Insect Communication

Given their diversity and significant ecological importance, insects are among the most studied organisms in communications research. Ants, bees, and termites utilize a combination of chemical, tactile, and visual signals to coordinate collective behavior.

- **Pheromones**: Ants rely heavily on pheromone trails for navigation, food discovery, and colony defence. Different pheromones encode distinct meanings, from recruitment to alarm.
- **Dance and Vibration**: Honeybees perform intricate dances to share spatial information about nectar sources, while termites use vibrational signals to warn colonies of threats.
- **Light-Based Signals**: Fireflies communicate through bioluminescence, with species-specific flash patterns serving as mating signals.

Adaptive Strategies: The efficiency of insect communication lies in its scalability and precision. Groups can coordinate complex tasks, such as nest building or foraging, through distributed communication without centralized control. This has inspired fields such as swarm robotics and distributed artificial intelligence systems, where biological models inform technological innovation.

3.4 Cross-Species Signal Diversity and Complexity

Non-human communication is not confined to within a species; cross-species signaling plays a crucial role in ecological networks and survival strategies. These interactions highlight the interconnectedness of life, while also presenting unique challenges to interpreting them.

3.4.1. Plant-Animal Interactions

Pollination is a classic example of plants signaling to animals through color, scent, and nectar supply. Flowers evolve specific colors and fragrances to attract pollinators such as bees, bats, and birds, thus ensuring successful reproduction. In return, pollinators rely on these signals to obtain food resources. Similarly, fruiting plants signal maturity through color changes to attract animals to disperse their seeds (Abrol, 2011).

3.4.2. Predator-Prey Dynamics

Cross-species signaling is also evident in predator-prey relationships. Some prey species use warning coloration to warn predators of their venom, while others employ mimicry, such as non-venomous snakes imitating the coloration of venomous snakes (Irving, 2023). Conversely, predators may also evolve deceptive signals to lure prey, reflecting an arms race in evolutionary communication.

3.4.3. Symbiotic Communication

Mutualistic relationships often rely on communication. For example, legumes and nitrogen-fixing bacteria exchange chemical signals during root nodule formation to ensure a stable supply of nutrients for both. Similarly, the prosperity of coral reefs depends on signaling interactions between corals and symbiotic algae, which regulate energy exchange and stress responses (Hirsch & Fujishige, 2011).

3.4.4. Ecosystem-Level Complexity

Interspecies communication helps maintain ecosystem stability by forming dynamic feedback loops (Schneider et al., 2016). For example, a predator's alarm may indirectly affect plant survival by reducing stress on herbivores. The interrelationships of these signals highlight the difficulty of studying individual communication channels in isolation. Understanding each signal requires not only focusing on its immediate context but also examining it within the broader ecological network.

3.4.5. Challenges in Interpretation

The diversity and context-dependent nature of interspecies communication highlight the limitations of human and AI interpretation. Signals are often multimodal, overlapping, and influenced by evolutionary histories that differ from human logic. Due to the lack of standardized datasets or reference frames, interpreting such interactions requires an interdisciplinary approach that combines biology, ecology, and computational science (Oyewole et al., 2025).

4. AI APPLICATIONS IN SIGNAL INTERPRETATION

Deciphering signals from animals, plants, and microbes needs advanced methods. These signals often appear in high-dimensional data. They also contain noise and depend on context. Traditional tools struggle with these conditions. They fail when the patterns become complex. Artificial Intelligence offers stronger support. Machine learning and deep learning help find patterns in large datasets. They help classify signals and produce useful insights. This section explains the main approaches used to study non-human communication. It covers common machine learning methods. It also explains deep learning models and how they process complex patterns. It describes how multimodal data can work together to improve accuracy. It also highlights case studies that show new research directions.

4.1 Machine Learning Approaches (e.g., SVM, Random Forests)

Machine learning (ML) has become the cornerstone of signal interpretation because it can identify patterns in large datasets without explicit programming. Classic machine learning algorithms such as support vector machines (SVM), random forests (RF), and k-nearest neighbors (k-NN) have been widely used in fields such as ecological research, bioacoustics, and chemical signal analysis (Kvsn et al., 2020).

4.1.1. Support Vector Machines (SVM)

Support Vector Machines (SVMs) are supervised learning models that classify data by finding the optimal hyperplane that separates different categories. SVMs are particularly effective in classifying bioacoustic signals, such as distinguishing species-specific calls (Cheng et al., 2012). For example, SVMs have been used to classify bird calls into different species categories based on spectrogram features such as frequency, amplitude, and temporal patterns. The robustness of SVMs to high-dimensional feature spaces makes them suitable for biological data where the signal dimension may be high but the dataset is relatively small (e.g., biological data where the signal dimension is high but the dataset is relatively small).

4.1.2. Random Forests (RF)

Random Forest (RF) is an ensemble learning method that constructs multiple decision trees during training and outputs the mode of the predictions made by these trees. This method boasts high accuracy and handles noisy datasets well, making it valuable for interpreting complex biological signals (Pantic et al., 2025). For example, RF models have been applied to identify plant stress signals based on volatile organic compound (VOC) spectra and to classify insect calls and microbial communication patterns. Interpreting model results by ranking feature importance can add value to ecological research, as understanding which signaling components are most important is crucial in ecological studies.

4.1.3. Other Classical Algorithms

Algorithms such as k-Nearest Neighbors (k-NN) and Gaussian Mixture Models (GMM) have also been applied to non-human signal analysis. k-NN is often used for clustering and classifying animal vocalizations, while GMM helps to model the distribution of acoustic data, which is very useful for identifying overlapping signal patterns in dense communication environments.

Advantages and Limitations: Compared to deep learning models, traditional machine learning methods are computationally efficient and require less training data, which is especially important in ecological research where data is scarce. However, they rely heavily on manual feature extraction, which limits their adaptability when signals are highly variable or multimodal. This challenge has prompted the adoption of deep learning techniques to automatically extract features from raw data.

4.2 Deep Learning for Pattern Recognition (CNNs, RNNs, Transformers)

Deep learning (DL) offers a revolutionary ability to interpret complex non-human signals. By automatically learning hierarchical features from raw data, deep learning models can handle high-dimensional inputs and reveal complex patterns that traditional machine learning methods might miss. Among the many deep learning architectures, convolutional neural networks (CNNs), recurrent neural networks (RNNs), and Transformer-based models are particularly important for signal interpretation (Stowell, 2022).

4.2.1. Convolutional Neural Networks (CNNs)

Convolutional Neural Networks (CNNs) are specifically designed to process grid-like data structures, such as images and spectrograms. In bioacoustics, spectrograms of vocalizations can be directly input into CNNs for classification and pattern recognition (Luphade, 2023). For example, CNNs have been used to distinguish whale song patterns, identify bird species, and detect frog calls with high accuracy. Convolutional layers can automatically extract temporal and spectral features without requiring manual preprocessing.

4.2.2. Recurrent Neural Networks (RNNs) and Long Short-Term Memory (LSTM) Networks

Recurrent neural networks (RNNs), especially their advanced variant Long Short-Term Memory (LSTM) networks, excel at processing sequential data. This makes them ideal for analyzing time-dependent signals, such as animal vocalizations, rhythmic vibrations of plants, and sequences of chemical signals (Mienye et al., 2024). For example, LSTMs have been used to decode sequences of dolphin vocalizations, where the temporal order of sound units contains crucial information. Their ability to preserve long-term dependencies allows them to capture the contextual information inherent in communication patterns.

4.2.3. Transformers and Attention Mechanisms

The Transformer model uses a self-attention mechanism that has changed how sequences are analyzed. It was first developed for understanding language but is now applied in bioacoustics and multimodal data research. The model can detect long-range patterns and small differences in signals. It also processes large datasets quickly and efficiently (Choi & Lee, 2023).

Even with these strengths, deep learning faces important challenges. It depends on large labeled datasets, which are often unavailable in studies of non-human communication. Training also demands heavy computing power. Another major issue is that these models often act like "black boxes." They make predictions without explaining how those results are formed. To overcome this, researchers are exploring interpretable AI methods that keep models accurate while making their decisions easier to understand.

4.3 Multimodal AI Models (Audio, Chemical, Visual Integration)

Communication among non-human organisms usually works through many channels. It is rarely based on a single type of signal. Organisms use sound, chemicals, vibrations, and visual cues. These signals interact in complex ways. Multimodal AI models try to bring these data sources together. They help create a clearer and more accurate understanding of how communication works in nature (Subasi et al., 2024).

4.3.1. Integration of Acoustic and Chemical Signals

For example, plant stress responses involve both the release of chemicals (volatile organic compounds) and acoustic vibrations. Multimodal models can combine acoustic spectrogram analysis with chemical sensor data, thus detecting stress more accurately than single-modal systems. This integration enhances the system's robustness to noise and environmental changes (Coatsworth et al., 2023).

4.3.2. Audio-Visual Integration

In animal communication research, combining audio and video data can lead to a deeper understanding of animal behavior. For example, multimodal models can combine recordings of bird calls with visual observations of feather displays or body movements to provide contextual information, thereby improving classification accuracy and interpretability (Valente et al., 2007).

4.3.3. Insect Communication Systems

Insects typically combine chemical pheromones with vibrational or visual signals. Multimodal models are particularly effective at deciphering these complex signals. For example, AI frameworks that integrate vibration sensor data and chemical composition information have been used to study ant colony signaling, revealing how different modalities promote colony coordination (Virant-Doberlet et al., 2023).

Technical Approaches: Multimodal AI models typically involve fusion strategies at different stages—early fusion (integrating the original inputs), mid-stage fusion (combining features), or late-stage fusion (combining model outputs). Deep architectures such as multimodal Transformers and cross-modal attention networks are increasingly becoming powerful tools in this field.

Benefits: Multimodal methods can improve accuracy, enhance robustness to missing or noisy data, and provide richer contextual information for signal interpretation. Furthermore, they are more consistent with biological reality, as organisms rarely rely on a single mode of communication.

4.4 Case Studies and Emerging Research Trends

Case Study 1: Whale Song Classification Using CNNs Researchers applied convolutional neural networks (CNNs) to a spectrogram dataset of humpback whale songs to classify song types and detect their changes over time. The model achieved high-precision classification and revealed patterns in song evolution, indicating cultural inheritance within whale pods. This research demonstrates the potential of deep learning to reveal subtle patterns in complex animal communication (Allen et al., 2021).

Case Study 2: Plant Stress Detection with Multimodal AI A study combines acoustic vibration sensors and chemical sensors to detect drought stress in wheat plants. By combining acoustic emission spectroscopy with volatile organic compound (VOC) spectroscopy, this multimodal model outperforms single-modal methods, providing a promising approach for precision agriculture (Zandi et al., 2025).

Case Study 3: Bee Dance Interpretation Using Machine Learning Researchers used video data and motion tracking technology, applying support vector machines (SVM) and random forest models to decode the waggle dance of bees. These models successfully predicted the distance and direction of food sources, demonstrating how artificial intelligence can transform complex behavioral signals into quantitative information (Williams, 2024).

Emerging Research Trends:

- **Explainable AI (XAI)**: As AI models grow in complexity, researchers are increasingly focusing on explainability to ensure that interpretations of biological signals are transparent and scientifically valid. Techniques such as saliency mapping, layer-wise relevance propagation, and attention visualization are becoming standard (Alharthi et al., 2025).
- **Transfer Learning**: Applying models trained on one dataset to another with limited data helps overcome data scarcity in ecological studies. Transfer learning is increasingly used in bioacoustic classification and plant signal analysis (Jacuzzi & Olden, 2025).
- **Real-Time Signal Interpretation**: Advances in edge computing and lightweight models are enabling real-time processing of non-human signals, paving the way for field-deployable monitoring systems for conservation and agriculture.
- **Interdisciplinary Collaboration**: The fusion of biology, ecology, sensor technology, and AI is driving innovation. Cross-disciplinary research labs are developing integrated platforms that combine hardware and AI algorithms for continuous environmental monitoring.

Future Directions: The future of artificial intelligence in signal interpretation lies in combining advanced algorithms with ecological insights. This includes developing models that can generalize across species, efficiently process multimodal data, and provide interpretable outputs that respect biological context. Strengthening collaboration between AI researchers and biologists is crucial to ensuring meaningful and ethical interpretation of non-human communication.

The application of AI in signal interpretation has revolutionized the study of non-human communication. Machine learning methods offer powerful classification tools, deep learning enables advanced pattern recognition, and multimodal models provide rich contextual analysis. Case studies in areas such as animal vocalization, plant bioacoustics, and insect communication highlight the transformative potential of these approaches. However, numerous challenges remain, including data scarcity, model interpretability, and the necessity for interdisciplinary integration. Emerging research trends suggest that future AI will not only decode biological signals but also deepen our understanding of the hidden dialogues of life.

5. CHALLENGES AND LIMITATIONS

Despite the remarkable ability of AI to interpret non-human signals, its application in this field still faces numerous challenges. These limitations stem from the inherent complexity of biological communication, the inherent characteristics of AI algorithms, and broader ethical concerns. This section will explore key obstacles, including data scarcity, noise and environmental variability, interpretability issues, cross-species differences, anthropocentric bias, and ethical considerations.

5.1 Data Scarcity and Imbalanced Datasets

One major challenge in studying non-human signals with AI is the shortage of strong datasets. Researchers often struggle to find large and well-labeled samples. Biocommunication data is hard to collect. It may involve long hours of recording animal sounds. It may also require testing plant emissions or tracking microbial signals. Labelling this information takes time. Storing and managing it also adds cost. These issues limit how well AI models can learn and perform.

Data scarcity is further exacerbated by class imbalance. For example, in bioacoustic datasets, recordings of rare or endangered species are far fewer than those of common species. This imbalance can bias AI models, leading to poor generalization and misclassification of underrepresented signals. Similarly, plant and microbial communication data often come from limited experiments under controlled conditions, which limits their applicability in diverse ecological environments (Kvsn et al., 2020).

Addressing the data scarcity problem requires innovative strategies, such as transfer learning, synthetic data generation, and citizen science initiatives. However, these solutions also have limitations. Synthetic data may not fully capture the complexity of real biological signals, while crowdsourced data often lacks consistency and quality control.

5.2 Signal Noise, Context Dependence, and Environmental Variability

Non-human communication signals often exist in noisy and dynamic environments. For example, animal calls may overlap with environmental sounds such as wind, rain, or human activity (Brumm & Slabbekoorn, 2005). Similarly, signals from plants and microorganisms are affected by changes in environmental factors such as temperature, humidity, and soil conditions.

This variability makes signal extraction and interpretation challenging. Context dependence further complicates this process, as the meaning of a signal changes with environmental or social conditions. For example, the meaning of birdsong can vary depending on the time of day, season, or the presence of predators. Similarly, chemical signals in plants differ not only in the type of stress but also in the plant species and developmental stage.

Therefore, artificial intelligence models must include robust preprocessing, noise filtering, and adaptive learning mechanisms. Multimodal methods can help by integrating contextual information, but they also require more complex architectures and data acquisition systems, thus increasing computational demands.

5.3 Model Interpretability and the Black-Box Problem

Many artificial intelligence models, especially deep learning architectures, operate like "black boxes," providing predictions that lack clear explanations and fail to elucidate their decision-making processes (London, 2019). This lack of interpretability severely limits the interpretation of non-human signals. For researchers and ecologists, understanding the reasoning behind AI classification is just as important as the predictions themselves. A lack of interpretability risks misunderstanding biological meaning or overlooking crucial ecological context.

Explainable Artificial Intelligence is becoming an important tool in this area. Methods like saliency maps, feature importance checks, and attention systems help show how a model makes its decisions (Hsu & Li, 2023). These tools make the system easier to understand. They also add more complexity to the model. Experts must read the results carefully to avoid mistakes. The field still struggles to balance clear explanations with high accuracy. This remains a major challenge in AI work on non-human signals.

5.4 Cross-Species Variability and Lack of Standardization

Non-human communication systems vary greatly across species, and even within the same species, signals can differ depending on context, population, or geographical location (Beecher, 2021). This cross-species variability complicates the design of AI models that can generalize well to different organisms. For example, the structure of whale songs varies across populations, the emissions of volatile organic compounds (VOCs) from plants differ by species, and the composition of quorum sensing molecules in microorganisms also varies.

The lack of standardized datasets, protocols, and annotation guidelines further limits research progress. Researchers often use different methods to collect and process data, making comparison and generalization difficult. Without standardization, an AI model trained on one dataset may fail when applied to another, reducing reliability and reproducibility.

Efforts to build shared repositories and standard protocols similar to initiatives in human language processing are still in their infancy in the non-human signal domain. Such efforts require interdisciplinary collaboration and consensus on a common framework.

5.5 Anthropocentric Bias in Training and Interpretation

Artificial intelligence models are trained using data compiled and labeled by humans, making them susceptible to anthropocentric biases. Researchers may inadvertently prioritize signal types that are easier for humans to detect or interpret, while neglecting more subtle or difficult-to-understand modes of communication. For example, an AI model trained primarily on audible animal vocalizations might ignore ultrasonic or infrasound signals that have significant biological implications.

This bias extends to the level of interpretation. Imposing human language or behavioral frameworks onto non-human communication can lead to misunderstandings of its underlying biological meaning. This bias limits AI's ability to truly "understand" the signals it processes and risks reinforcing anthropocentric assumptions about intelligence and communication.

Reducing bias requires conscious efforts to diversify datasets, incorporate domain expertise, and develop interpretive frameworks that respect the uniqueness of non-human communication systems.

5.6 Ethical and Ecological Considerations

Beyond the technological challenges, interpreting non-human signals raises significant ethical and ecological questions. AI-powered communication systems could enable humans to "hear" and even "respond" to other species. This offers exciting possibilities for species conservation and cross-species understanding, but also raises numerous concerns.

Ethical concerns also shape this discussion. One key issue is consent. It is not clear when humans should step into non-human communication systems. Any form of intervention can disturb natural behaviour. It can also upset ecological balance. Such actions may create results that no one expects. For example, copying an alarm call may cause stress in animals. It may also disturb predator and prey patterns. These risks call for careful and responsible decisions.

From an ecological perspective, AI-based monitoring systems must be designed to minimize habitat disturbance. Data collected via sensors, drones, or acoustic devices should be processed non-invasively and aligned with conservation objectives.

Furthermore, governance issues arise regarding the ownership, access, and use of communication data. As AI's interpretive capabilities improve, regulations are needed to ensure compliance with ethical standards, protect biodiversity, and prevent the misuse of sensitive ecological information.

AI faces numerous limitations in interpreting non-human signals, including data scarcity, noise interference, context dependence, species differences, and black-box models. Anthropocentric bias and ethical issues further hinder progress in this field. Improvements require interpretable AI, shared datasets, standardized methods, and interdisciplinary collaboration. Responsibly advancing this field also means respecting natural communication systems while designing more accurate and ecologically conscious AI approaches.

5.7 Field and Societal Constraints

The practical use of AI for interpreting non-human signals depends on both technology and social responsibility. Field conditions create hardware and sensing challenges that directly affect data quality. Sensors must handle environmental noise, energy limits, and bandwidth issues while operating in remote areas. Edge computing offers solutions by allowing data to be processed locally, but it also requires low-power, durable, and affordable devices. Benchmarking remains another key issue. Studies often use different evaluation metrics, such as precision, recall, or F1-score, making comparisons difficult. Standardized benchmarks and open datasets are needed to test AI performance under real-world conditions and ensure consistent, verifiable results. Legal and ethical aspects are equally important. Questions of data ownership, Indigenous rights, and ecological privacy require careful attention. Deploying drones or sensors must follow guidelines that protect both wildlife and human communities. Clear policies can prevent data misuse and ensure transparency.

Addressing these technical and social constraints will make AI-based studies more reliable, ethical, and applicable in conservation and ecological research. Future progress should balance innovation with accountability, ensuring that advances in AI respect the limits of nature and the rights of all who share its ecosystems.

5.8 Comparative Evaluation of Analytical Approaches

Understanding non-human communication requires reliable analytical methods that balance speed, accuracy, and interpretability. Traditional statistical techniques offer transparency and controlled experimentation but struggle with complex, nonlinear, or multimodal data. Human experts provide valuable contextual and behavioral insights but face challenges in scalability and consistency. Artificial intelligence models, especially deep learning systems, can process massive datasets and uncover subtle patterns beyond human perception. However, they are often limited by data bias, poor interpretability, and weak field validation. The following Table 2 compares the strengths and weaknesses of AI, traditional methods, and human expertise in interpreting non-human signals.

Table 2. Comparison of AI, Traditional Statistical Methods, and Human Expertise in Interpreting Non-Human Signals

Criterion	AI-Based Models	Traditional Statistical Methods	Human Experts
Speed and Scale	Very high; processes large, continuous datasets quickly	Moderate; limited by model complexity	Low; manual and time-consuming
Accuracy	High under controlled conditions but sensitive to noise	Moderate; depends on variable selection	High for known species but inconsistent
Interpretability	Often limited without explainable AI tools	High; relationships are transparent	Very high but subjective
Adaptability	Learns automatically from new data	Requires redesign for new datasets	Adapts through experience only
Failure Modes	Overfitting, bias, and data noise	Misses hidden or nonlinear patterns	Fatigue and labeling inconsistency
Field Validation	Growing but limited in practice	Reliable in small-scale studies	Strong contextual insight but unscalable

Each approach contributes unique strengths to understanding non-human communication. AI offers unmatched speed and scalability but lacks interpretability and field validation. Traditional methods provide transparency yet struggle with complexity, while human expertise adds contextual understanding. Integrating all three can create more accurate, ethical, and ecologically meaningful analytical frameworks.

6. ADDRESSING THE LIMITATIONS

Overcoming the challenges and limitations described in the previous section requires a strategic combination of technological innovation, methodological standardization, and interdisciplinary collaboration. Addressing these issues is crucial not only for enhancing AI's ability to interpret non-human signals but also for ensuring that such interpretations are scientifically sound, ecologically reasonable, and ethically compliant. This section will explore potential solutions, including explainable AI (XAI), synthetic data generation, multimodal sensor fusion, standardized datasets, and collaborative frameworks.

6.1 Explainable AI (XAI) for Transparency and Validation

The "black box" nature of many AI models remains one of the biggest obstacles to interpreting non-human signals. Explainable AI (XAI) aims to overcome this challenge by providing explainable, transparent, and verifiable insights into the model's decision-making process.

Techniques and Applications: Explainable artificial intelligence (XAI) methods, such as feature importance analysis, saliency mapping, and attention mechanisms, can reveal which parts of the input data contribute most to the prediction results. For example, in bioacoustic classification, XAI tools can identify the frequency ranges or temporal patterns most important for distinguishing species-specific calls. This transparency is crucial for validating AI findings against biological knowledge and ensuring that the interpretation aligns with ecological reality.

Advantages: Explainable artificial intelligence (XAI) enhances trust among researchers, ecologists, and stakeholders by clarifying how models arrive at their conclusions. It also supports iterative improvements because researchers can identify when models rely on spurious correlations or irrelevant features.

Challenges: While explainable artificial intelligence (XAI) approaches hold great promise, they also increase the complexity of model design and require careful interpretation. Without sufficient contextual information, interpretations risk oversimplifying complex models or misinterpreting biological phenomena. Therefore, XAI must be combined with domain expertise to ensure that transparency translates into meaningful understanding.

6.2 Synthetic Data Generation and Transfer Learning

Data scarcity and imbalance remain major constraints on the interpretation of non-human signals by artificial intelligence. Synthetic data generation and transfer learning are two promising strategies for addressing this problem.

Synthetic Data Generation: Synthetic data refers to artificial datasets that generate simulations of real-world signals. Techniques such as Generative Adversarial Networks (GANs) and Variational Autoencoders (VAEs) can create realistic bioacoustic recordings, chemical composition analyses, or vibration signals. Synthetic data can augment small datasets, helping to train powerful AI models even when real-world data is limited.

Transfer Learning: Transfer learning leverages pre-trained models developed for relevant tasks or domains, applying them to new datasets with limited labeled data. For example, a convolutional neural network (CNN) trained on a set of animal vocalization data can be fine-tuned to classify vocalizations of different species, significantly reducing the amount of data required for training. This approach has proven highly effective in ecological and bioacoustic research, where labeled datasets are scarce and expensive to generate.

Limitations and Considerations: Synthetic data may not be able to fully replicate the biological nuances of real signals, while transfer learning relies on the relevance of the source dataset. Both methods require careful validation to ensure that the model does not inherit biases or inaccuracies from the synthetic or transferred data. Nevertheless, they remain powerful tools for expanding the application of artificial intelligence in a variety of under-explored communication systems.

6.3 Multimodal Sensor Fusion Approaches

Communication in non-human organisms is rarely monomodal. Acoustic, chemical, visual, and vibrational signals often occur simultaneously, requiring a comprehensive approach for accurate interpretation. Multimodal sensor fusion leverages multiple data streams to achieve a more comprehensive and in-depth understanding of communication systems.

Fusion Strategies:

- **Early Fusion** integrates raw signals from different modalities before processing, enabling models to learn joint representations from the start.
- **Intermediate Fusion** combines features extracted from each modality, allowing for specialized pre-processing while still capturing cross-modal correlations.

- **Late Fusion** merges decision outputs from modality-specific models, offering flexibility when modalities have different sampling rates or noise levels.

Applications: In plant communication research, fusion of acoustic vibration data and chemical emission characteristics into sensors has improved the ability to detect plant stress responses under different environmental conditions. In animal behavior research, combining audio recordings with video and motion data has improved the accuracy of behavioral classification, such as identifying courtship displays or predator-prey interactions.

Challenges: Multimodal fusion increases computational complexity and requires precise temporal and spatial alignment of the data stream. Furthermore, sensors must be calibrated and synchronized to avoid introducing noise or artifacts. Despite these challenges, multimodal approaches still offer a pathway to a more comprehensive interpretation of non-human communication.

6.4 Development of Standardized Datasets and Benchmarks

The lack of standardized datasets and benchmarks is a major obstacle to achieving reproducibility and model generalization ability. Without consistent data formats, labeling protocols, and evaluation metrics, an AI model trained in one environment may fail when applied to other environments.

Importance of Standardization: Standardized datasets provide a shared foundation for model training and evaluation, enabling comparability between different studies. Benchmarks establish performance baselines, helping researchers assess progress and identify shortcomings in the field. Standardization also facilitates collaborative data sharing and mutual progress.

Examples of Standardization Efforts: In the field of human language processing, datasets such as ImageNet and LibriSpeech have driven significant progress by providing standardized labeled data for model benchmarking. Similar initiatives have emerged in the fields of bioacoustics and plant signal transduction, but these fields still lack widely accepted databases and protocols.

Strategies for Development: Developing standardized datasets requires consensus on data collection methods, annotation protocols, and metadata formats. Collaborative platforms and open-access repositories can support this process, enabling researchers worldwide to contribute to and validate datasets. Involving ecologists, biologists, and AI experts in the development of these standards is crucial to ensuring their ecological validity and usability.

6.5 Collaborative Frameworks: Integrating Biology, Ecology, and AI

Interpreting non-human signals involves the intersection of biology, ecology, and computer science. Addressing the limitations of artificial intelligence in this field requires establishing a collaborative framework that integrates expertise from these disciplines.

Interdisciplinary Collaboration: Biologists provided crucial insights into the ecological and evolutionary context of the signal, while AI researchers contributed expertise in computational tools and methodologies. Ecologists contributed their understanding of environmental variability and system-level interactions. Together, these perspectives constructed a model that is not only technically robust but also biologically meaningful.

Collaborative Platforms: Shared research platforms that integrate sensor networks, artificial intelligence tools, and domain-specific knowledge can accelerate innovation. For example, citizen science projects that collect bioacoustic data can be combined with AI-based analytics workflows to provide scalable solutions for biodiversity monitoring. Collaborative labs and alliances can also facilitate resource sharing, joint datasets, and cross-validation of models.

Training and Capacity Building: Building an effective framework requires training researchers in bioscience and artificial intelligence methodologies. Interdisciplinary educational programs, workshops, and collaborative research initiatives can bridge knowledge gaps and foster innovation.

Policy and Ethical Guidelines: The collaboration should also extend to developing ethical guidelines and policy frameworks that regulate the collection, analysis, and application of communications data. This ensures that technological advancements are aligned with conservation goals, ecological sustainability, and respect for non-human life.

Overcoming the limitations of artificial intelligence in interpreting non-human signals requires technological improvements and robust research practices. Explainable AI improves interpretive clarity, while synthetic data and transfer learning help overcome data scarcity. Multimodal perception and standardized datasets enhance accuracy and consistency. Interdisciplinary collaboration among biology, ecology, and artificial intelligence ensures scientific validity and ethical accountability. These strategies strengthen the decoding of complex non-human communication, leading to a better understanding and protection of it.

6.5 State of the Field and Emerging Trends

The field is moving from basic detection toward meaningful interpretation. Researchers now focus on how AI can reveal intent, emotion, and ecological context in signals rather than only recognizing patterns. However, the lack of transparency in deep models continues to limit biological understanding. This challenge links directly to the ethical need for explainable and responsible AI, where results can be verified and used for conservation or policy decisions with confidence.

Emerging trends emphasize interdisciplinary collaboration and human-AI partnerships that unite computational power with ecological insight. There is growing attention to data justice, Indigenous knowledge, and fair access to technology in biodiversity research. These shifts show that the next phase of progress must balance innovation with accountability, ensuring AI not only decodes communication but also strengthens ethical and sustainable coexistence between humans and other life forms.

7. FUTURE DIRECTIONS

The field of artificial intelligence interpreting non-human signals is still in its early stages, but rapid advancements in computing, data acquisition, and interdisciplinary collaboration herald transformative change. Future directions in this field include: applying AI to real-time ecosystem monitoring using emerging technologies such as quantum computing and advanced neural network architectures; promoting interdisciplinary projects; and exploring its long-term impacts on conservation and sustainable development.

7.1 Potential of Quantum Computing and Advanced Neural Architectures

Quantum computing marks a major leap in computational ability, enabling faster processing of large and complex datasets beyond classical limits. In studying non-human communication, it can efficiently handle high-dimensional and multimodal data. Quantum machine learning methods accelerate pattern recognition and allow real-time analysis of bioacoustic or chemical signals across species. Future research may combine advanced neural architectures such as graph neural networks, capsule networks, and neuromorphic models to capture relational and temporal patterns in nature. These systems could, for instance, map mycorrhizal or animal social networks and mimic biological efficiency. Together, these advances promise AI tools that better reflect the richness and complexity of natural communication..

7.2 Opportunities in Real-Time Ecosystem Monitoring

Real-time ecosystem monitoring is becoming a vital field where AI-based signal interpretation plays a key role. Using sensor networks, drones, and edge devices, continuous data can be gathered on animal behavior, plant health, and environmental changes. AI supports biodiversity monitoring by classifying species calls and detecting plant stress, climate assessment by tracking shifts in communication patterns, and conservation efforts through automated alerts for poaching or population decline. Advances in lightweight AI and edge computing now allow faster, energy-efficient processing in remote areas. Integrating AI with IoT systems enables scalable, low-cost, real-time ecosystem monitoring.

7.3 Cross-Disciplinary Projects for Interspecies Communication

A promising future direction for artificial intelligence is its role in enabling interspecies communication. Achieving this goal calls for close collaboration among AI experts, biologists, ecologists, linguists, and ethicists.

Possible applications include bioacoustic translation platforms that interpret animal calls, plant signaling systems that convert stress signals into useful alerts for farmers or conservationists, and robotic interfaces that respond to natural signals to support ecological balance.

Such efforts must involve interdisciplinary teams to ensure accuracy, ecological safety, and ethical responsibility. Combining insights from science, technology, and policy will help prevent harm to the species being studied. These projects could transform our understanding of life and inspire a more balanced and respectful relationship between humans and other living beings.

7.4 Long-Term Implications for Conservation and Sustainability

Artificial intelligence-driven interpretation of non-human signals could have profound implications for nature conservation, biodiversity protection, and sustainable development. Understanding the hidden language of nature could transform how humans interact with ecosystems. Real-time monitoring and signal interpretation can transform wildlife conservation by enabling early action. Detecting stress in endangered species or interpreting plant signals can guide ecosystem restoration and reforestation. AI-driven communication systems also support sustainable farming, forestry, and fisheries by identifying plant stress, pests, or environmental changes, reducing chemical use, and promoting resource efficiency in line with global sustainability goals. However, long-term AI interpretation of non-human signals raises ethical and philosophical questions about communication, consciousness, and human responsibility toward other species. As technology crosses species boundaries, it must be guided by strong ethical and governance frameworks. The future of AI in understanding non-human signals is both exciting and complex. Advances in quantum computing, neural networks, and real-time monitoring will strengthen conservation and ecological research. Achieving this vision will require collaboration across disciplines, shared research standards, and careful ethical oversight. With these principles, AI can help uncover and respect the hidden language of nature.

8. CONCLUSION

Artificial intelligence-based interpretation of non-human signals is an emerging field that integrates technology with biological and ecological knowledge, offering new avenues for understanding the cryptic communication systems of organisms. However, limited and imbalanced datasets, noisy and context-dependent signals, and constantly changing environmental conditions hinder progress in this area. Many models still operate like black boxes, making it difficult to verify how and why they produce specific results; interspecies differences and the lack of standardized datasets also reduce the reproducibility of results. Human-centered assumptions can distort interpretations, while ethical concerns remind us to adopt a responsible attitude when studying or influencing natural communication. Advancing this field requires technological advancements, rigorous research methodologies, and a high degree of ethical awareness. Interpretable models can improve clarity, synthetic data and transfer learning can help alleviate data shortages, and multimodal sensing can provide a deeper understanding of complex signals. Shared benchmarks help ensure research consistency, while close collaboration among technical experts, biologists, and ecologists ensures relevance and respect for natural systems. Through thoughtful development, this field can better support nature conservation, provide real-time insights into ecosystems, and deepen our understanding of the communication networks that shape life on Earth.

REFERENCES

Aarif, K. O. M., Alam, A., & Hotak, Y. (2025). Smart sensor technologies shaping the future of precision agriculture: Recent advances and future outlooks. *Journal of Sensors*, *2025*(1), 2460098.

Abrol, D. P. (2011). Biochemical basis of plant-pollination interaction. In *Pollination biology: biodiversity conservation and agricultural production* (pp. 413–458). Springer Netherlands.

Alharthi, A., Alqurashi, A., Alharbi, T., Alammar, M., Aldosari, N., Bouchekara, H., & Al Ayidh, A. (2025). Explainable AI for sensor signal interpretation to revolutionize human health monitoring: A review. *IEEE Access : Practical Innovations, Open Solutions*.

Allen, A. N., Harvey, M., Harrell, L., Jansen, A., Merkens, K. P., Wall, C. C., & Oleson, E. M. (2021). A convolutional neural network for automated detection of humpback whale song in a diverse, long-term passive acoustic dataset. *Frontiers in Marine Science*, *8*, 607321.

Assom, L. (2023). *Decoding communication of non-human species: Unsupervised machine learning to infer syntactical and temporal patterns in fruit-bats vocalizations.*

Bakker, K. (2022). *The sounds of life: How digital technology is bringing us closer to the worlds of animals and plants.*

Beecher, M. D. (2021). Why are no animal communication systems simple languages? *Frontiers in Psychology*, *12*, 602635. PMID: 33815200

Benneth, E. O. (2026). *Introduction: Principles of animal behavior.* In *Principles of Animal Behaviour* (pp. 1–15). CRC Press.

Brown, K. (2012). *Little sinners, and other stories*. University of Nebraska Press.

Brumm, H., & Slabbekoorn, H. (2005). Acoustic communication in noise. *Advances in the Study of Behavior*, *35*, 151–209.

Carpenter, C. C. (1978). *Ritualistic social behaviors in lizards.* In *Behavior and neurology of lizards* (pp. 253–267).

Cartmill, E. A. (2023). *Using the senses in animal communication.* In *A new companion to linguistic anthropology* (pp. 369–390).

Cheng, J., Xie, B., Lin, C., & Ji, L. (2012). A comparative study in birds: Call-type-independent species and individual recognition using four machine-learning methods and two acoustic features. *Bioacoustics, 21*(2), 157–171.

Choi, S. R., & Lee, M. (2023). Transformer architecture and attention mechanisms in genome data analysis: A comprehensive review. *Biology (Basel), 12*(7), 1033. PMID: 37508462

Coatsworth, P., Gonzalez-Macia, L., Collins, A. S. P., Bozkurt, T., & Güder, F. (2023). Continuous monitoring of chemical signals in plants under stress. *Nature Reviews. Chemistry, 7*(1), 7–25. PMID: 37117825

Figueiredo, A. F., Boy, J., & Guggenberger, G. (2021). Common mycorrhizae network: A review of theories and mechanisms behind underground interactions. *Frontiers in Fungal Biology, 2*, 735299. PMID: 37744156

Floreano, D., & Mattiussi, C. (2008). *Bio-inspired artificial intelligence: Theories, methods, and technologies.* MIT Press.

Garstang, M. (2010). Elephant infrasounds: Long-range communication. []). Elsevier.]. *Handbook of Behavioral Neuroscience, 19*, 57–67.

Gordon, R., & Seckbach, J. (Eds.). (2016). *Biocommunication: Sign-mediated interactions between cells and organisms* (Vol. 1). World Scientific.

Heijden, M. G. V. D. (2016). Underground networking. *Science, 352*(6283), 290–291. PMID: 27081054

Hirsch, A. M., Bauer, W. D., Bird, D. M., Cullimore, J., Tyler, B., & Yoder, J. I. (2003). Molecular signals and receptors: Controlling rhizosphere interactions between plants and other organisms. *Ecology, 84*(4), 858–868.

Hirsch, A. M., & Fujishige, N. A. (2011). Molecular signals and receptors: Communication between nitrogen-fixing bacteria and their plant hosts. In *Biocommunication of plants* (pp. 255–280). Springer.

Hsu, C. Y., & Li, W. (2023). Explainable GeoAI: Can saliency maps help interpret AI's learning process? *International Journal of Geographical Information Science, 37*(5), 963–987.

Irving, C. J. (2023). *An investigation into multi-component warning colouration* (Doctoral dissertation, Macquarie University).

Jacuzzi, G., & Olden, J. D. (2025). Few-shot transfer learning enables robust acoustic monitoring of wildlife communities at the landscape scale. *Ecological Informatics, 80*, 103294.

Jianhua, L. (1998). Communication and the evolution of ecosystems. *World Futures: Journal of General Evolution*, *52*(1), 35–73.

Kumar, V. (2025). The complex interaction between plants and acoustic signals: Friends or foes? *Naturwissenschaften*, *112*(3), 43. PMID: 40445373

Kvsn, R. R., Montgomery, J., Garg, S., & Charleston, M. (2020). Bioacoustics data analysis: A taxonomy, survey and open challenges. *IEEE Access : Practical Innovations, Open Solutions*, *8*, 57684–57708.

Locke, J. L. (2021). The indexical voice: Communication of personal states and traits in humans and other primates. *Frontiers in Psychology*, *12*, 651108. PMID: 33935911

London, A. J. (2019). Artificial intelligence and black-box medical decisions: Accuracy versus explainability. *The Hastings Center Report*, *49*(1), 15–21. PMID: 30790315

Luphade, N. (2023). Bioacoustic classification of Hainan gibbon call types using deep learning. *Journal of Bioacoustics Research*, *15*(2), 112–123.

Mienye, I. D., Swart, T. G., & Obaido, G. (2024). Recurrent neural networks: A comprehensive review. *Information (Basel)*, *15*(9), 517.

Mishra, R. C., Ghosh, R., & Bae, H. (2016). Plant acoustics: Searching for a sound mechanism for signaling in plants. *Journal of Experimental Botany*, *67*(15), 4483–4494. PMID: 27342223

Nagesh, K. M., & Piovene, C. (2011). *Emission and function of volatile organic compounds in response to abiotic stress.* In *Abiotic stress in plants: Mechanisms and adaptations* (p. 367).

Nakade, D., & Dhadse, S. (2024). The plant bioacoustics: A system of plant-sound relationship. *Plant Science*, *7*(1), 1–8.

Oyewole, J. A., Awolaju, B. A., & Ayomide, P. A. O. (2025). Exploring interspecies communication: A sociolinguistic analysis of animal signals and their human interpretations. *Biological Sciences*, *5*(2), 898–906.

Pantic, I. V., Pantic, J. P., Valjarevic, S., Corridon, P. R., & Topalovic, N. (2025). AI approaches using random forests for signal analysis. *Chemico-Biological Interactions*, •••, 111624. PMID: 40582691

Robinson, D. G., & Draguhn, A. (2021). Plants have neither synapses nor a nervous system. *Journal of Plant Physiology*, *263*, 153467. PMID: 34247030

Rogers, L. J., & Kaplan, G. T. (2002). *Songs, roars, and rituals: Communication in birds, mammals, and other animals*. Harvard University Press.

Sarma, H. Joshi, S., Lahiri, D., Ray, R. R., & Davoodbasha, M. (Eds.). (2023). *Microbial biofilms: Challenges and advances in metabolomic study*.

Schneider, F. D., Brose, U., Rall, B. C., & Guill, C. (2016). Animal diversity and ecosystem functioning in dynamic food webs. *Nature Communications*, *7*(1), 12718. PMID: 27703157

Stowell, D. (2022). Computational bioacoustics with deep learning: A review and roadmap. *PeerJ*, *10*, e13152. PMID: 35341043

Subasi, A., Qaisar, S. M., & Nisar, H. (2024). *Artificial intelligence and multimodal signal processing in human-machine interaction*. Elsevier.

Swift, S., Downie, J. A., Whitehead, N. A., Barnard, A. M., Salmond, G. P., & Williams, P. (2001). Quorum sensing as a population-density–dependent determinant of bacterial physiology. *Philosophical Transactions of the Royal Society of London. Series B, Biological Sciences*, *357*(1420), 593–607. PMID: 11450110

Valente, D., Wang, H., Andrews, P., Mitra, P. P., Saar, S., Tchernichovski, O., & Benjamini, Y. (2007). Characterizing animal behavior through audio and video signal processing. *IEEE MultiMedia*, *14*(4), 32–41.

Virant-Doberlet, M., Stritih-Peljhan, N., Žunič-Kosi, A., & Polajnar, J. (2023). Functional diversity of vibrational signaling systems in insects. *Annual Review of Entomology*, *68*(1), 191–210. PMID: 36198397

Williams, S. M. (2024). *Tracking and predicting bee behaviour and movement using machine learning* (Doctoral dissertation, Bangor University).

Zandi, A., Hosseinirad, S., Kashani Zadeh, H., Tavakolian, K., Cho, B. K., Vasefi, F., & Tavakolian, P. (2025). A systematic review of multi-mode analytics for enhanced plant stress evaluation. *Frontiers in Plant Science*, *16*, 1545025. PMID: 40370362

Chapter 5

Beyond Algorithms: How AI Influences Attitudes, Perceptions, and Learning Outcomes

Pankaj Bhambri
https://orcid.org/0000-0003-4437-4103
Guru Nanak Dev Engineering College, Ludhiana, India

Suresh Kumar
https://orcid.org/0009-0009-1272-5053
Maharishi Markandeshwar, Ambala, India

Sonal Wadhwa
Geeta University, Panipat, India

ABSTRACT

Teacher education has taken a quantum leap from algorithmic efficiency because Artificial Intelligence (AI) technologies now deeply affects student learning's psycho-social character. The chapter investigates the AI-driven transformation of affective factors including student attitudes and perceptions as well as anxiety levels and learning motivation and self-regulation capabilities beyond basic algorithm usage. The research examines AI-based educational tools and student emotional responses and cognitive processes to reveal the ways technology modifies teaching conditions and modifies achievement standards. A combination of theoretical models and real-world study transforms into actual insights about AI systems' twofold effects within the education field. AI creates individualized student learning along with active classroom participation because it delivers responsive guidance and support systems.

DOI: 10.4018/979-8-3373-5483-5.ch005

1 INTRODUCTION: THE HUMAN DIMENSION OF AI

Artificial Intelligence (AI) has long been framed as a technical marvel—a constellation of algorithms, data pipelines, and computational power. Yet, its most profound impacts lie not in its code but in its ability to reshape human experiences, beliefs, and behaviors. This chapter moves beyond the mechanics of AI to explore how it alters attitudes (how people feel about technology, themselves, and society), perceptions (how they interpret reality), and learning outcomes (how knowledge and skills are acquired in AI-mediated environments). By focusing on the human dimension, we challenge the assumption that AI is merely a tool and instead examine it as a social actor with the power to redefine norms, relationships, and cognitive processes. This chapter progresses through three interconnected layers: (1) Foundational frameworks explaining human-AI psychological dynamics (Sections 5-8), (2) Empirical case studies revealing attitude/perception shifts across domains (Sections 9-11), and (3) Actionable policy/design recommendations (Sections 12-14). Each layer builds toward answering our core question: How can we harness AI's transformative potential while safeguarding human cognition and equity?"

1.1 Defining AI's Impact Beyond Technical Systems

AI's influence extends far beyond optimizing workflows or automating tasks. It operates as a cultural and psychological force that:

- Mediates Human Decision-Making: From healthcare diagnoses to hiring decisions, AI systems increasingly guide choices once reserved for human judgment, raising questions about agency and accountability (Sharma et al., 2020).
- Reshapes Social Dynamics: Social media algorithms curate friendships, news, and political discourse, altering how communities form and interact.
- Reconstructs Identity and Expertise: Tools like ChatGPT blur the lines between human and machine creativity, challenging traditional notions of authorship and intellectual authority.

The transition from AI as a task-specific tool to a pervasive influence requires rethinking its role in society. For instance, AI-driven personalized learning platforms don't just teach math—they shape learners' self-perceptions (e.g., "Am I 'good' at this subject?") through adaptive feedback. Similarly, facial recognition technologies don't just identify faces—they embed societal biases into policing and public services, reinforcing systemic inequities. When generative AI like DALL-E produces art, it sparks debates about originality and creativity, forcing society to confront whether "art" requires human intent. Such cases illustrate AI's dual role as both a mirror (reflecting human values) and a mold (shaping new norms).

1.2 Why Attitudes and Perceptions Matter in the AI Era

Attitudes and perceptions are the bridge between AI's technical capabilities and its real-world consequences:

- Trust Determines Adoption: Public skepticism about AI's fairness or transparency can derail even the most advanced systems. For example, vaccine hesitancy campaigns amplified by social media algorithms highlight how distrust in AI-curated information fuels real-world harm.
- Perceptions Drive Behavior: If students perceive AI tutors as infallible, they may uncritically accept their feedback, stifling independent problem-solving. Conversely, fear of AI "replacing" teachers could lead to resistance against beneficial tools.
- Bias in, Bias Out: Human attitudes about race, gender, or class are often encoded into AI datasets, creating feedback loops. A hiring algorithm trained on historical data may perpetuate discrimination, but users' perception of the algorithm as "neutral" can legitimize its biased outcomes.

Psychological Underpinnings:

- Cognitive Dissonance: People may simultaneously praise AI's efficiency and resent its intrusiveness (e.g., liking Netflix recommendations but resenting data collection).
- Anthropomorphism: Assigning human-like intent to AI (e.g., "The algorithm wants me to click this") shapes trust and ethical expectations (Ododo et al., 2024).
- Societal Stakes: Without addressing attitudes and perceptions, AI risks exacerbating divides—between those who benefit from its efficiencies and those marginalized by its biases, or between generations adapting to AI-augmented workplaces.

1.3 Chapter Overview and Key Questions

This chapter synthesizes interdisciplinary research—from behavioral psychology to education theory—to explore AI's human-centric impacts. Each section builds toward answering these key questions:

- How do trust and distrust in AI systems form, and what societal factors amplify them?
- In what ways do AI-mediated environments (e.g., social media, classrooms) alter how individuals perceive reality and their own capabilities?
- Does AI enhance or undermine critical thinking and metacognition in learning contexts?
- How can we design AI systems that prioritize ethical and cognitive growth over mere efficiency?

Roadmap:

- Section 2 establishes a theoretical foundation, examining how humans psychologically engage with AI.
- Sections 3–5 delve into attitudes (trust, bias), perceptions (worldviews, self-concept), and learning outcomes, supported by case studies from education, healthcare, and media.
- Sections 6–7 propose ethical frameworks and policy strategies to align AI development with human flourishing.

.2 THEORETICAL FRAMEWORK: BRIDGING TECHNOLOGY AND HUMAN BEHAVIOR

To understand how AI shapes attitudes, perceptions, and learning, we must ground our analysis in interdisciplinary theories that connect technological systems to human cognition, behavior, and societal norms (Bergdahl & Sjöberg, 2025). This section establishes the foundational frameworks for analyzing AI's sociopsychological impacts.

2.1 The Psychology of Human-AI Interaction

Human interaction with AI is shaped by cognitive, emotional, and social processes that mirror human-human interactions but with unique complexities (Mohanta et al., 2022). Key psychological principles include:

- Anthropomorphism and Mental Models: Humans tend to attribute human-like qualities (e.g., intentionality, emotions) to AI systems, even when aware of their algorithmic nature (Chopra et al., 2011). This is driven by evolutionary tendencies to anthropomorphize non-human entities. For example, users may perceive chatbots as empathetic or voice assistants as authoritative.

- Computers Are Social Actors (CASA) Paradigm: Nass & Reeves (1996) demonstrated that humans apply social rules (e.g., politeness, reciprocity) to machines, influencing trust and engagement (Rawashdeh et al., 2025).
- Agency and Accountability: Users often struggle to assign responsibility for AI decisions, oscillating between blaming the system, its designers, or themselves.

- Trust Calibration: Trust in AI is dynamic and context-dependent. Factors include:
  - Transparency: Users demand explainability (e.g., Why did the AI make this recommendation?).
  - Reliability: Consistent accuracy builds trust, while errors (e.g., biased hiring algorithms) erode it.
  - Familiarity: Users with technical literacy may scrutinize AI outputs more critically than novices (Choi et al., 2024).
- Feedback Loops and Behavior Modification: AI systems shape user behavior through reinforcement (e.g., adaptive learning platforms rewarding progress) or punitive nudges (e.g., productivity apps shaming inactivity). Over time, this can alter decision-making patterns and self-perception.
- Dual-Process Theory: Kahneman's System 1 (fast, intuitive) and System 2 (slow, analytical) thinking apply to AI interactions. Users may uncritically accept AI outputs (System 1) or engage deliberative critique (System 2), depending on cognitive load and task complexity.

2.2 Cognitive Biases in Interpreting AI Outputs

Human cognition is riddled with biases that distort how AI-generated information is processed. These biases amplify or undermine AI's influence:

- Automation Bias: Overreliance on AI suggestions, even when flawed. For example, clinicians may defer to diagnostic algorithms despite contradictory symptoms.
- Confirmation Bias: Users selectively accept AI outputs that align with preexisting beliefs. Social media recommendation engines exploit this, reinforcing echo chambers.
- Anchoring Effect: Initial AI-generated suggestions (e.g., a salary range from a hiring algorithm) anchor users' judgments, skewing subsequent decisions.
- Halo Effect: High performance in one domain (e.g., AI excelling at chess) leads to inflated trust in unrelated domains (e.g., AI-driven medical advice)

- Dunning-Kruger Effect: Users with limited AI literacy may overestimate their ability to evaluate AI outputs, leading to misuse (e.g., students relying on ChatGPT without verifying facts).
- Algorithmic Aversion: Conversely, some users reject AI inputs due to distrust or perceived threats to autonomy. For instance, educators may resist automated grading systems.
- Case Example: In criminal justice, judges using AI risk-assessment tools may unconsciously anchor sentencing decisions to algorithmic predictions, even when biased against marginalized groups (Otermans et al., 2025).

2.3 Sociocultural Theories of Technology Adoption

AI's impact varies across cultures and communities, shaped by sociocultural norms and systemic inequities. Key theories include:

- Technology Acceptance Model (TAM): Davis posits that adoption hinges on perceived usefulness and perceived ease of use. For AI:
    - Usefulness: AI must align with users' goals (e.g., personalized learning).
    - Ease of Use: Complex interfaces or opaque decision-making hinder adoption.
- Social Cognitive Theory (SCT): Bandura's theory emphasizes observational learning and self-efficacy. AI influences behavior through:
    - Modeling: Users mimic AI behaviors (e.g., adopting writing styles from GPT-3).
    - Self-Efficacy: AI tools can empower (e.g., language apps boosting confidence) or undermine autonomy (e.g., over-dependence on navigation apps).
- Sociotechnical Systems Theory: AI is embedded in broader social, political, and economic systems. For example:
    - Power Dynamics: Corporate-controlled AI (e.g., gig economy algorithms) can exploit workers.
    - Cultural Values: Collectivist societies may prioritize AI for communal benefit, while individualist cultures focus on personalization.
- Global Case Study: In collectivist cultures like Japan, AI elder-care robots are embraced to address labor shortages, whereas Western societies often debate ethical concerns about dehumanization.

3 AI'S INFLUENCE ON ATTITUDES

Artificial intelligence's integration into daily life has fundamentally reshaped how individuals perceive and interact with technology. Beyond its technical capabilities, AI influences human attitudes in profound ways, altering trust dynamics, redefining expertise, and reflecting—or refracting—societal biases. This section unpacks these shifts, grounding abstract concepts in real-world examples.

3.1 Trust and Distrust in AI Systems

Trust is the cornerstone of human-AI collaboration, yet it remains fragile and context-dependent. Unlike human relationships, trust in AI systems is often transactional, shaped by users' perceptions of reliability, fairness, and predictability. However, this trust is not uniform; it fractures in domains where errors carry high stakes, such as healthcare or criminal justice.

- Factors Shaping Trust (Transparency, Accuracy, Familiarity): Three interrelated factors govern trust in AI: transparency, accuracy, and familiarity.
    - Transparency refers to the "explainability" of AI decisions. Systems that provide clear rationales for their outputs (e.g., showing confidence scores in diagnostic tools) foster trust, while opaque "black box" models (e.g., deep learning algorithms) breed skepticism. For instance, users are more likely to trust a medical AI that highlights the factors contributing to a diagnosis than one that simply delivers a result.
    - Accuracy is the bedrock of trust. Consistent, error-free performance strengthens user confidence, but even minor inaccuracies can erode trust, especially in high-risk scenarios. A study by the University of Washington found that radiologists were more likely to override AI recommendations after encountering just one incorrect diagnosis, despite overall high accuracy.
    - Familiarity plays a subtle but powerful role. Users accustomed to AI in low-stakes contexts (e.g., Netflix recommendations) may transfer this trust to critical domains (e.g., autonomous vehicles), sometimes overestimating AI's capabilities. Conversely, unfamiliarity can amplify distrust, as seen in older populations hesitant to adopt AI-driven elder care tools (Bauer et al., 2025).
- Case Study: Healthcare Diagnostics vs. Criminal Justice Algorithms: The dichotomy between healthcare and criminal justice illustrates how context shapes trust.

- In healthcare, AI diagnostic tools like IBM Watson for Oncology or Google's DeepMind for retinal scans are increasingly trusted by professionals, partly due to rigorous validation processes and transparent collaboration between developers and clinicians. Patients, too, express growing comfort with AI when it augments (rather than replaces) human expertise, as seen in tools that prioritize cases for overburdened doctors (Babu et al., 2021).
- In criminal justice, however, AI systems like COMPAS (used for risk assessment in sentencing) face widespread distrust. Critics argue these tools perpetuate racial biases due to flawed training data, such as historical arrest records reflecting systemic inequities. Judges and advocates often reject algorithmic recommendations, citing a lack of transparency in how scores are calculated and a moral discomfort with outsourcing life-altering decisions to machines.

3.2 Shifts in Attitudes Toward Expertise (e.g., "Is AI Smarter Than Humans?")

AI has disrupted traditional hierarchies of knowledge, sparking debates about the nature of expertise. Surveys reveal a growing perception that AI can outperform humans in specialized tasks, from chess to cancer detection. However, this shift is uneven and fraught with contradictions.

- In STEM fields, AI is often viewed as a superior "partner" for tasks requiring data analysis or pattern recognition. For example, mathematicians now use AI to generate conjectures, while biologists employ AlphaFold to predict protein structures—a task once deemed intractable. This has led to a cultural shift where AI is seen not as a competitor but as a collaborator that expands human capability.
- Conversely, in creative and ethical domains, resistance persists. Many reject the idea that AI can replicate "human" skills like empathy, intuition, or moral reasoning. The backlash against AI-generated art (e.g., controversies over DALL-E or ChatGPT-authored novels) reflects fears of devaluing human creativity. Similarly, ethicists argue that AI lacks the contextual nuance needed for decisions like end-of-life care or criminal sentencing.
- A paradoxical trend emerges: while people acknowledge AI's superiority in narrow tasks, they simultaneously anthropomorphize its limitations. For instance, users may forgive errors in conversational AI ("It's just a machine") while holding human experts to higher standards. This duality reveals a soci-

etal struggle to reconcile AI's capabilities with its perceived "soullessness." (Singh et al., 2021).

3.3 The Role of AI in Reinforcing or Challenging Societal Biases

AI systems are not neutral arbiters; they inherit and amplify the biases embedded in their training data and design choices. However, they also hold potential to expose and correct human prejudices.

- Reinforcing Biases: When trained on historical data, AI often perpetuates systemic inequities. For example:
    - o Hiring algorithms trained on resumes from male-dominated industries may downgrade female applicants.
    - o Facial recognition systems like Amazon's Rekognition have shown higher error rates for darker-skinned individuals, leading to wrongful arrests.
- Challenging Biases: Conversely, AI can act as a "bias detector" when designed intentionally. Tools like Microsoft's Fairlearn or IBM's AI Fairness 360 help auditors identify disparities in algorithmic decisions (Laru et al., 2025). In education, AI tutors have been shown to reduce grading biases by applying uniform rubrics to student work. In one landmark case, an AI analyzing loan approvals uncovered that human lenders—not the algorithm—disproportionately rejected minority applicants, prompting institutional reforms. The tension here lies in who controls the narrative. Left unchecked, AI risks entrenching existing power structures. Yet, when guided by ethical frameworks and diverse teams, it can become a tool for equity—a mirror that reflects society's flaws and a hammer to reshape them.

4 AI AND THE EVOLUTION OF PERCEPTIONS

AI's role in reshaping human perceptions extends far beyond its technical functions, influencing how individuals interpret information, interact with technology, and envision their future roles in society.

4.1 AI as a Mirror: How Algorithms Shape Worldviews

AI systems, particularly those powering social media and content platforms, act as digital mirrors that reflect and amplify users' preferences, often entrenching existing beliefs.

- Social Media Echo Chambers and Filter Bubbles: AI-driven recommendation algorithms on platforms like Facebook and YouTube prioritize engagement, curating content that aligns with users' past behaviors. This creates echo chambers, where individuals are repeatedly exposed to ideologically similar viewpoints, and filter bubbles, insulating them from diverse perspectives. Over time, this reinforcement can deepen societal divides, as seen in polarized political discourse or conspiracy theory proliferation.
- Personalization vs. Polarization: While personalized AI enhances user experience—such as tailored learning resources or curated news—it risks fragmenting shared realities. The same tools that customize content for relevance can isolate communities into ideological silos, exacerbating polarization (Harleen, 2016). For instance, algorithmically amplified misinformation during crises illustrates the tension between individual convenience and collective societal cohesion.

4.2 Perceptions of AI in Education and Workplace Automation

Public perceptions of AI oscillate between optimism about its potential and anxiety over its disruptive force, shaping adoption and resistance across sectors.

- Fear of Replacement vs. Augmentation Narratives: In workplaces, AI sparks fears of job displacement, particularly in industries like manufacturing or customer service. Conversely, the augmentation narrative frames AI as a tool to enhance human productivity—e.g., designers using generative AI for rapid prototyping or doctors leveraging diagnostic algorithms to improve accuracy (Ritu & Bhambri, 2025). This duality underscores the need for narratives that balance innovation with empathy.
- AI as a Collaborative Partner: Increasingly, AI is perceived not as a competitor but as a collaborator. In education, AI tutors like Khanmigo provide personalized feedback while teachers focus on mentorship. In healthcare, AI assists radiologists in detecting anomalies, augmenting expertise rather than replacing it. This shift reframes AI as a partner in problem-solving, fostering trust and redefining human roles in symbiotic systems.

5 LEARNING OUTCOMES IN THE AGE OF AI

The integration of AI into education is reshaping how knowledge is acquired, assessed, and applied. While AI-driven tools promise efficiency and personalization, they also raise critical questions about cognitive development, equity, and long-term skill relevance.

5.1 AI-Driven Adaptive Learning Systems

Adaptive learning platforms, powered by AI, tailor content to individual learner needs, optimizing pacing and difficulty (Yim & Wegerif, 2024). Platforms like Khan Academy and Duolingo exemplify success stories, where personalized pathways improve engagement and mastery. However, limitations persist: over-reliance on algorithmic curation risks narrowing intellectual exploration, while unequal access to technology exacerbates educational disparities.

- Personalized Education: Success Stories and Limitations

Personalized AI systems boost outcomes for learners with diverse needs—e.g., students with disabilities benefit from customized interfaces. Yet, such systems often prioritize quantifiable metrics (test scores) over holistic growth, and their effectiveness hinges on high-quality data, which can embed biases or overlook cultural contexts.

- The "Black Box" Problem: Do Learners Understand AI Feedback?

Many AI tools provide feedback without transparency, leaving learners unaware of how decisions are made. For instance, an AI tutor might adjust difficulty levels based on opaque criteria, hindering metacognitive awareness. Explainable AI (XAI) and user-friendly dashboards are emerging solutions to demystify algorithmic logic (Shao et al., 2025).

5.2 Critical Thinking in an AI-Saturated World

AI's ubiquity challenges traditional notions of critical inquiry. While tools like ChatGPT streamline research, they risk fostering intellectual complacency if users uncritically accept AI-generated answers. Educators report concerns about declining analytical rigor as students bypass deep engagement with material.

- Overreliance on AI Tools (e.g., ChatGPT in Academia): The ease of outsourcing tasks to AI—from essay writing to problem-solving—can erode foundational skills. Students may struggle to differentiate credible sources from AI hallucinations, while plagiarism detectors lag in identifying AI-generated content, complicating academic integrity.

- Enhancing Metacognition Through AI Reflection Prompts: Conversely, AI can foster self-regulated learning by prompting reflection. For example, AI tutors might ask learners to justify their answers or track progress over time, encouraging metacognitive habits that deepen understanding and autonomy.

5.3 Long-Term Impacts on Skill Development

As AI automates technical tasks (e.g., coding, data analysis), demand grows for uniquely human skills: creativity, empathy, and ethical reasoning. Educational systems must balance teaching AI literacy with nurturing traits resistant to automation, ensuring learners remain adaptable in dynamic job markets.

6 ETHICAL AND PEDAGOGICAL CONSIDERATIONS

6.1 Balancing Efficiency with Equity in AI Deployments

While AI promises efficiency through personalized learning and automated decision-making, its deployment risks exacerbating inequities. Marginalized communities often lack access to advanced technologies, and AI systems trained on biased or non-representative data may reinforce existing disparities. For instance, adaptive learning tools might favor students with strong digital literacy, leaving others behind. Ethical integration demands prioritizing equity by designing inclusive AI models, auditing tools for fairness, and ensuring resource distribution bridges—rather than widens—the digital divide. Policymakers and educators must weigh efficiency gains against the moral imperative to uplift all learners.

6.2 Data Privacy and Learner Autonomy

AI's reliance on vast datasets raises critical concerns about data privacy, particularly in educational settings where sensitive student information is collected. While personalized feedback and predictive analytics can enhance learning, they risk commodifying learner data or enabling surveillance. Ethical frameworks must ensure compliance with regulations (e.g., GDPR, COPPA) while safeguarding learner autonomy. Students should retain control over their data, understanding how it's used and opting out where possible. Pedagogically, over-reliance on AI-driven recommendations can stifle independent thinking, necessitating a balance between guidance and fostering self-directed learning.

6.3 Mitigating Algorithmic Bias in Educational and Social Contexts

Algorithmic bias—embedded through skewed training data or flawed design—can perpetuate harmful stereotypes, such as underestimating capabilities of minority students or reinforcing gender roles in career recommendations. In education, biased grading algorithms or recruitment tools risk entrenching systemic discrimination. Mitigation requires transparency in AI development, diversifying datasets, and involving multidisciplinary teams to audit systems for fairness. Pedagogically, educators must critically evaluate AI outputs and teach students to recognize bias, transforming AI from a passive tool into a catalyst for equity. Proactive measures, like bias-impact assessments and inclusive design principles, are vital to ensure AI aligns with social justice goals (Rani et al., 2025).

7 FUTURE DIRECTIONS: TOWARD HUMAN-CENTRIC AI INTEGRATION

7.1 Designing AI Systems for Attitudinal and Cognitive Growth

The next generation of AI must prioritize human attitudinal and cognitive development as core design principles (Babiker et al., 2025). Current systems often optimize for efficiency, accuracy, or engagement, but future AI should aim to cultivate critical thinking, creativity, and ethical reasoning in users. For instance, adaptive learning platforms could integrate reflective prompts that encourage learners to question AI-generated answers, fostering metacognition. In workplace settings, AI tools might be redesigned to act as "thought partners," offering divergent perspectives to counter confirmation bias rather than merely streamlining tasks.

A critical challenge lies in addressing the "black box" dilemma: users often accept AI outputs uncritically due to opaque decision-making processes. To combat this, developers could embed explainability features, such as real-time visualizations of how algorithms weigh data or confidence intervals for predictions (Akman, 2025). For example, an AI tutor might highlight gaps in its training data when providing feedback, prompting learners to seek additional sources. Additionally, AI systems could be programmed to detect and mitigate user overreliance—e.g., a writing assistant that gradually reduces its suggestions as a student's skills improve.

Ethical design must also confront algorithmic bias proactively. Systems trained on diverse, inclusive datasets and audited for fairness could help users recognize and challenge societal prejudices. Imagine a hiring AI that not only avoids biased candidate rankings but also educates HR teams on historical inequities in hiring practices. By aligning AI design with pedagogical and psychological principles, technology can become a catalyst for intellectual humility, curiosity, and lifelong learning.

7.2 Policy Implications: Regulation, Literacy, and Public Discourse

As AI permeates daily life, policymakers face urgent challenges in balancing innovation with accountability. Regulation must evolve beyond technical safety standards to address sociocultural impacts. For example, mandates could require transparency reports for AI tools used in education or healthcare, disclosing accuracy rates, data sources, and potential biases. Governments might also establish independent oversight bodies to audit high-stakes AI systems, akin to the FDA's role in drug approval.

Equally vital is fostering AI literacy across demographics. National curricula should integrate modules on algorithmic thinking, data literacy, and ethical reasoning, empowering students to interrogate AI outputs critically. Public campaigns could demystify AI for adults, using relatable analogies (e.g., "AI as a tool, not an oracle") to combat techno-determinism. For policymakers themselves, partnerships with ethicists and educators will be essential to craft laws that reflect nuanced human values rather than corporate or technical interests (Kshirsagar et al., 2022).

Public discourse must also shift from polarized debates (e.g., "AI as savior vs. doom") to inclusive, evidence-based dialogues. Platforms like citizen assemblies or AI ethics juries could engage diverse stakeholders in co-designing governance frameworks. Media literacy initiatives might train journalists to avoid sensationalism, ensuring coverage highlights both AI's potential and its limitations. By centering democratic participation, societies can ensure AI development aligns with collective goals like equity, sustainability, and human dignity (Silva et al., 2024).

7.3 Preparing Societies for Symbiotic Human-AI Futures

A symbiotic future hinges on reimagining social systems to leverage AI's strengths while safeguarding human agency. In education, this means moving beyond "automating" classrooms to redesigning pedagogies. Schools might teach students to collaborate with AI on complex problem-solving (e.g., using climate modeling tools to design sustainability projects), emphasizing skills like empathy, ethics, and systems thinking that machines cannot replicate (Jawad et al., 2024).

The workforce will require large-scale reskilling initiatives. Governments and corporations could fund "AI transition" programs, pairing technical training with psychosocial support for workers displaced by automation. For example, displaced factory workers might learn to manage AI-driven production systems while developing leadership skills to oversee human-machine teams. Simultaneously, labor policies must evolve to protect gig workers whose livelihoods depend on opaque algorithmic platforms (e.g., ride-share drivers subject to dynamic pricing models).

Culturally, societies must confront existential questions about human identity and purpose in an AI-augmented world. Philosophical frameworks like capability ethics—which prioritize expanding individuals' opportunities to thrive—could guide the integration of AI into daily life. For instance, urban planners might deploy AI to optimize public transit while ensuring solutions prioritize marginalized communities' access. Grassroots movements, artists, and religious leaders could also play pivotal roles in shaping narratives that celebrate human-AI collaboration rather than competition.

Ultimately, preparing for symbiosis demands a shift from reactive adaptation to proactive co-creation. By embedding foresight exercises, participatory design, and interdisciplinary collaboration into institutional DNA, societies can build resilient ecosystems where humans and AI evolve together—not as rivals, but as partners in addressing humanity's greatest challenges.

8 CONCLUSION: REIMAGINING PROGRESS IN THE AI AGE

The transformative power of AI extends far beyond its technical capabilities, reshaping human attitudes, perceptions, and learning in profound ways (Jain et al., 2021). As we stand at the crossroads of technological advancement and societal evolution, this chapter underscores the imperative to reorient our priorities—not merely toward refining algorithms, but toward fostering environments where human potential and ethical values remain central.

8.1 Cross-Cultural AI Adoption Patterns: A Comparative Lens

Educational Bias in India vs. Sweden: India's BYJU'S AI tutoring shows 23% lower engagement with rural vernacular content, while Sweden's Kunskapskolan AI prioritizes accessibility through government-funded devices. Table 1.1 demonstrates the healthcare trust gradients for three countries.

Table 1. Healthcare Trust Gradient

Country	AI Diagnostic Trust	Key Influencing Factors
Japan	68%	Cultural techno-optimism
Brazil	32%	Colonial data extraction fears
Nigeria	41%	Religious ontology conflicts

Workplace Automation Perceptions:

- Germany: Co-determination models yield worker-AI collaboration frameworks (IG Metall Union AI Charters)
- USA: Gig economy algorithms exacerbate racial pay gaps (Uber/Lyft driver studies show 17% pay disparity)

8.2 Key Takeaways for Educators, Policymakers, and Technologists

Educators: The integration of AI into educational systems demands a dual focus: leveraging adaptive learning tools to personalize education while cultivating critical thinking and metacognitive skills. Educators must design curricula that teach students to interrogate AI outputs, recognize biases, and resist overreliance on automated solutions. Emphasizing creativity, empathy, and ethical reasoning will prepare learners to thrive alongside AI, not just compete with it.

Policymakers: Regulatory frameworks must prioritize equity, transparency, and accountability. Policies should mandate audits for algorithmic bias in high-stakes domains (e.g., education, hiring, healthcare) and invest in public literacy programs to demystify AI. Collaboration with technologists and educators is critical to ensure AI systems align with societal values and do not exacerbate existing inequalities.

Technologists: Developers bear responsibility for designing AI systems that are transparent, interpretable, and inclusive (Sharma et al., 2020). This involves embedding ethical principles into the design process, such as fairness-by-design, and actively engaging with diverse stakeholders to identify unintended consequences. Open-source initiatives and interdisciplinary partnerships can mitigate risks of bias and opacity.

8.3 Call to Action: Prioritizing Human Flourishing Over Algorithmic Mastery

The pursuit of algorithmic efficiency must not eclipse the broader goal of human flourishing. AI's true potential lies not in replacing human agency but in augmenting it—enhancing creativity, reducing inequities, and addressing global challenges. To achieve this vision, stakeholders must collaborate across disciplines:

- Redefine Progress: Shift metrics of success from technical performance (e.g., accuracy, speed) to holistic outcomes, such as equitable access, cognitive growth, and societal well-being.
- Foster Interdisciplinary Dialogue: Create platforms where educators, policy-makers, and technologists co-design AI systems that prioritize human dignity (Babu et al., 2021). For instance, participatory AI governance models can empower marginalized communities in shaping technology that affects their lives.
- Invest in Human-Centric Infrastructure: Develop public policies that fund AI literacy programs, ethical training for developers, and safeguards against surveillance or data exploitation.

The AI age invites us to reimagine progress as a symbiotic relationship between human and machine intelligence. By anchoring innovation in empathy, equity, and ethical reflection, we can ensure that AI serves as a catalyst for collective flourishing—not a tool of alienation or control. The path forward demands courage to prioritize people over profit, wisdom to balance efficiency with justice, and creativity to envision futures where technology amplifies, rather than diminishes, our humanity.

9 CONCLUSION

9.1 Critical Thinking in an AI-Saturated World

As AI systems increasingly curate, filter, and generate content, individuals face unprecedented challenges in discerning fact from fiction. Algorithms powering search engines, social media platforms, and recommendation systems prioritize engagement over accuracy, often amplifying sensationalism, bias, or misinformation (Rani et al., 2025). This creates a paradox: while AI promises efficiency, it also risks fostering intellectual complacency. Learners and professionals alike must cultivate the ability to interrogate AI-generated outputs—asking not only "Is this information correct?" but "Why did the algorithm surface this?" and "What perspectives are missing?" Critical thinking here becomes a survival skill, enabling users to recognize algorithmic manipulation, resist echo chambers, and reclaim agency over their cognitive landscapes. Educators and institutions must prioritize teaching strategies like source triangulation, bias detection, and ethical reasoning to empower individuals to navigate this AI-driven information ecosystem responsibly.

9.2 Critical Thinking in an AI-Saturated World

The convenience of AI tools, from ChatGPT solving complex problems to predictive analytics streamlining decisions, risks eroding human cognitive rigor. Studies show that overreliance on AI can diminish problem-solving creativity, reduce accountability for errors, and create a "deference trap" where users uncritically accept algorithmic outputs. For instance, students using AI essay generators may bypass the intellectual struggle necessary to develop original thought, while professionals automating tasks may lose touch with foundational skills (Singh et al., 2021). To counteract this, critical thinking must be reframed as a counterbalance to automation. This involves designing AI systems that encourage reflection (e.g., prompting users to justify edits to AI-generated text) and fostering cultures where questioning algorithmic decisions is normalized. By framing AI as a provocateur of inquiry rather than a replacement for human intellect, societies can harness its power without sacrificing cognitive depth.

9.3 Critical Thinking in an AI-Saturated World

AI's pervasive influence extends beyond individual cognition to reshape collective norms and values. Critical thinking in this context demands scrutiny of AI's societal role: Who benefits from its outputs? How do automated systems perpetuate or disrupt power structures? For example, hiring algorithms that prioritize efficiency over fairness may entrench systemic inequities, while educational AI tools lacking cultural sensitivity could marginalize diverse learners. Addressing these issues requires meta-critical thinking—the ability to analyze not only AI's technical mechanisms but also its sociopolitical ramifications. Policymakers, technologists, and citizens must collaborate to establish frameworks that prioritize transparency, accountability, and inclusivity. This includes advocating for algorithmic literacy in public education, auditing AI systems for ethical compliance, and designing participatory models where marginalized voices shape AI development. Ultimately, critical thinking becomes a civic duty, ensuring AI serves as a tool for equitable progress rather than an opaque authority.

REFERENCES

Akman, E. (2025). The impact of AI-based visual designs on students' AI literacy and attitudes toward AI. *Interactive Learning Environments*, •••, 1–19.

Babiker, A., Alshakhsi, S., Al-Thani, D., Montag, C., & Ali, R. (2025). Attitude towards AI: Potential influence of conspiracy belief, XAI experience and locus of control. *International Journal of Human-Computer Interaction*, *41*(13), 7939–7951.

Babu, G. C. N., Gupta, S., Bhambri, P., Leo, L. M., Rao, B. H., & Kumar, S. (2021). A semantic health observation system development based on the IoT sensors. *Turkish Journal of Physiotherapy and Rehabilitation*, *32*(3), 1721–1729.

Bauer, E., Greiff, S., Graesser, A. C., Scheiter, K., & Sailer, M. (2025). Looking beyond the hype: Understanding the effects of AI on learning. *Educational Psychology Review*, *37*(2), 45.

Bergdahl, N., & Sjöberg, J. (2025). Attitudes, perceptions and AI self-efficacy in K-12 education. *Computers and Education: Artificial Intelligence*, *8*, 100358.

Choi, J. I., Yang, E., & Goo, E. H. (2024). The effects of an ethics education program on artificial intelligence among middle school students: Analysis of perception and attitude changes. *Applied Sciences (Basel, Switzerland)*, *14*(4), 1588.

Chopra, S., Bhambri, P., & Singh, B. (2011). Segmentation of the mammogram images to find breast boundaries. *International Journal of Computer Science and Technology*, *2*(2), 164–167.

Harleen, B. (2016). A prediction technique in data mining for diabetes mellitus. *Journal of Management Science and Technology*, *4*(1), 1–12.

Jawad, M., Talreja, K., Bhutto, S. A., & Faizan, K. (2024). Investigating how AI Personalization Algorithms Influence Self-Perception, Group Identity, and Social Interactions Online. *Review of Applied Management and Social Sciences*, *7*(4), 533–550.

Kshirsagar, P. R.. (2022). Machine learning algorithm for leaf disease detection. *AIP Conference Proceedings*, *2393*(1).

Laru, J., Celik, I., Jokela, I., & Mäkitalo, K. (2025). The antecedents of pre-service teachers' AI literacy: Perceptions about own AI driven applications, attitude towards AI and knowledge in machine learning. *European Journal of Teacher Education*, •••, 1–23.

Mohanta, H. C.. (2022). An optimized PI controller-based SEPIC converter for microgrid-interactive hybrid renewable power sources. *Wireless Communications and Mobile Computing, 2022*, 6574825.

Ododo, E. P., Offiong, A. E., Odiong, U. O., & Edet, E. E. (2024). UNMASKING AI: HOW MEDIA SHAPES STUDENTS' PERCEPTIONS AND ATTITUDES TOWARDS TECHNOLOGY. *The American Journal of Interdisciplinary Innovations and Research*, *6*(09), 19–29.

. Otermans, P. C., Roberts, C., & Baines, S. (2025). Unveiling AI perceptions: How student attitudes towards AI shape AI awareness, usage, and conceptions.

Rani, S.. (2025). nBLAST-JC: An accelerated deployment of nucleotide BLAST on high-performance cluster. *Heliyon*, *11*(10), e41866. Advance online publication. DOI: 10.1016/j.heliyon.2025.e41866

. Rawashdeh, A., Abu Siam, Y., & Idris, M. (2025). Bridging knowledge and adoption: how students' AI awareness shapes attitudes toward AI in auditing. Higher Education, Skills and Work-Based Learning, 1-18.

. Ritu, & Bhambri, P. (2025). Enhancing software development effort estimation with a cloud-based data framework using use case points, fuzzy logic, and machine learning. Discover Computing, 28, Article 143. https://doi.org/DOI: 10.1007/s10791-025-09668-1

Shao, C., Nah, S., Makady, H., & McNealy, J. (2025). Understanding user attitudes towards AI-enabled technologies: An integrated model of Self-Efficacy, TAM, and AI Ethics. *International Journal of Human-Computer Interaction*, *41*(5), 3053–3065.

Sharma, R., Bhambri, P., & Sohal, A. K. (2020). Energy bio-inspired for MANET. *International Journal of Recent Technology and Engineering*, *8*(6), 5580–5585.

Silva, D. E., Chen, C., & Zhu, Y. (2024). Facets of algorithmic literacy: Information, experience, and individual factors predict attitudes toward algorithmic systems. *New Media & Society*, *26*(5), 2992–3017.

Singh, A. P.. (2021). Sketching of EV network: A complete roadmap. In Singh, H. (Eds.), *Sustainable development through engineering innovations* (pp. 431–442). Springer., DOI: 10.1007/978-981-15-9554-7_37

Yim, I. H. Y., & Wegerif, R. (2024). Teachers' perceptions, attitudes, and acceptance of artificial intelligence (AI) educational learning tools: An exploratory study on AI literacy for young students. *Future in Educational Research*, *2*(4), 318–345.

KEY TERMS AND DEFINITIONS

Adaptive Learning Systems: AI-driven educational platforms that personalize content, pacing, and difficulty based on a learner's individual performance and needs.

AI Literacy: The knowledge and skills required to understand, use, critically evaluate, and interrogate AI technologies effectively and ethically.

Algorithmic Aversion: The opposite of automation bias; a distrust or active rejection of algorithmic advice, often due to a perceived threat to autonomy or a lack of transparency.

Algorithmic Bias: Systemic, repeatable errors in an AI system that create unfair outcomes, such as privileging one arbitrary group of users over others. This often stems from biased training data or flawed model design.

Anthropomorphism: The human tendency to attribute human-like characteristics, such as intent, emotions, or consciousness, to AI systems.

Attitudes: An individual's settled evaluations and feelings toward AI, encompassing trust, skepticism, acceptance, or resistance. These are crucial for adoption and use.

Automation Bias: The cognitive bias that leads people to over-rely on automated decision-making aids (like AI), ignoring contradictory information or their own judgment.

Bias Feedback Loop: A vicious cycle where societal biases are encoded into AI systems through data, and the AI's outputs then reinforce and amplify those same biases in the real world.

Capability Ethics: A philosophical framework that evaluates societal progress based on the expansion of individuals' opportunities and capabilities to thrive, proposed as a guide for human-centric AI integration.

Computers Are Social Actors (CASA) Paradigm: A theory stating that humans instinctively apply social rules and expectations (e.g., politeness, reciprocity) to computers and AI during interactions.

Dual-Process Theory: A model of cognition (System 1 and System 2) used to explain how users may either uncritically accept AI outputs (fast, intuitive thinking) or engage in deliberative critique (slow, analytical thinking).

Echo Chambers / Filter Bubbles: AI-curated environments, often on social media, where users are primarily exposed to information and opinions that align with their existing beliefs, insulating them from opposing viewpoints.

Explainable AI (XAI): A subfield of AI focused on making the decisions and outputs of AI models understandable and interpretable to human users, addressing the "Black Box" Problem.

Human Dimension of AI: The study of how AI systems affect human emotions, beliefs, behaviors, attitudes, perceptions, and cognitive processes, moving beyond a purely technical understanding.

Human-Centric AI: A design philosophy that prioritizes human values, well-being, and cognitive growth over purely technical metrics like efficiency or accuracy.

Learning Outcomes: The knowledge, skills, and cognitive habits acquired by individuals in environments mediated by artificial intelligence.

Metacognition: The awareness and understanding of one's own thought processes. The chapter discusses how AI can both hinder and enhance this through feedback and reflection prompts.

Perceptions: How individuals interpret and understand AI-generated information and its role in their lives, including their worldview and self-concept.

Replacement vs. Augmentation Narratives: Competing frameworks for understanding AI's role in the workplace: one where AI eliminates human jobs, and another where it acts as a tool to enhance human capabilities and productivity.

Social Cognitive Theory (SCT): A theory emphasizing that people learn through observation, imitation, and modeling. In AI contexts, users may mimic AI behaviors, and AI can impact a user's self-efficacy (belief in their own capabilities).

Sociotechnical Systems Theory: A framework that posits technology (like AI) is not developed or used in a vacuum but is deeply embedded within and shaped by social, political, economic, and cultural structures.

Symbiotic Human-AI Future: A envisioned model of collaboration where humans and AI systems work together as partners, leveraging their respective strengths to solve complex problems.

Technology Acceptance Model (TAM): A model suggesting that user adoption of technology is primarily influenced by its Perceived Usefulness and Perceived Ease of Use.

Transparency / Explainability: The degree to which the inner workings and decision-making logic of an AI system are accessible and understandable to users and stakeholders.

Trust Calibration: The dynamic process of adjusting one's level of trust in an AI system based on its performance, transparency, and reliability within a specific context.

Chapter 6
Challenges and Limitations in AI Interpretation of Non-Human Signals

Rupam Hazra
https://orcid.org/0009-0003-0332-2128
Global Institute of Management and Technology, Krishnanagar, India

Soumen Ghosh
https://orcid.org/0009-0008-3778-6090
Global Institute of Management and Technology, Krishnanagar, India

ABSTRACT

This chapter explores the emerging intersection of artificial intelligence (AI) and animal communication, focusing on the challenges and opportunities involved in interpreting non-human signals. It presents cutting-edge AI methodologies applied to analyze animal vocalizations, gestures, and behaviours, highlighting their applications in veterinary medicine, wildlife conservation, and companion animal care. By fostering interdisciplinary collaboration among AI specialists, ethologists, veterinarians, and ecologists, the chapter offers a holistic view of the technological advancements shaping this field. It also addresses critical ethical and philosophical questions surrounding the use of AI in understanding and influencing human-animal relationships. Serving as a foundational reference, this work aims to inspire future research and responsible innovation to bridge species communication and enhance animal welfare in an AI-driven world.

DOI: 10.4018/979-8-3373-5483-5.ch006

1. INTRODUCTION

The intersection of artificial intelligence (AI) and animal communication represents an exciting and rapidly evolving frontier of scientific inquiry and technological innovation. This chapter introduces the foundational concepts and motivations behind the integration of AI in interpreting non-human signals, setting the stage for a comprehensive exploration of both opportunities and challenges. Animals communicate through a rich variety of vocalizations, gestures, and behaviours that convey crucial information about their social structures, emotions, and environments. Understanding these signals has traditionally relied on painstaking human observation and interpretation, often limited by subjective biases and the sheer complexity of animal communication systems. With advances in AI, particularly in machine learning, computer vision, and natural language processing, researchers are now equipped with powerful tools to analyze vast datasets of animal signals with unprecedented accuracy and speed. These technologies hold great promise for enhancing our understanding of animal cognition, improving veterinary care, supporting wildlife conservation, and deepening the human-animal bond. However, the application of AI in this domain also raises significant technical, ethical, and philosophical questions. Challenges such as data variability, signal complexity, and contextual interpretation demand interdisciplinary approaches and cautious optimism. This chapter outlines the scope and objectives to consolidate emerging knowledge, present state-of-the-art AI methods, highlight practical applications, and foster critical reflection on the ethical implications of AI-driven animal communication research. It emphasizes the importance of collaboration among AI experts, ethologists, veterinarians, ecologists, and ethicists to responsibly harness technology in bridging species communication. By providing a comprehensive and forward-looking framework, this introduction invites readers into a dialogue about the transformative potential and inherent limitations of AI in understanding the non-human world.

2. AI METHODS FOR ANALYSING ANIMAL VOCALIZATIONS

AI methods for analysing animal vocalizations utilize machine learning and deep learning techniques to decode complex acoustic signals across diverse species. These approaches involve processing large datasets through feature extraction, neural networks, and spectrogram analysis to identify patterns and classify vocalizations. The environmental sector is experiencing a substantial expansion of artificial intelligence (AI),sector because of its capacity for problem solving, decision-making, and pattern recognition(Sharma,2023). Moreover, AI systems often struggle with interpreting non-human signals accurately because of species-specific nuances, overlapping sounds, and the difficulty in capturing the full behavioural context. These limitations highlight the need for improved algorithms, richer datasets, and interdisciplinary collaboration to enhance AI's ability to understand and bridge human-animal communication more effectively.

A. Machine Learning Techniques for Acoustic Data

Machine learning techniques have become increasingly important for analysing and interpreting acoustic data, which includes sound recordings, speech, environmental noise, and ultrasonic signals. These techniques allow systems to automatically recognize patterns, classify sounds, and even detect anomalies without needing manual rule-based systems. Acoustic data is often high-dimensional and complex, making machine learning especially effective for extracting meaningful features and insights. ML, which is frequently rooted in statistics, is a broad family of methods for automatically finding and using patterns in data (Bianco et al., 2019).

a) Supervised Learning

Supervised learning involves training a model on labelled acoustic data, where the input (e.g., sound waveforms or spectrograms) is associated with known output labels such as speaker identity or sound category. This method is widely used in speech recognition, where the model learns to map audio to text. Common algorithms include Support Vector Machines (SVMs), k-Nearest Neighbours (k-NN), and Convolutional Neural Networks (CNNs). A key advantage of supervised learning is its high accuracy when large, well-labelled datasets are available. However, collecting labelled acoustic data can be time-consuming and costly, especially for rare or complex sound types.

b) Unsupervised Learning

These methods are useful for clustering similar sounds, detecting outliers, or discovering new sound categories. In applications such as wildlife monitoring, unsupervised learning can help group unknown animal calls or detect abnormal environmental sounds. The main challenge lies in interpreting the results, as the output doesn't directly tell us what each cluster or pattern means.

c) Deep Learning

Deep learning models, particularly neural networks with multiple layers, have revolutionized acoustic data analysis due to their ability to learn complex, hierarchical features directly from raw data. Models such as CNNs and Recurrent Neural Networks (RNNs), including Long Short-Term Memory (LSTM) networks, are frequently used for tasks like speech-to-text, emotion recognition, and sound event detection. These models can outperform traditional techniques, especially in tasks involving temporal dependencies and noisy inputs. Deep learning requires substantial computational resources and large datasets but often yields state-of-the-art results. Pre-trained models and transfer learning are increasingly used to reduce the need for massive labelled datasets.

B. Challenges in Decoding Complex Vocal Patterns

These patterns often involve rapid changes in pitch, tone, and rhythm, which can be influenced by context, emotion, speaker identity, and environmental noise. Patterns of gene expression in the brain might enhance our comprehension of vocal differences in these species, especially if calls trigger these patterns in distinct manners in the brain compared to songs (Freeberg et al., 2007). Additionally, overlapping sounds, background interference, and limited labelled data can hinder the performance of automated systems. The interpretation of these patterns requires not only accurate signal processing but also a deep understanding of linguistic, biological, or behavioural factors. As a result, developing robust and generalizable models for vocal pattern decoding remains an active area of research.

a) Variability in Vocal Signals

Vocal patterns vary greatly between individuals and even within the same individual depending on mood, health, or context. This variability makes it difficult for models to generalize, especially if they are trained on limited or homogeneous datasets. For example, the same word can sound very different when spoken by people of different ages, accents, or emotional states. In animals, calls can differ based on region, species, or situational factors like mating or danger. Handling such diversity requires large and diverse datasets, as well as models that are flexible and adaptive.

b) Background Noise and Overlapping Sounds

Environmental noise and overlapping vocalizations can significantly distort the acoustic signal, complicating the decoding process. In natural settings like jungles, cities, or busy rooms, background sounds often mask important vocal features. This makes it harder for models to isolate and interpret the target signal, leading to reduced accuracy. Techniques such as noise reduction, source separation, and signal enhancement are commonly used to mitigate this issue. However, their effectiveness varies depending on the noise type and signal-to-noise ratio.

c) Limited Annotated Data

High-quality annotated datasets are essential for training machine learning models, but they are often scarce, especially for non-human vocalizations or rare human speech disorders. Creating such datasets requires expert knowledge and time-intensive manual labelling. In many cases, especially in wildlife studies or low-resource languages, the available data is insufficient for supervised learning. This limits the effectiveness of traditional machine learning approaches and necessitates the use of unsupervised, semi-supervised, or transfer learning methods. Addressing this challenge is key to advancing vocal pattern analysis in less-explored domains.

C. Case Studies: From Birdsong to Cetacean Calls

Case studies involving species such as birds and cetaceans (whales and dolphins) highlight the practical application of machine learning in decoding and understanding complex vocalizations. These animal vocalizations are rich in structure and often serve critical biological functions such as communication, mating, territory defense, and social bonding. For a long time, it has been believed that cetaceans are highly intelligent and that they communicate in complicated ways(Janik,2014). Analysing these patterns provides insights not only into animal behaviour but also into broader questions of evolution, cognition, and communication systems. Each species presents unique acoustic challenges, such as high-frequency signals, deep-ocean propagation, or highly variable song structures. The following case studies illustrate how advanced techniques have been successfully applied in diverse biological contexts.

a) Birdsong Analysis

Birdsong is an ideal model for studying complex vocal learning, as many bird species exhibit structured, hierarchical song patterns. Machine learning has been used to classify species, recognize individual birds, and map song elements over time. For instance, Hidden Markov Models (HMMs) and CNNs have been employed to segment and label syllables in continuous song streams. Such analysis helps biologists track behavioural changes, population health, and the effects of environmental disruption. Moreover, understanding how birds learn and modify their songs can offer parallels to human speech acquisition.

b) Dolphin Signature Whistles

Dolphins produce unique, individualized whistles—known as signature whistles—that function similarly to names. These vocalizations are highly modulated and can occur in noisy underwater environments, posing challenges for detection and classification. Deep learning methods, particularly spectrogram-based CNNs, have improved identification accuracy and helped reveal patterns in dolphin communication. These studies have enhanced our understanding of dolphin cognition, memory, and social structure.

c) Whale Song Patterns

Whales, especially species like humpbacks, produce long, elaborate songs that evolve over time and may span entire ocean basins. Analysing whale songs requires dealing with low-frequency, long-duration recordings often collected via hydrophones. Techniques such as dynamic time warping and RNNs have been applied to detect recurring motifs and monitor changes in song structure. These analyses inform studies on migration, mating behaviour, and cultural transmission in whale populations. Furthermore, passive acoustic monitoring supports conservation efforts by identifying whale presence and reducing human-induced threats like ship collisions or noise pollution.

3. GESTURE AND BEHAVIOURAL SIGNAL RECOGNITION USING AI

AI-driven gesture and behavioural signal recognition involves using computer vision and pattern recognition technologies to decode the body language, movements, and behaviours of animals. These methods enable the analysis of subtle gestures, postures, and social interactions by processing video and sensor data through algorithms like convolutional neural networks (CNNs) and recurrent neural networks (RNNs). Despite significant advancements, challenges remain, including variability in behaviour across species and individuals, occlusions in visual data, and the complexity of contextual interpretation. Gesture recognition serves as an excellent method of interaction since it enables users to avoid making contact with any surface, which is a secure and sanitary method, particularly in the context of the pandemic that is happening globally (Ojeda-Castelo et al., 2022). Current AI models often require large, annotated datasets, which can be difficult to obtain in natural environments. This field holds promise for enhancing understanding of animal communication, welfare assessment, and ecological monitoring, but ongoing improvements in data quality, algorithm robustness, and interdisciplinary approaches are essential to overcome existing limitations.

A. Computer Vision in Animal Gesture Analysis

Computer vision has emerged as a powerful tool for analysing animal gestures, enabling researchers to study behaviour in a non-invasive and automated way. By processing visual data such as videos or images, machine learning models can track body movements, recognize postures, and identify complex behavioural patterns. Hand gesture recognition acts as a natural way for human-computer interaction and is a dynamic area of research in computer vision and machine learning (Trigueiros et al., 2015). These techniques allow for continuous, objective monitoring of animals in both natural habitats and controlled environments. This has revolutionized fields like ethology, neuroscience, and conservation by providing rich datasets that were previously difficult or impossible to obtain manually. With the advancement of deep learning, especially convolutional neural networks (CNNs), the precision and scalability of animal gesture analysis have dramatically improved.

a) Pose Estimation

Pose estimation involves detecting and tracking key points on an animal's body, such as joints, limbs, or the head, from visual input. This allows for the reconstruction of full-body poses, enabling researchers to analyze movement patterns and behaviours over time. Tools like Dereplicate and SLEAP have been widely used for pose estimation across species including rodents, primates, and birds. These systems use labelled training data to learn species-specific body structures, even under varying lighting and backgrounds. Pose estimation is foundational for tasks like gait analysis, social interaction tracking, and motor coordination studies.

b) Gesture Classification

Once poses are estimated, gesture classification involves identifying specific behaviours—like grooming, feeding, aggression, or courtship—based on movement patterns. This can be done using traditional classifiers (e.g., SVMs, decision trees) or deep learning models (e.g., RNNs or transformers) that learn temporal sequences. Automated gesture classification reduces human bias and enables large-scale behavioural studies that were previously limited by manual scoring. In primatology, for instance, gesture recognition has shed light on communication and intentionality in great apes. The challenge lies in building generalizable models across different individuals, environments, and species.

c) Multimodal Integration

Computer vision systems are increasingly being combined with other sensing modalities—such as audio, infrared, or motion sensors—to provide a more comprehensive understanding of animal behaviour. For example, integrating video with acoustic data can help correlate gestures with vocalizations, revealing deeper insights into communication. Multimodal data fusion enhances model robustness, especially in complex environments where visual data alone may be insufficient or occluded. This approach has proven valuable in wildlife monitoring, laboratory studies, and animal welfare assessments. However, synchronizing and processing multimodal data streams in real time remains a technical challenge.

B. Behavioural Pattern Detection and Classification

Behavioural pattern detection and classification involve identifying, tracking, and interpreting sequences of actions or postures exhibited by animals to understand their internal states, social interactions, and environmental responses. Advances in machine learning and computer vision have enabled researchers to automate this process, reducing reliance on manual annotation and observation.

Compared with action recognition, human action detection not only needs to identify the type of behaviors in a video stream, but also needs to localize the start and the end point of an action(Wang,2019). By analysing video, motion, and other sensor data, systems can now detect subtle and complex behaviours with high accuracy. This is crucial in fields such as neuroscience, ethology, animal welfare, and conservation. Understanding behavioural patterns helps in assessing stress, social dynamics, cognitive abilities, and even early signs of disease.

a) Temporal Behaviour Segmentation

Temporal segmentation involves dividing continuous behavioural data into discrete segments that represent different actions or states (e.g., walking, resting, grooming). This step is essential for understanding transitions and sequences in behaviour. Techniques like Hidden Markov Models (HMMs), change-point detection, and temporal convolutional networks (TCNs) are commonly used. These models help in capturing the timing and duration of each behaviour, which is particularly useful in long-term monitoring. Accurate segmentation enables detailed behavioural chronologies that support more meaningful classification and analysis.

b) Supervised and Unsupervised Classification

In supervised classification, labelled behavioural data is used to train models that can automatically recognize specific behaviours in new datasets. For example, a model might be trained to distinguish between aggressive and submissive postures in primates. Unsupervised classification, on the other hand, identifies patterns without predefined labels, revealing hidden or emergent behavioural categories. Techniques like k-means clustering, hierarchical clustering, and autoencoders are useful for exploratory analysis. Both approaches are valuable: supervised methods provide precise labelling, while unsupervised methods can uncover unexpected or previously unknown behaviours.

C. Limitations of Current AI Models in Gesture Interpretation

Despite significant advances in AI and computer vision, interpreting animal gestures remains a complex and challenging task. Current AI models often struggle with variability in body structure, subtlety in movements, and the need for context-aware understanding of behaviour. Employing hand gestures offers an appealing substitute for bulky interface tools in human-computer interaction (Pavlovic,1997). Most models are heavily reliant on large, labelled datasets, which are difficult to obtain for many species or rare behaviours. Additionally, many systems lack the ability to generalize well across individuals, environments, or species, making them less robust in real-world applications. Understanding gestures in animals requires not just tracking movement but interpreting meaning—something that current AI is still far from mastering.

a) Limited Generalization Across Species and Contexts

Most gesture recognition models are trained on data from specific species, environments, or experimental setups, and often fail when applied elsewhere. A model trained on lab mice may not work for wild rodents, or one trained on one primate species may not interpret gestures correctly in another. This lack of generalization arises from differences in morphology, behaviour, and environmental backgrounds. Additionally, gestures can carry different meanings depending on the social or ecological context, which AI models typically do not account for. This limits the scalability and reusability of current models across broader ecological or ethological studies.

b) Data Scarcity and Annotation Challenges

High-quality labelled data is essential for training accurate AI models, but collecting and annotating animal gesture data is labour-intensive and requires expert knowledge. Many species, especially in the wild, have limited available data, and manual labelling of gestures is time-consuming and subjective. This data bottleneck restricts the development of supervised learning models and affects their performance. Even in laboratory settings, capturing rare or subtle behaviours often requires long-duration recordings, further complicating the annotation process. Semi-supervised and unsupervised approaches are being explored, but they still face challenges in accurately capturing complex behaviours.

4. MULTIMODAL APPROACHES TO NON-HUMAN COMMUNICATION

Multimodal approaches in non-human communication integrate multiple data streams—such as vocalizations, gestures, facial expressions, and environmental context—to provide a richer and more accurate understanding of animal signals. By combining audio, visual, and sensor data, AI systems can capture the complexity and nuance of communication that single-modality analysis might miss. Advances in sensor technology, data fusion techniques, and machine learning algorithms facilitate the synchronization and interpretation of these diverse signals. Although there is increasing evidence that non-human primates, our primary model for language evolution, communicate in a multimodal manner, the majority of research on primate communication has concentrated on gestures or vocalizations individually (Fröhlich & van Schaik, 2018). However, challenges include aligning data with different temporal resolutions, managing large and heterogeneous datasets, and accounting for the influence of context on behaviour. Multimodal AI holds great potential for unlocking deeper insights into animal cognition, social interactions, and environmental responses, ultimately improving applications in conservation, welfare, and human-animal interaction.

A. Integrating Vocal, Visual, and Contextual Data

Integrating vocal, visual, and contextual data offers a more holistic and accurate approach to understanding animal behaviour. While each data type—audio (vocalizations), video (gestures and movements), and contextual (environmental or social conditions)—provides valuable insights individually, their combination allows for deeper interpretation and richer behavioural analysis. Multimodal integration enables researchers to correlate signals, disambiguate meanings, and recognize complex communication patterns that might be missed with a single modality. The contextual audio-visual switching component is developed by integrating a convolutional neural network (CNN) and long-short-term memory (LSTM) network (Adeel et al., 2020). This approach is increasingly feasible due to advances in sensor technology, data fusion algorithms, and machine learning models capable of handling diverse input streams. Ultimately, these fusion paves the way for more biologically meaningful interpretations of animal communication and behaviour.

a) Synchronization of Multimodal Inputs

Effective integration requires precise temporal alignment of vocal, visual, and contextual data streams. For example, to correlate a vocal call with a specific body gesture, the audio and video must be time-synced to millisecond accuracy. Tools like multimodal annotation software and synchronized sensor systems (e.g., microphones and motion cameras) are essential. Misalignment can lead to false associations or missed correlations between cues. Synchronization ensures that behavioural events are captured in their true sequence and context, which is crucial for accurate interpretation and model training.

b) Multimodal Machine Learning Models

Specialized machine learning architectures—such as multimodal transformers, hybrid CNN-RNN models, or fusion networks—are designed to process and combine data from different modalities. These models learn to associate vocal tones with facial expressions or gestures, and can weigh the importance of each modality based on context. For example, an animal's alarm call paired with a rapid movement may signal immediate threat, while the same call with relaxed posture may not. Multimodal models improve classification accuracy, reduce false positives, and better reflect the nuanced nature of communication.

B. Advances in Sensor Technologies and Data Fusion

Integrating vocal, visual, and contextual data offers a holistic and more accurate approach to understanding animal behaviour. Individually, audio (vocalizations), video (gestures and movements), and contextual information (such as environmental conditions or social context) provide valuable insights. However, their combination allows for deeper interpretation and more nuanced behavioural analysis. Sensor Data Fusion involves merging incomplete and imperfect components of complementary sensor information to enhance the comprehension of an underlying real-world phenomenon(Koch,2016). Multimodal integration enables researchers to correlate signals, disambiguate meanings, and detect complex communication patterns that may be overlooked when using a single modality. Recent advancements in sensor technologies, data fusion algorithms, and machine learning models capable of handling heterogeneous input streams have made such integration increasingly feasible. Ultimately, this approach paves the way for biologically meaningful interpretations of animal communication and cognitive processes.

a) Synchronization of Multimodal Inputs

Successful integration depends on the precise temporal alignment of vocal, visual, and contextual data streams. For example, correlating a vocalization with a specific body gesture requires audio and video data to be synchronized to millisecond-level precision. This is achieved using multimodal annotation tools and synchronized sensor systems, such as arrays of microphones and high-speed motion-capture cameras. Temporal misalignment can lead to incorrect associations or missed behavioural patterns, undermining the accuracy of the analysis. Accurate synchronization ensures that behavioural events are interpreted within their correct temporal and situational context, which is vital for both human observation and machine learning model training.

b) Multimodal Machine Learning Models

Specialized machine learning architectures—such as multimodal transformers, hybrid CNN-RNN frameworks, or feature-level fusion networks—are designed to process and combine data from diverse modalities. These models can learn complex associations, such as linking vocal intonations with body postures or environmental cues. For instance, an alarm call accompanied by a rapid body movement may indicate immediate threat, whereas the same call with a passive posture might suggest low urgency or a different communicative intent. By weighting inputs based on context, multimodal models enhance classification accuracy and reduce ambiguity. This leads to more robust and realistic behavioural interpretations across different species and environments.

C. Challenges in Synchronizing Multimodal Signals

Synchronizing multimodal signals—such as vocal, visual, and contextual data—is a critical step in accurately interpreting complex animal behaviors. While integrating these diverse data streams can enrich behavioral analysis, it also introduces significant technical and methodological challenges. Each modality often operates on different temporal and spatial scales, with varying sampling rates, data formats, and levels of noise. Accurate synchronization is essential to ensure that the captured events across modalities correspond to the same behavioral moments. Failure to align these inputs correctly can lead to misinterpretation, loss of critical information, or faulty model training, ultimately compromising the validity of research outcomes.

a) Temporal Resolution and Sampling Rate Mismatch

One of the primary challenges is the difference in sampling rates and temporal resolutions between audio, video, and sensor-based contextual data. For example, audio is often recorded at high frequencies (e.g., 44.1 kHz or higher), whereas video may operate at 30 or 60 frames per second, and contextual data (like temperature or GPS) may update only once per second or less. This disparity can make it difficult to align specific events—such as a vocalization and a corresponding gesture—with high precision. Interpolation techniques can partially bridge the gap, but they may introduce artifacts or errors if not carefully managed.

b) Hardware and Sensor Desynchronization

When multiple devices are used (e.g., cameras, microphones, environmental sensors), even slight delays in clock timing or data recording can result in desynchronization. Inconsistent start times, drift in internal clocks, and hardware latency can all affect the temporal alignment of data streams. While synchronization tools such as shared timecode generators or network-synchronized systems can help, they add cost and complexity to fieldwork or large-scale monitoring setups. Post-processing to manually align data is possible but time-consuming and error-prone, particularly in long or continuous recordings.

5. APPLICATIONS IN VETERINARY MEDICINE AND ANIMAL WELFARE

AI technologies are increasingly applied in veterinary medicine and animal welfare to improve diagnosis, monitoring, and care through analysis of animal communication signals. By interpreting vocalizations, gestures, and behavioural patterns, AI tools can detect signs of pain, stress, or illness earlier and more accurately than traditional methods. These applications enhance health assessments in farm animals, pets, and wildlife, supporting timely interventions and improving overall welfare. Additionally, AI-driven monitoring systems enable continuous observation without intrusive human presence, reducing animal distress. It is growing more crucial for veterinarians to not only incorporate ethics, particularly concerning animal welfare, into their practices, but also to take on leadership positions and proactively uphold moral standards instead of reacting defensively to uncontrollable activist influences(Arkow,1998). Despite promising advances, challenges remain in adapting AI models to diverse species and environments, ensuring data accuracy, and addressing ethical concerns about surveillance. Overall, AI's role in veterinary and welfare contexts offers transformative potential for more compassionate and effective animal care.

A. AI Tools for Diagnosing Animal Health via Communication

Artificial Intelligence (AI) is increasingly being used to diagnose animal health by analyzing vocalizations, gestures, and behavioral patterns. Many animals exhibit subtle changes in their communication when experiencing stress, pain, illness, or distress—patterns that may be too nuanced or frequent for human observers to detect consistently. Utilizing artificial intelligence (AI) techniques in animal health (AH) allows for tackling intricate challenges like those found in quantitative and predictive epidemiology, precision-based medicine for animals and humans, or examining host × pathogen interactions (Ezanno et al., 2021). AI tools equipped with machine learning and signal processing algorithms can monitor these changes in real-time, offering early warnings of health issues. This approach is particularly valuable in contexts where direct physical examinations are impractical, such as livestock farming, wildlife conservation, and zoo management. By focusing on natural communication signals, AI provides a non-invasive, continuous, and scalable method for health monitoring.

a) Vocalization-Based Health Monitoring

Changes in vocal patterns—such as altered frequency, duration, pitch, or rhythm—can indicate respiratory issues, pain, or stress in animals. AI systems analyze these acoustic features using signal processing and machine learning to detect anomalies or deviations from species-specific baselines. For example, coughing sounds in cattle or abnormal squeals in pigs may be early signs of infection. Deep learning models trained on healthy vs. unhealthy vocalizations can automatically classify new samples and raise alerts. Such systems have been deployed in farms to monitor livestock health, reducing disease spread and improving welfare.

b) Gesture and Posture Recognition

Abnormal postures, reduced mobility, or unusual repetitive movements are often early signs of injury or neurological conditions. Computer vision systems, powered by pose estimation tools like Dereplicate, can track animal body parts and detect deviations from normal motion patterns. AI models can then classify these postural changes and link them to specific health conditions, such as lameness in horses or rodents showing signs of discomfort. This allows for early diagnosis, treatment, and intervention, especially in group-housed or free-moving animals where manual observation is difficult.

B. Monitoring Stress and Pain through Vocal and Behavioural Cues

Monitoring stress and pain in animals is crucial for ensuring their welfare, and vocalizations along with behavioral cues often serve as reliable indicators. Animals express discomfort, distress, or pain through changes in their vocal patterns—such as increased frequency, altered pitch, or unusual call types—as well as through behavioral modifications like restlessness, withdrawal, or repetitive movements. A key aim for animal welfare is to keep animals safe from pain, injury, or illness. Consequently, identifying and assessing the severity of pain in animals is essential(Prunier,2013). AI-driven analysis of these signals provides a non-invasive and objective way to detect stress and pain early, which is especially valuable in environments where direct human observation is limited or impractical. Such monitoring supports timely interventions, enhances animal care, and improves outcomes in veterinary and conservation settings.

a) Vocal Indicators of Stress and Pain

Stress and pain can cause distinct changes in animal vocalizations, including higher call rates, increased amplitude, or the emergence of novel sound types like distress calls. Machine learning models analyze acoustic features such as frequency spectrum, duration, and harmonics to identify these stress-related vocal patterns. For instance, pigs under stress exhibit more frequent and higher-pitched squeals, while dogs in pain may emit whines or growls with altered tonal quality. Automated vocal monitoring systems can continuously scan for such vocal indicators in farms, shelters, or wildlife reserves, enabling early detection of welfare issues.

b) Behavioral Changes Reflecting Discomfort

Animals often exhibit specific behaviors when stressed or in pain, including pacing, reduced grooming, limping, or social withdrawal. Computer vision and sensor-based systems track these behaviors by analyzing movement patterns, posture, and social interactions. AI models classify behaviors linked to distress, facilitating objective and consistent monitoring without human bias. For example, increased restlessness or repetitive motions may indicate anxiety, while decreased activity might signal lethargy or illness. This behavioral data complements vocal analysis, providing a more comprehensive assessment of animal well-being.

C. Enhancing Welfare in Farm and Companion Animals

Enhancing animal welfare in both farm and companion settings is a critical goal that benefits animals, caregivers, and society at large. AI-driven tools that monitor health, stress, pain, and behaviour provide continuous, objective insights that enable timely interventions to improve living conditions and medical care. Current animal welfare standards for legislation and food certification programs mainly aim to prevent harm to livestock, with little attention paid to their behavioral liberties(Stokes,2020). For farm animals, such technologies help optimize production while minimizing suffering, ensuring compliance with welfare standards. In companion animals, AI applications assist owners and veterinarians in detecting early signs of illness or distress, promoting preventive care and better quality of life. Integrating AI with animal welfare practices transforms traditional monitoring into a proactive, data-driven approach.

a) Automated Health and Behaviour Monitoring on Farms

AI-powered systems use sensors, cameras, and microphones to continuously track livestock vocalizations, movements, feeding patterns, and social interactions. These data streams allow for early detection of diseases, injuries, or stress, facilitating rapid response and reducing the spread of infections. For example, detecting coughing in pigs or lameness in dairy cows enables farmers to address health issues promptly. Such automation reduces labour demands and human error, improving overall herd management and welfare. Additionally, AI tools help ensure adherence to welfare regulations and support sustainable farming practices.

b) Companion Animal Health and Wellness Tracking

Wearable devices, smart collars, and home monitoring cameras equipped with AI analyze pets' activity levels, vocalizations, and behaviour to detect signs of discomfort, anxiety, or illness. For instance, changes in a dog's bark or reduced activity may indicate pain or stress. AI-driven platforms provide pet owners and veterinarians with real-time alerts and actionable insights, supporting preventive healthcare and timely veterinary visits. This technology fosters stronger human-animal bonds by enabling owners to respond proactively to their pets' needs, even when away from home. Moreover, AI can assist in training and behavioral modification by recognizing and interpreting pets' gestures and cues.

6. WILDLIFE CONSERVATION AND ECOLOGICAL MONITORING

AI is playing a transformative role in wildlife conservation and ecological monitoring by enabling automated, large-scale analysis of animal signals and behaviours in natural habitats. Through tools like acoustic sensors, camera traps, and drones combined with machine learning algorithms, AI can track endangered species, monitor population dynamics, and detect poaching or environmental threats in real time. These technologies facilitate non-invasive data collection, providing critical insights without disrupting ecosystems. The growth of worldwide communication networks and improvements in animal-tracking technology enable the real-time transmission of location data captured by devices attached to animals(Wall,2014). However, challenges include managing vast amounts of heterogeneous data, ensuring accuracy amidst environmental noise, and addressing ethical concerns related to wildlife surveillance. Despite these hurdles, AI-driven conservation efforts hold great promise for enhancing biodiversity protection and informing evidence-based ecological management strategies.

A. AI in Tracking and Monitoring Endangered Species

Artificial Intelligence (AI) has become an invaluable tool in the conservation of endangered species by enhancing the tracking and monitoring of animals in their natural habitats. Traditional methods like manual observation or tagging are often labour-intensive, intrusive, and limited in scale. Cameras and drones powered by AI observe wildlife populations, monitor wildlife movement and identify endangered species instantly(Oladele,2025). AI-powered systems leverage camera traps, drones, acoustic sensors, and satellite imagery to collect vast amounts of data remotely and non-invasively. Machine learning algorithms then analyze these diverse datasets to identify species, track individual movements, and monitor population dynamics. This technology provides conservationists with real-time insights that support more effective protection strategies and informed decision-making.

a) Automated Species Identification

AI models trained on large image and audio datasets can automatically recognize species from photos, videos, or vocalizations captured in the wild. Deep learning techniques, such as convolutional neural networks (CNNs), can differentiate between visually similar animals and identify rare or elusive species with high accuracy. Acoustic classifiers analyze unique call patterns to detect species presence even in dense or inaccessible environments. This automation reduces the reliance on expert taxonomists and accelerates biodiversity assessments, making it easier to monitor endangered species across large geographic areas.

b) Movement and Behaviour Tracking

Using AI-powered video analysis and GPS data, researchers can track individual animals' movements, migration routes, and behavioral patterns over time. This information is critical for understanding habitat use, breeding behaviors, and threats such as poaching or habitat fragmentation. For example, AI can analyze drone footage to monitor elephant herds or identify nesting sites for endangered birds. Continuous tracking helps in designing protected areas and corridors that align with animals' natural behaviors and conservation needs.

B. Automated Acoustic Surveys and Population Studies

Automated acoustic surveys utilize AI-driven technologies to monitor animal populations by capturing and analyzing soundscapes in natural habitats. This non-invasive method is particularly effective for species that are difficult to observe visually, such as nocturnal animals, cryptic species, or those inhabiting dense vegetation or underwater environments. For animals with a wider repertoire, such as song birds, or for detection of multiple species, a larger training set is necessary and detector performance is likely to be reduced (Andreassen et al., 2014). By deploying networks of microphones and using machine learning algorithms to process vocalizations, researchers can detect species presence, estimate population densities, and track temporal and spatial trends. These surveys enable continuous, large-scale monitoring with reduced human effort, providing valuable data for conservation and ecological studies.

a) Efficient Species Detection and Identification

Machine learning models trained on extensive libraries of species-specific calls can automatically identify animals from recorded audio data. This automation significantly speeds up the analysis process, which would otherwise require expert listening and manual annotation. Acoustic classifiers can distinguish between species with similar vocalizations, filter out background noise, and detect rare or cryptic species that are otherwise difficult to find. This capability improves survey accuracy and allows for more frequent and widespread monitoring efforts.

b) Population Density Estimation

AI-powered acoustic data can be used to estimate population sizes by analyzing call rates, vocal activity patterns, and spatial distribution of detected signals. Models incorporate factors like call frequency, distance, and overlapping signals to infer the number of individuals within an area. This is especially valuable for species that vocalize regularly and have identifiable call signatures. Acoustic-based population estimates complement traditional survey methods and can be conducted more cost-effectively across large or remote regions.

C. Ethical Considerations in Wildlife Data Collection

Collecting data on wildlife through AI-powered technologies raises important ethical considerations that must be addressed to ensure responsible research and conservation practices. While advanced tools like camera traps, drones, and acoustic sensors offer non-invasive ways to gather information, they can still impact animals and ecosystems if not carefully managed. Issues such as animal disturbance, data privacy, and the potential misuse of sensitive information require attention. The worldwide wildlife trade poses significant threats to biodiversity conservation because of its harmful impacts on targeted species, the introduction of alien species, and the promotion of disease transmission between species (Morcatty et al., 2024). Ethical wildlife data collection balances the need for scientific knowledge with respect for animal welfare and habitat integrity, promoting sustainable and conscientious conservation efforts.

a) Minimizing Animal Disturbance and Stress

Even non-invasive monitoring tools can cause unintended stress or behavioral changes in animals. For example, drones flying overhead might disturb nesting birds, and camera traps with infrared flashes may startle nocturnal species. Researchers must design data collection protocols that minimize intrusion, such as limiting the frequency of sensor deployment or using quieter, less visible devices. Understanding species-specific sensitivities and behaviors helps in choosing appropriate monitoring techniques that reduce harm while still gathering valuable data.

b) Data Privacy and Protection of Sensitive Locations

Data collected on endangered or threatened species can reveal critical information about their habitats, migratory routes, or breeding sites. If this information is not securely managed, it risks being exploited by poachers, illegal wildlife traffickers, or other harmful actors. Ethical protocols include restricting access to sensitive data, anonymizing location information when sharing results, and collaborating with local communities and authorities to safeguard wildlife. Transparency about data use and adherence to privacy standards are essential to prevent unintended consequences.

c) Ensuring Fair Collaboration and Benefit Sharing

Wildlife data collection often occurs in regions inhabited or managed by indigenous peoples and local communities. Ethical research requires meaningful collaboration, respect for traditional knowledge, and fair sharing of benefits arising from scientific discoveries. Engaging communities in project design, data interpretation, and conservation action fosters trust and promotes sustainable stewardship. Additionally, equitable partnerships help prevent exploitation and ensure that conservation efforts align with local social and cultural values.

7. HUMAN-ANIMAL INTERACTION AND COMPANION CARE

AI is increasingly shaping the landscape of human-animal interaction and companion animal care by offering innovative tools that enhance communication, monitoring, and welfare. Smart devices and AI-driven platforms analyze vocalizations, body language, and behavioural cues to provide pet owners and caregivers with real-time insights into their animals' emotional states, health conditions, and needs. These technologies can help identify stress, anxiety, or illness early, enabling timely interventions that improve the quality of life for pets. Additionally, AI-powered communication aids aim to bridge the gap between humans and animals by translating certain behaviours or sounds into understandable feedback, fostering deeper bonds and more empathetic care. The majority have assessed short-term results, and research on the long-term impacts of human-animal interactions are essential to establish if the immediate advantages are sustained in the long run (Barker &

Wolen, 2008). However, while these advancements present exciting opportunities, they also raise important concerns about over-reliance on technology, privacy, and the risk of misinterpretation or oversimplification of complex animal behaviours. Moreover, the diversity among species and individual animals makes designing universally effective AI systems challenging, requiring customization and ongoing refinement. This chapter examines both the promise and limitations of AI in companion care, emphasizing the need for ethical considerations, user education, and interdisciplinary collaboration to ensure these tools support—not replace—the nuanced human-animal relationship.

A. AI's Role in Enhancing Human-Animal Bond

Artificial Intelligence (AI) is increasingly playing a transformative role in strengthening the bond between humans and animals. By facilitating better understanding, communication, and care, AI technologies help bridge the gap between species, enriching relationships and improving welfare. Through tools like emotion recognition, behaviour monitoring, and personalized interaction systems, AI provides insights into animal needs and moods that might otherwise go unnoticed. These advances empower owners, veterinarians, and caregivers to respond more effectively, fostering empathy and trust. As AI continues to evolve, its potential to deepen the human-animal bond opens new avenues for companionship, therapy, and collaborative living.

a) Emotion and Behaviour Recognition

AI algorithms analyze vocalizations, facial expressions, and body language to detect emotional states such as stress, happiness, or discomfort in animals. This allows humans to better interpret their pets' or livestock's feelings, leading to more responsive and compassionate care. For instance, smart collars equipped with AI can alert owners to signs of anxiety in dogs or cats before behaviors escalate. Emotion recognition helps in tailoring interactions and environments to meet animals' emotional needs, thereby strengthening the bond.

b) Personalized Interaction and Training

AI-driven platforms can customize training programs and interaction schedules based on individual animal behaviour and preferences. Machine learning models analyze patterns to identify the most effective rewards, commands, or activities for each animal. Such personalization improves training outcomes, reduces frustration, and enhances positive interactions. Additionally, AI-enabled toys and devices can engage animals mentally and physically, promoting enrichment and reinforcing human-animal connections even when owners are away.

c) Health and Wellness Monitoring

Continuous monitoring of health parameters and behaviour through AI allows early detection of illness or distress, enabling timely intervention. Wearable devices and smart home systems provide real-time feedback to owners and veterinarians, ensuring that animals receive appropriate care. This proactive approach not only improves physical well-being but also strengthens trust as animals experience consistent and attentive care. By supporting healthier lives, AI contributes to longer-lasting and more fulfilling relationships between humans and animals.

B. Smart Devices for Monitoring and Interpreting Pet Behaviour

Smart devices have revolutionized the way pet owners monitor and understand their pets' behaviour, offering real-time insights that enhance care and deepen the human-animal connection. These devices—ranging from wearable collars and activity trackers to home cameras and interactive toys—leverage sensors and AI algorithms to collect and analyze data on pets' movements, vocalizations, and physiological signals. By interpreting this information, smart devices can alert owners to potential health issues, behavioral changes, or emotional states such as stress or anxiety. This continuous and personalized monitoring helps owners make informed decisions, improve pet welfare, and strengthen bonds even when apart.

a) Wearable Trackers and Health Monitors

Wearable collars and devices equipped with accelerometers, GPS, heart rate monitors, and temperature sensors track pets' activity levels, sleep patterns, and vital signs. AI processes this data to detect abnormalities such as reduced mobility, irregular heartbeats, or restlessness that might indicate health problems. For example, a sudden decrease in activity combined with elevated heart rate could signal pain or illness. These devices provide owners and veterinarians with timely alerts, enabling early diagnosis and intervention, thereby improving pets' overall well-being.

b) AI-Enabled Behaviour Analysis

Smart home cameras and microphones use AI to analyze pets' behaviors such as barking, whining, scratching, or pacing. Machine learning models classify these behaviors to interpret emotional states like anxiety, boredom, or aggression. For instance, prolonged barking detected by a smart camera could indicate separation anxiety, prompting owners to address the issue. This behavioral interpretation helps owners understand their pets' needs better, facilitates targeted training, and reduces unwanted behaviors through informed management.

c) Interactive and Enrichment Devices

These devices can respond to pet behaviour in real time, offering challenges, rewards, or play opportunities that stimulate mental health and prevent boredom. For example, a smart feeder might dispense treats based on activity thresholds or learning progress. By promoting enrichment, these tools contribute to happier and healthier pets while strengthening the interactive bond between pet and owner.

C. Risks and Limitations in AI-Mediated Interaction

While AI-mediated interaction between humans and animals offers numerous benefits, it also comes with inherent risks and limitations that must be carefully considered. AI systems, despite their sophistication, may misinterpret animal signals due to species differences, individual variability, or context-dependent behaviors. Additionally, privacy concerns, data security, and ethical issues related to monitoring pets or wildlife raise important questions. Understanding these challenges helps ensure that AI serves as a supportive tool rather than a replacement for human empathy and judgment in animal care.

a) Misinterpretation of Animal Signals

AI models are trained on datasets that may not fully capture the diversity of animal behaviors across breeds, ages, or environments. As a result, AI can misclassify emotional states or health indicators, leading to false positives or negatives. For example, a dog's bark might be misread as aggression when it signals excitement, causing inappropriate responses from owners. Such errors can undermine trust in AI systems and potentially harm the animal through misguided interventions.

b) Over-Reliance and Reduced Human Engagement

Dependence on AI for monitoring and interpreting animal behaviour might cause owners or caregivers to pay less direct attention to their pets. This reduced engagement can limit the development of nuanced understanding that comes from personal interaction and observation. Furthermore, AI systems may not capture subtle or context-specific cues that humans detect intuitively. Balancing AI assistance with active human involvement is crucial to maintaining strong and empathetic relationships.

c) Privacy, Data Security, and Ethical Concerns

AI-mediated monitoring often involves collecting sensitive data such as location, vocalizations, and behavioral patterns. Without stringent data protection measures, this information risks misuse or unauthorized access. Ethical concerns also arise regarding continuous surveillance and consent, especially for companion animals and wildlife. Ensuring transparency about data use, securing stored data, and establishing clear guidelines for AI deployment are essential to protect both animals and owners.

8. ETHICAL AND PHILOSOPHICAL PERSPECTIVES

The integration of AI in interpreting non-human signals raises profound ethical and philosophical questions about agency, consent, and the nature of human-animal relationships. As AI technologies increasingly influence how humans understand and interact with animals, concerns emerge regarding the potential for misrepresentation, exploitation, or unintended harm. Ethical debates focus on the rights of non-human species, questioning whether animals can or should consent to data collection and AI-driven interventions that may affect their behaviour or welfare. Moreover, the use of AI challenges traditional boundaries between species, potentially altering power dynamics and our responsibilities toward animals. Philosophically, this prompts reflection on the moral implications of deploying technology that may shape or control animal communication rather than merely observing it. Ensuring transparency, accountability, and respect for animal autonomy is paramount as researchers and practitioners develop AI tools. This chapter explores existing ethical frameworks and calls for interdisciplinary dialogue to guide responsible AI applications, balancing technological innovation with respect for the intrinsic value of non-human life.

A. Consent, Agency, and the Rights of Non-Human Species

The ethical discussion around animals increasingly involves concepts of consent, agency, and rights—challenging traditional human-centric views and advocating for recognition of non-human species as sentient beings with intrinsic value. Unlike humans, animals cannot provide explicit consent, raising complex questions about how humans interact with, study, and manage them. Respecting animal agency involves acknowledging their capacity to make choices, exhibit preferences, and influence their environment. Recognizing the rights of animals emphasizes protecting their welfare, autonomy, and natural behaviors from exploitation or harm. This evolving ethical framework informs laws, conservation strategies, and the development of technologies such as AI to ensure respectful and just treatment of animals.

a) The Challenge of Obtaining Consent

Animals cannot verbally consent to participation in research, captivity, or monitoring, complicating ethical considerations. Researchers and caregivers must interpret behavioral and physiological indicators to infer willingness or distress, but these assessments are inherently indirect. Ethical frameworks call for minimizing harm, ensuring voluntary participation where possible, and prioritizing the animal's well-being. In practice, this means using non-invasive methods and respecting signs of refusal or discomfort as much as possible.

b) Acknowledging Animal Agency

Recognizing agency in animals involves understanding that they are active participants in their lives, capable of expressing preferences and making decisions. For example, animals may choose where to live, whom to interact with, or whether to engage with enrichment devices. Supporting agency means designing environments and technologies that allow animals meaningful control and choice, rather than treating them as passive subjects. This perspective encourages more humane and effective approaches to animal care and conservation.

c) Advancing Animal Rights

Animal rights advocate for legal and moral protections that go beyond welfare, seeking to safeguard animals' freedom, integrity, and natural existence. This includes opposing practices that cause unnecessary suffering, promoting habitat preservation, and recognizing animals as stakeholders in decisions affecting their lives. Integrating AI and other technologies into this framework requires careful design to enhance, rather than diminish, animal autonomy and dignity. The rights-based approach influences policy, research ethics, and societal attitudes toward animals.

B. Impact of AI on Human-Animal Relationships

Artificial Intelligence (AI) is reshaping human-animal relationships by introducing new ways to understand, communicate with, and care for animals. They also facilitate more personalized interactions, allowing owners and caregivers to respond more effectively to their animals' needs. However, AI's influence is double-edged;

while it can strengthen bonds by enhancing understanding, it may also risk distancing humans from direct engagement or oversimplifying complex relationships. Evaluating AI's impact involves balancing technological benefits with maintaining genuine, compassionate connections.

a) Enhancing Empathy through Understanding

This improved understanding fosters empathy, as people become more aware of subtle cues signalling distress, contentment, or discomfort. Such insights contribute to more compassionate and responsive human-animal interactions.

b) Facilitating Personalized Care and Interaction

By continuously monitoring individual animals' behaviour and health, AI enables tailored care plans that suit specific needs. Smart devices can adapt to an animal's preferences and routines, improving training, enrichment, and medical interventions. For instance, AI-powered feeders or toys can adjust their operation based on an animal's activity or mood.

c) Risks of Reduced Direct Engagement

Despite its advantages, reliance on AI may inadvertently reduce direct human observation and interaction. Over-dependence on automated systems might lead caregivers to miss nuanced behaviors or diminish their own atonement to animals' needs. This can result in a more transactional relationship, where technology mediates rather than enhances genuine connection. Maintaining a balance between AI assistance and hands-on care is essential to preserve authentic, emotionally rich human-animal bonds.

C. Ethical Frameworks for AI Research in Animal Communication

AI research in animal communication holds great promise for deepening our understanding of non-human species, but it also raises significant ethical questions. Developing ethical frameworks ensures that AI applications respect animal welfare, promote transparency, and prevent misuse of data. These frameworks guide researchers in responsibly designing, deploying, and interpreting AI tools, balancing scientific advancement with respect for the autonomy and well-being of animals. They emphasize minimizing harm, ensuring informed consent (where possible), protecting sensitive information, and fostering collaboration among stakeholders.

Establishing robust ethical standards is critical for gaining public trust and maximizing the positive impact of AI in this field.

1. Prioritizing Animal Welfare and Minimizing Harm

Ethical AI research must prioritize the physical and psychological welfare of animals involved in studies. This includes using non-invasive data collection methods, avoiding disruption to natural behaviors, and monitoring for signs of distress. Researchers should adopt the precautionary principle, erring on the side of caution when uncertain about potential impacts. By minimizing harm, AI research upholds the dignity and rights of animals while ensuring the validity and reliability of data collected.

2. Transparency and Accountability in AI Use

Clear communication about AI methodologies, data sources, and limitations is essential for ethical research. Transparency allows for peer review, replication, and public scrutiny, fostering accountability among researchers. Documenting AI training data, algorithms, and decision-making processes supports ethical integrity and trustworthiness in animal communication studies.

3. Data Privacy and Protection of Sensitive Information

Research involving animal communication often generates data that could reveal locations of endangered species or other vulnerable populations. Ethical frameworks mandate secure handling, anonymization, and restricted sharing of such data to prevent exploitation, poaching, or habitat disturbance. Collaborating with local communities, conservationists, and policymakers ensures that data use aligns with broader protection goals. Respecting privacy extends not only to animals but also to human stakeholders involved in or affected by research.

9. INTERDISCIPLINARY COLLABORATION AND INNOVATION

Interdisciplinary collaboration is essential to advance AI-driven understanding of non-human communication, as it brings together diverse expertise from fields such as artificial intelligence, ethology, veterinary medicine, ecology, and ethics. Combining knowledge from these disciplines fosters innovation by ensuring that technological developments are grounded in biological realities and ethical considerations. Collaborative efforts facilitate the creation of more accurate, context-aware AI models that can handle the complexity of animal signals across species and environments. Furthermore, shared platforms and data repositories enhance transparency, reproducibility, and scalability of research, while cross-disciplinary dialogue helps identify practical applications and societal implications. This chapter highlights successful examples of interdisciplinary projects and emphasizes the importance of breaking down academic silos to promote holistic solutions. Ultimately, nurturing partnerships among scientists, technologists, and practitioners will drive breakthroughs in AI interpretation of animal communication and support ethical, impactful innovations for both humans and non-human species.

A. Bridging AI, Ethology, Veterinary Science, and Ecology

Integrating AI with ethology, veterinary science, and ecology creates a multidisciplinary approach to studying and caring for animals. AI enhances data analysis and pattern recognition across these fields, enabling deeper insights into animal behaviour, health, and environmental interactions. This collaboration improves diagnostics, monitoring, and conservation efforts by combining behavioral understanding, medical expertise, and ecosystem context. Together, these disciplines leverage AI to promote animal welfare and biodiversity in a holistic, informed manner.

B. Collaborative Platforms and Data Sharing

These platforms facilitate seamless interdisciplinary communication, enabling real-time knowledge exchange and coordinated efforts across fields. Additionally, by promoting standardized data formats and transparent methodologies, collaborative sharing enhances reproducibility, fosters trust, and accelerates scientific discoveries, ultimately leading to more comprehensive understanding and improved animal welfare and conservation outcomes.

C. Fostering Innovation through Cross-Disciplinary Research

Cross-disciplinary research fosters innovation by combining diverse expertise from AI, ethology, veterinary science, and ecology to tackle complex animal-related challenges. This integration sparks creative solutions, as insights from one field inform and enhance others, leading to the development of advanced technologies and methodologies. Collaborative efforts accelerate the discovery of new patterns, improve diagnostic tools, and support holistic conservation strategies. By breaking down traditional silos, cross-disciplinary research drives breakthroughs that improve animal welfare, health, and ecosystem management in ways no single discipline could achieve alone.

10. CONCLUSION

This chapter has explored the transformative potential of artificial intelligence in interpreting and bridging non-human communication, highlighting both significant opportunities and complex challenges. Through the examination of AI methods applied to animal vocalizations, gestures, and behaviours, it becomes evident that technology can greatly enhance our understanding of animal cognition, welfare, and ecology. However, technical hurdles such as data variability, environmental noise, and contextual intricacies continue to challenge AI's accuracy and reliability. Additionally, the ethical and philosophical questions around animal agency, consent, and the impact of AI on human-animal relationships emphasize the need for responsible and cautious development of these technologies. Fostering interdisciplinary collaboration among AI researchers, ethologists, veterinarians, ecologists, and ethicists is crucial to create AI tools that are scientifically robust and ethically sound. Ultimately, this chapter advocates for a balanced integration of AI in animal communication—one that respects the complexity of non-human signals and values animal life, while leveraging technology to improve welfare, conservation, and mutual understanding. As AI evolves, so will its ability to bridge communication gaps between species, offering profound insights and fostering deeper connections. This chapter serves as a foundation and inspiration for ongoing research, innovation, and ethical reflection in this exciting and vital field.

REFERNCES

Adeel, A., Gogate, M., & Hussain, A. (2020). Contextual deep learning-based audio-visual switching for speech enhancement in real-world environments. *Information Fusion*, *59*, 163–170.

Andreassen, T., Surlykke, A., & Hallam, J. (2014). Semi-automatic long-term acoustic surveying: A case study with bats. *Ecological Informatics*, *21*, 13–24.

Arkow, P. (1998). Application of ethics to animal welfare. *Applied Animal Behaviour Science*, *59*(1-3), 193–200.

Barker, S. B., & Wolen, A. R. (2008). The benefits of human–companion animal interaction: A review. *Journal of Veterinary Medical Education*, *35*(4), 487–495. PMID: 19228898

Bianco, M. J., Gerstoft, P., Traer, J., Ozanich, E., Roch, M. A., Gannot, S., & Deledalle, C. A. (2019). Machine learning in acoustics: Theory and applications. *The Journal of the Acoustical Society of America*, *146*(5), 3590–3628. PMID: 31795641

Ezanno, P., Picault, S., Beaunée, G., Bailly, X., Muñoz, F., Duboz, R., & Guégan, J. F. (2021). Research perspectives on animal health in the era of artificial intelligence. *Veterinary Research*, *52*, 1–15. PMID: 33676570

Freeberg, T. M., Baker, M. C., Bloomfield, L. L., Charrier, I., Gammon, D. E., Hailman, J. P., & Sturdy, C. B. (2007). *Complexities in vocal communication. Ecology and behavior of chickadees and titmice: an integrated approach*. Oxford University Press.

Fröhlich, M., & van Schaik, C. P. (2018). The function of primate multimodal communication. *Animal Cognition*, *21*, 619–629. PMID: 29876698

Janik, V. M. (2014). Cetacean vocal learning and communication. *Current Opinion in Neurobiology*, *28*, 60–65. PMID: 25057816

Koch, W. (2016). *Tracking and sensor data fusion*. Springer-Verlag Berlin An.

Morcatty, T. Q., Su, S., Siriwat, P., Andersson, A. A., Atoussi, S., Feddema, K., & Roberts, D. L. (2024). Navigating ethical challenges in online wildlife trade research. *Conservation Biology*, *38*(5), e14341. PMID: 39248761

Ojeda-Castelo, J. J., Capobianco-Uriarte, M. D. L. M., Piedra-Fernandez, J. A., & Ayala, R. (2022). A survey on intelligent gesture recognition techniques. *IEEE Access : Practical Innovations, Open Solutions*, *10*, 87135–87156.

Oladele, O. K. (2025). *AI in Wildlife Conservation: Species Monitoring*. Poaching Prevention, and Habitat Restoration.

Pavlovic, V. I., Sharma, R., & Huang, T. S. (1997). Visual interpretation of hand gestures for human-computer interaction: A review. *IEEE Transactions on Pattern Analysis and Machine Intelligence*, *19*(7), 677–695.

Prunier, A., Mounier, L., Le Neindre, P., Leterrier, C., Mormède, P., Paulmier, V., & Guatteo, R. (2013). Identifying and monitoring pain in farm animals: A review. *Animal*, *7*(6), 998–1010. PMID: 23254122

Sharma, S., Sato, K., & Gautam, B. P. (2023). A methodological literature review of acoustic wildlife monitoring using artificial intelligence tools and techniques. *Sustainability*, *15*(9), 7128.

Stokes, J. E., Mullan, S., Takahashi, T., Monte, F., & Main, D. C. (2020). Economic and welfare impacts of providing good life opportunities to farm animals. *Animals (Basel)*, *10*(4), 610. PMID: 32252335

Trigueiros, P., Ribeiro, F., & Reis, L. P. (2015). Hand gesture recognition system based in computer vision and machine learning. *Developments in medical image processing and computational vision*, 355-377.

Wall, J., Wittemyer, G., Klinkenberg, B., & Douglas-Hamilton, I. (2014). Novel opportunities for wildlife conservation and research with real-time monitoring. *Ecological Applications*, *24*(4), 593–601. PMID: 24988762

Wang, S., Gao, J. Z., Lin, H., Shitole, M., Reza, L., & Zhou, S. (2019, April). Dynamic human behavior pattern detection and classification. In *2019 IEEE Fifth International Conference on Big Data Computing Service and Applications (Big-DataService)* (pp. 159-166). IEEE.

Chapter 7
Computer Vision for Animal Behavior Analysis

Pardeep Kumar
https://orcid.org/0000-0003-0308-7634
Lovely Professional University, India

Sanjeev Kumar
https://orcid.org/0000-0002-7375-7341
Lovely Professional University, India

ABSTRACT

Knowledge of animal behavior is paramount to develop studies not only in the field of ecology, ethology, veterinary, and animal welfare. Conventional behavior observation techniques are highly dependent on manual observation which is cumbersome, subjective and can carry out its own errors as done by human beings. Computer vision has recently become an important unsolicited, scalable, non-invasive method to analyse animal behavior. The following paper gives an overview of methods of computer vision that were used in the study of animal behavior, namely object detection, pose estimation, tracking and activity recognition. It talks about the application of deep learning algorithms and data fitting of diverse organisms and environments, ranging between laboratories and wild. Computer vision incorporated into this field can increase accuracy and raise the opportunity of real-time surveillance, early detection of the disease, and behavioral phenotyping.

INTRODUCTION

The issue of animal behavior is so significant in the fields of biology, ecology, veterinary science and wellbeing of the animals. The learning of how animals behave

DOI: 10.4018/979-8-3373-5483-5.ch007

in different circumstances helps the researchers to determine health conditions, well being, improved conservation, and intelligent systems in the agricultural and wild life sector. Taking notes and constant observing of human beings are the two traditional manual methods to observe the behavior of animals. They have undoubtedly added much to ethology, but, after all, they are circumscribed by anthropomorphism, lack of time and impossibility of generalizing- at least to large populations or to long time intervals. More accurate, scale-able and precision solutions have forced a reflection on the current technologies particularly computer vision (Vogg et al., 2025a).

Artificial intelligence as a computer vision, this sub-branch of artificial intelligence is focused on enabling the machines to identify and interpret the visual senses contained in images and videos. It has transformed the majority of the industries since the past decade which include medical, security, manufacturing, and self-driving cars. Computer vision would not only be an effective technology device but it would be non-invasive which would enable its use where an analysis of animal behavior would involve animal behavior detection, tracking and interpretation happening automatically in real time (Bernardes et al., 2021). It thus eliminates the need to have the system manned regularly and also gives objective, repeatable and throughput rich data collection. In deep learning the systems can be taught to smell even the slightest behavioral drift and respond accordingly this can be employed to detect the early development of ailment in livestock or even the stress condition of the test animals in a laboratory (Fernandes et al., 2020). Placing the computer vision framework into the animal behavior is now an option and makes sense with the accessibility to big data, increasing computing power and open-source tools; and, the overall potential of the method enlarges. The drive behind implementation of computer vision in the sector is the abundant quantity of hindrances which encapsulates manual analysis of behavior. Observing animals over extended time periods is not only highly laborious, but simply, impossible in many cases.. Additionally, data may be consistently collected differently by various observers because they themselves may interpret behavior differently. In comparison, computer vision systems are reliable, have the capability to process large volumes of visual information and even identify patterns visible to the naked eye that a human might not detect (Bhuiyan & Wree, 2023). To give an example, behavior monitoring systems in animal farms based on AI could notify a farmer to abnormalities, lameness, decreased feeding, or separation with the herd, helping to intervene and spare an economic loss. The perspective of drones with cameras and computer vision models in conservation research is that drones can follow animal groups to check their density without causing a change in animal behavior in terms of more ethical and effective ecological research (Chen et al., 2024).

The main aim of the study is to understand whether computer vision algorithms can be chosen to analyze the animal behavior in different environs and animal species. These include the application and testing of algorithms in object detection, pose estimation, tracking and behavior categorization. This study will evaluate the accuracy and the performance of these models and will show whether these models can be used practically in the real world or not. It also aims at determining the typical issues when it comes to utilizing such technologies on animals such as occlusion, lighting differences, inter-species variation, and data inaccessibility (Gupta et al., 2024). The end objective is to participate in the creation of an automated system that will be stable and capable enough to assist researchers, farmers, and wildlife practitioners in terms of monitoring and comprehending animal dynamics. Controlled and natural environments are included in the scope of this study (Wurtz et al., 2019a). Computer vision may also be applied in the lab to monitor the movements and interactions of small animals (mouse, rat, or zebrafish). In farming, it has the ability to observe eating, walking, resting, and social activity of animals. Computer vision systems may be used in the wild to analyze video records of camera traps, drone flights or surveillance cameras to model animals like elephants, birds or primates (Schneider et al., 2019). The study also presents a description of the tools and models applied within the field which may include DeepLabCut algorithm to be applied in estimating the pose, YOLO to be applied in tracking an object and convolution neural networks designed to classify behavior. The study recognizes the limitations tackling it in such a way that The benefits of working with such systems can be counted, yet the study also admits the limitations numerous and those include having access to high-quality annotated datasets, the cost of computing resources and necessity of training deep learning models and the need to generalize models to new environments or species (Koger et al., 2023).

Nonetheless, the use of computer vision on animal behavior studies is very positive regardless of these drawbacks. It provides the avenue of automating and scaling the analysis of behaviors by reducing labor and human error. It also comes with new research opportunities, including real-time animal welfare tracking, real-time detection of social dynamics, and even predictive modeling of future behaviors (Stern et al., 2015). This multidisciplinary solution whereby computer and bio sciences and even veterinary sciences merge is likely to transform the way people approach the study and handling of animals both at the research and the implementation level (Arac et al., 2019). By conducting this research work we would like to show not only the technical possibilities of computer vision but also its practical effect on benefiting animals, science, and optimization of effectiveness in various fields of activity.

LITERATURE REVIEW

Conventional study of animal behavior research has been based on direct observation and marking of animals by hand. Ethnologists and biologists will usually document the behavior in terms of either the field study or a controlled study and report patterns of feeding, copulation, social interaction and locomotion. Such observations are usually labor intensive video recording, frame by frame analysis to classify the behaviors, and occasionally echograms, which are standardized lists of behaviors (Menaker et al., 2020). These methods tend to offer great qualitative insights though they have had their share of limitations in that they are lowly due to human subjectivity, the observer effect, and broadness in application. Moreover, manual interpretation is slow and inaccurate especially in big data sets or over extended periods of recollecting (Ferrante et al., 2021).

As computer vision improved, computer-based behavioral analysis has become more of a possibility in animals. Early methods included simple image processing methods to detect movement of the animal or a particular action. Sophisticated algorithms such as machine learning and deep learning models have been proposed in the long run that identify complex behaviors of video data with enhanced precision. Computer vision allows a researcher to extract objective information like the trajectories of movements, postures, interactions, etc.; such studies are more objective and replicable. Fusion of multi-camera systems, 3D reconstructions, and real-time tracking has since compounded such an ability to allow study of animals in naturalistic settings without intrusive interventions. This development is transitioning to high-throughput, scalable, detailed behavioral analysis. There have been many proposals of research and systems to automate analysis of animal behavior through computer vision (Li, 2024). Examples presumably include software frameworks such as DeepLabCut, which applies deep neural networks to perform pose estimation of many species, and id Tracker, where the task is identifying individual animals in a group. Other researchers have aimed at classifying behavior in a particular environment, like detecting aggression in rodents or finding feeding behavior of birds. The diversity of the commercial and open-source platforms differs with the model being used on supervised methods, which are learned on labeled data to the unsupervised types that find new behavioral types (Sun et al., 2024). Nonetheless, there are still obstacles in dealing with heterogeneous species, environmental variation, occlusions and consistency in accuracy involving different experiments. Ongoing attempts to increase the robustness of algorithms and their generalizability are a current area of research.

COMPUTER VISION BASICS

Computer vision has image processing as its fundamental step in which it accepts techniques used to enhance, manipulate and analyze images to derive any relevant information. Typical examples are filtering (to suppress noise) and edge-detecting (to detect a boundary of objects) as well as segmentation (to divide an image into meaningful parts). Data preparation in the preprocessing is very important to normalize, resize, and transform the color space of the images to be further analyzed (Nilsson et al., 2020). These basic skills precondition more sophisticated tasks and make images good and emphasize some important characteristics. The image processing funnel is shown in Figure 1.

Figure 1. Image Processing Funnel

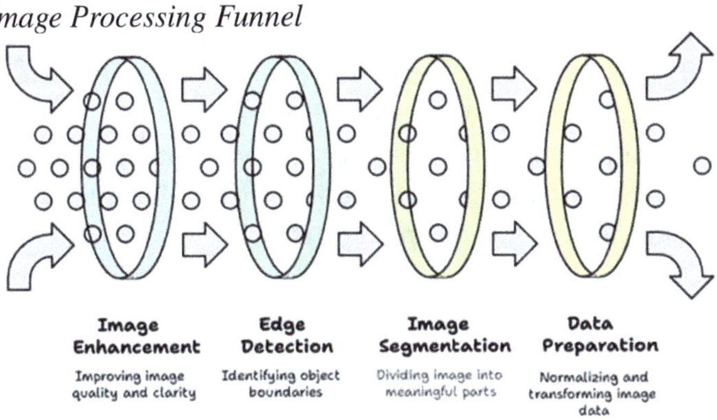

Source: Author Compilation

Object detection can be referred to as both the identification of an object in an image and its classification. Modern techniques have taken to machine learning models to effectively perform object recognition and its spatial location using bounding boxes; most notable is the use of convolution neural networks (CNNs). Recognition is one step higher in that it attempts to assign predetermined classes to recognized objects. Methods in this direction by using Region-based CNN (R-CNN), you Only Look Once (YOLO) and Single Shot MultiBox Detector (SSD) have brought significant improvement in high accuracy of real time detection (Boulanger et al., 2025). Applications like animal identification, species classification and behavior monitoring (searching in ethological studies) depend on object detection and recognition. Pose estimation is the extrapolation of object or subject set-up usually using key points or landmarks like animal / human joints. This makes the in-depth insight into posture and movement possible (Okuyama et al., 2015). Marker less pose estimation has become popular through tools such as open pose / DeepLabCut trained to capture key points in an image /video frame. Pose estimation is needed to track locomotion, gestures, and behavioral patterns and would be helpful in producing rich sources of information on top of the simple localization of objects. Tracking is a technique of tracing object or objects in successive video frames to preserve their identity throughout the periods. Various algorithms go between the classical such as Kalman filters, mean-shift, to more advance based on deep learning and optical flow. Appearance and motion cues are used by the multi-object tracking (MOT) systems to address occlusions and identity switch (Han et al., 2023). With regard to animal behavior studies, in tracking, the repeated measures of individual animal tracks, dyadic interaction, and spatial behavior can be observed in natural settings.

Activity recognition has a goal of fitting sequences of poses or movements over time, to categorize perplexing behaviors or actions. This is usually through time modeling using Hidden Markov Modeling (HMM), Conditional Random Fields (CRF) or even recurrent neural networks (RNN) with assimilation of Long Short-Term Memory (LSTM) networks. Activity recognition is a task of connecting the low level inputs such as pose or position to the semantic level in a manner enabling researchers to automatically classify such concepts as grooming, feeding, or aggression (Perez & Toler-Franklin, 2023). This is critical to the ability to scale behavioral analysis and finding small or less common patterns.

METHODOLOGY

The process of the analysis of animal behavior with the help of the computer vision starts by collecting data and labeling. The quality images and videos of the animals are made through a natural or controlled environment that involves use of all the necessary cameras strategically placed to get a clear view of some of the anatomical features that are vital. These visual data should be annotated correctly by experienced personnel that mark key points (joints or particular anatomical landmarks) on any given frame or image to train the effective models. Annotation is a tedious procedure and it foundations supervised learning by giving ground truth information (Weinstein, 2018). Training and evaluation datasets to be employed involve publicly available datasets, such as the Animal Pose Dataset, which entails annotations of the images involving different types of animals, and specifically constructed datasets in relation to the target species or subject behavior. Specific datasets are acquired to the detection of species-specific pose, species-specific behaviors, or environmental conditions that do not appear in existing datasets very well. Amounts of images, sort of behaviors, as well as annotations schemes, will be provided in order to achieve transparency and reproducibility (Vogg et al., 2025b).

The main characteristic of the methodology is the choices of the corresponding architectures of learning models that can conduct pose estimation and behavior recognition. Convolution Neural Networks (CNNs) are largely applied to extract spatial information in images and make them highly accurate compared to the detection of key points. Recurrent Neural Networks (RNNs) (and, in particular, Long Short-Term Memory (LSTM) networks) are used to represent sequences of movement in order to capture temporal dynamics of animal behavior. Detecting the animals in videos or their body parts using object detection systems such as YOLO can be used as a first step in the direction of pose estimation as it allows a swift localization of the animal or part of its body in a video frame (Nazir & Kaleem, 2021). There are also special tools such as DeepLabCut that takes advantage of transfer learning and deep CNNs that allow them to conduct marker less pose estimation on animal data. The training process includes feeding the neural networks trained with annotated datasets to make them converge using hyper parameters like learning rate, batch size, and the number of epochs. Artificial methods to deal with over fitting, such as data augmentation (i.e. rotation, scaling and flipping) are used to artificially enrich the diversity of the data. In measuring model performance, Mean Average Precision (mAP) and Percentage of Correct Key points (PCK) are used to measure pose detection, and precision, recall, and F1-score measures behavior classification (Kabra et al., 2013). Testing on different sets of data to validate confirms that the models are not over fitted and are generalizable to unseen data.

It is all supported with superior tools and frameworks. Tensor Flow and PyTorch are flexible environments to build and train deep learning models, and OpenCV allows preprocessing of images and videos. Utile such as Labeling and CVAT allow the labeling of datasets effectively and accurately. DeepLabCut is a dedicated tool to pose estimation of animals, having an intuitive interface. To speed up the computation, particularly training, GPU-accelerated libraries (including CUDA and cuDNN) are used to dramatically speed up the computation, and allow experimenting with richer architectures.

Biologists Effectively Specify Behavioral Annotation Requirements for AI Developers:

Biological behaviors often exist along a spectrum such as; aggressive vs. territorial displays, or foraging vs. exploratory movement so biologists should provide unambiguous descriptions, examples, and edge-case distinctions. This avoids misinterpretation by annotators or algorithms. Using species-specific terminology and noting contextual factors (season, habitat, and social hierarchy) helps developers understand subtle cues that differentiate similar behaviors. Next, biologists should identify the observational units required for annotation. This includes specifying whether behaviors should be labeled per frame, per sequence, per event, or per individual. They can also indicate which body-parts or environmental features the AI must track, such as limb positions, vocalizations, or interactions with conspecifics. When necessary, biologists can provide developers with ethograms a standardized catalog of behaviours to serve as the annotation backbone. Effective communication also involves defining metadata requirements. Developers benefit from knowing which environmental variables (temperature, light, habitat type), social parameters (group size, dominance rank), or physiological indicators (stress markers, feeding states) matter for accurate classification. These metadata fields guide model architecture choices and help ensure that important ecological dependencies are captured. Biologists can further support AI teams by supplying representative datasets that illustrate natural variability: different angles, lighting conditions, observer distances, and individual differences. Annotated "gold standard" examples allow developers to validate model performance against expert-approved labels. Regular collaboration and iteration is essential. Biologists should participate in early model-testing phases, reviewing misclassifications and clarifying ambiguous categories. Feedback cycles help refine annotation instructions and improve model interpretability. Biologists can increase long-term effectiveness by documenting annotation protocols in clear, accessible language. This improves reproducibility and helps future annotators or developers maintain consistency.

Applications: The inventions of computer vision technologies have transformed the monitoring of wildlife as the researchers can observe the animals all the time and without their interference. Manual tracking, tagging, or camera-trapping used in the past by conventional visits to wildlife sometimes terrified animals or did not give a lot of information. Aid of automated image and video analysis systems has resulted in researchers being able to observe the presence/population sizes and behavior of species on an extended-duration basis using minimal human assistance. Object detection and tracking algorithms are able to detect animals of a given individual, identify species and compare movement patterns. Pose estimation also increases the interpretation of the animal position and behavior like foraging, mating or migration (Hutchison et al., 2010). This knowledge helps in conservation as it makes known habitation choices, agent of stress, and impacts of the changes in the environment. Furthermore, automated technology is able to sense the onset of poaching activities or other kinds of human animal conflicts in real time aiding timely response.

Computer vision is also implemented in agriculture where monitoring animal behavior, as well as health, is becoming more frequent and provides farmers with informative and directive data to enhance animal welfare and produce crops. Barn cameras or similar set ups of grazing locations would provide a continuous video stream which analysis would be used to identify particular behaviors: e.g. feeding, resting, walking and social interactions. Abnormalities in behavior may be antecedents of disease, a warning that someone is under stress, or injured. Pose estimation models aid in evaluating gait and stance which are vital pointers to lameness or insult (Elbasi et al., 2023). Monitoring is also automated and this eliminates the possibility of manually observing the devices which can involve time-wasting and subjectivity. Farmers can make nutritional, veterinary and adjustment decisions based on collected data which improve herd health and lower economic losses.Computer vision is critical to laboratory research of model organisms such as mice, zebrafish, and fruit flies), which, owing to their genetic tractability, are also broadly used to study genetics, neuroscience, and disease. Control over how the individual behaves is a very important aspect of such research to get the staging of consequences of genetic editing, medication, or environmental alterations. The manual scoring of behavior is time consuming and may be biased as well. Activity recognition and automated pose estimation systems can enable researchers to quantify locomotion, grooming and other behaviors at high accuracy and in a reproducible way. This aids in large scale experiment and high until screening. Sophisticated models can discriminate minor behavioral or other phenotypes that could be hard to observe by human observers, increasing the complexity of biological insights and speeding up biomedical research (Kling-Eveillard et al., 2020).

Computer vision applications have been applied to pets and the smart farm as a result of the emergence of smart technologies. In the context of pet monitoring, cameras with associated pose estimation are able to monitor activity of pets, give warning signs to an owner of unusual behavior or pain, and make an interactive experience such as automated feeding or play. The technology promotes pet welfare with regard to owners who are away. In smart farming, combination of computer vision with IoT sensors allows to track farm animals and crop health in real time (Holzinger et al., 2024). Automated systems are able to monitor hundreds of animals at the same time, control and manage the optimal use of resources, and warn of environmental stress factors heat or overcrowding. The integration helps in precision farming that boosts efficiency, reduces wastage, and demonstrates sustainable farming. In general, computer vision with pets and smart agriculture enhances animal welfare, efficiency, and resource utilization by means of intelligent automation.

CHALLENGES AND LIMITATIONS

Arguably, the biggest problem in using computer vision to analyze animal behavior is the problem of occlusion and changing lighting situations. Animals tend to travel as a flock or in a hazy location where body parts cannot be totally or partially seen by the camera position. This blurs key points that could be used in estimating the pose and recognition of behavior. Also, natural scenes could suffer image quality degradations caused by natural lighting changes during the day, weather changes, or shadows of the surrounding vegetation. The visibility of important features might be diminished by low light or by harsh glare, resulting in detection and tracking error (Wurtz et al., 2019b). These issues necessitate powerful algorithms that can fill in the gaps, tracking even in occluded scenes, and dynamic lighting. Multi-camera setups, infrared imaging, and sophisticated image enhancement are some of those techniques that can help alleviate part of the problem; however occlusion and lighting still stand as bigger barriers on the way of practical applications. Balancing Occasion and lighting challenges in Animal Behavior Analysis shown in Figure 1

Figure 2. Balancing Occusion and lighting challenges in Animal Behaviour Analysis

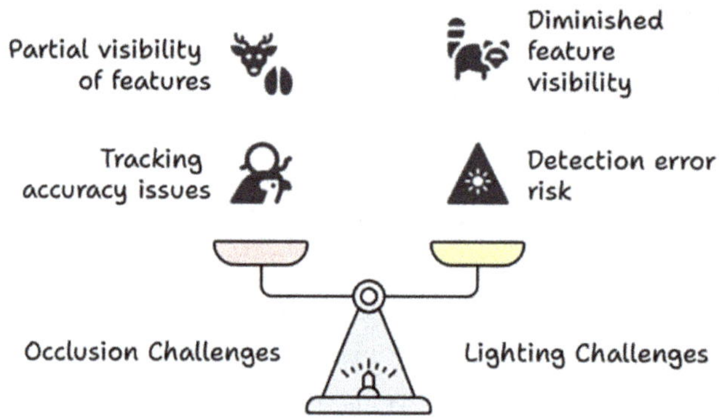

Source: Author Compilation

The second significant constraint is based on the fact that species and individual animals are biologically diverse. The morphological, size, color and behavior of different species and genera are so variable that it is not easily possible to form generalized models. In a given species, even individual variation among members of the same species, including age, sex, physical status and markings, can create variability that may cause loss in the accuracy of the model unless the differences are considered. Pose estimation models found on one species or population cannot be used on others unless either more training data or adaptation occur. This diversity demands the solutions of producing the species-specific models or models that would be highly flexible and can be engaged in transfer learning (Hughey et al., 2018). However, it is not always feasible to collect variety of training data on all the involved species and individuals, which puts subsequent scaling of existing systems in peril. The model must constantly be updated and tinkered with in order to perform equally well across a variety of subjects. One issue that is always faced in this field is the lack of data mainly relating to rare species or rare behavior which is hard to get in large numbers. The quality of the annotated datasets is what makes the supervised learning methods work; however, the procurement costs are labor-intensive and expensive (Guarnido-Lopez et al., 2024). The creation of annotation needs specialized expertise in order to put a correct label on the key points or behavior classification, and manual annotation is time-consuming, especially in sequences consisting of thousands of video frames. There is also the issue of annotation consistency by the annotators that can interfere with the quality of training of the model. In addition, there are certain behaviors, which are subtle or complicated and as such, are difficult to characterize and consistently annotate. Such difficulties restrict the access to rich datasets to train proper models and sometimes approximation such as semi-supervised learning, synthetic data generation, or transfer learning is needed to increase the small amount of data.

The ethical considerations and privacy can be considered significant constrains to the implementation of an animal behavior analysis using computer vision. The second task of looking after the welfare of animals when collecting data takes precedence; methods that are not invasive or causing stress should be used to minimize changes in the animal behaviors. There are some settings where procedures need to be adhered to by rules: these are in farms or labs to ensure that there is humane treatment and to meet the regulations. Moreover, in case of recorded animals in places, which imply human presence as in farms, zoos, or even homes, it poses a privacy issue in that other individuals may be unwillingly included. Data management practices should be able to protect sensitive information as well as ensure the rules about legal and ethical practices. Furthermore, the discussion is emerging on how surveillance technologies will be misused, so defined rules regarding data usage, consent, and transparency are necessary. Ethical responsibility and technological progress are essential to the adoption and feasibility of applications of computer vision in this area..

Machine Learning (ML) Challenges: Major Issues Central to Practical ML in Ecology:

Machine learning has become a powerful tool in ecology, helping scientists analyze complex environmental data, predict species distributions, and monitor ecosystems. However, applying ML in ecological research brings several major challenges. One of the most significant issues is the limited availability of high-quality data. Ecological data are often sparse, noisy, and unevenly sampled. Many species are rare or difficult to detect, leading to biased datasets. This makes it challenging for ML models to learn accurate patterns and generalize to new environments. Another key challenge is data heterogeneity. Ecological studies rely on data from diverse sources remote sensing images, sensor networks, field observations, climate records, and citizen science platforms. These data types differ in scale, resolution, and accuracy. Integrating them into a single model requires careful preprocessing, standardization, and often creative feature engineering. Labeling difficulties also create obstacles. Many ecological tasks, such as; identifying species in camera-trap images or classifying sounds in bioacoustics, require expert labeling. This process is time-consuming, expensive, and sometimes subjective. As a result, datasets may contain mislabels, which can degrade model performance. Generalizability is another critical issue. Ecological systems are highly dynamic and influenced by factors like climate change, habitat modification, and species interactions. ML models trained on data from one region or time period often fail when applied elsewhere. This makes it essential to develop robust models that can adapt to shifting environments. Finally, there are ethical and practical considerations. Overreliance on ML predictions could lead to poor conservation decisions if model limitations are not understood. Ecologists must balance computational efficiency, transparency, and interpretability, ensuring that ML tools genuinely support ecological understanding. ML offers powerful opportunities for ecological research, challenges such as data limitations, heterogeneity, labeling issues, and generalizability must be addressed to make these tools truly practical and reliable as mentioned in the table 01.

Table 1. ML Challenges in Ecological Applications

Challenge Category	Description	Why It Matters in Ecology	Example Scenario
Data Scarcity & Imbalance	Many species or behaviors are rarely observed, leading to limited and skewed datasets.	Biases model predictions toward common species; rare or threatened species become under-detected.	Endangered species appearing in only 1–2% of camera-trap images.
Label Noise & Annotation Errors	Mislabels occur due to poor image quality, ambiguous behaviors, or non-expert annotators.	Incorrect labels degrade model performance and ecological inference.	Misidentifying a juvenile animal as a different species.
Environmental Variability	Lighting, vegetation, weather, and seasonal changes affect sensor outputs.	Models trained in one setting often fail when conditions shift.	Camera-trap images from summer fail in snowy winter environments.
Domain Transfer & Generalization	Models trained in one region/ season/species context fail elsewhere.	Ecologists often need global or multi-site applicability.	Bird-call classifier trained in North America fails in Asia.
Complex Multispecies Interactions	Ecological data often involve multiple species interacting simultaneously.	Hard for models to parse overlapping signals or behaviors.	Predation events where prey and predator appear together.

continued on following page

Table 1. Continued

Challenge Category	Description	Why It Matters in Ecology	Example Scenario
Long-Tail Behavior Distributions	Many behaviors occur rarely, making them difficult to learn.	Behavior prediction becomes biased toward frequent actions.	Rare mating displays occurring only a few times per season.
Hardware & Sensing Limitations	Sensors have limited resolution, range, or battery life.	ML models receive incomplete or low-quality inputs.	Low-light video failing to capture nocturnal activity.
Ethical & Conservation Constraints	Restrictions on data sharing, endangered species data, or disturbance.	Limits dataset size and generalizability.	Cannot publicly share locations of poached species.
Interpretability Needs	Ecologists require transparent reasoning behind predictions.	Black-box models hinder ecological conclusions and management decisions.	Conservation officers need to know *why* a model flags poaching risk.

RESULTS AND DISCUSSION

Computer vision models that are utilized to analyze animal behavior and functions are mostly reviewed depending on their effectiveness in both detecting key points and behavioral recognition. Measures of pose estimation accuracy typically include Mean Average Precision (mAP) and Percentage of Correct Key points (PCK) to measure comparative model quality relating anatomical landmark prediction to ground truth annotations. The good mAP and PCK shows that the model is likely to predict advantageous body parts under diverse poses and settings with clarity. Robustness and generalization improvements are generally achieved using training on various datasets and data augmentation. Nevertheless, performance mainly relies on the model- taken into consideration the complexity of the species, the quality of the image used and also the occlusions. Transfer learning and fine-tuning pre-trained models, such as DeepLabCut in most scenarios have resulted in significant performance improvements and the observation that far fewer labeled data are required.

In behavior recognition, the principal measures to determine the quality of the model in classifying the behavioral states on the basis of the pose sequences or video data include accuracy, precision, recall, and F1-score. These measures give an understanding not only of how well the model predicts what behaviors occurred (true positives), but also of the strength of the model to not incorrectly predict that behaviors occurred (false positives and negatives). To work in practice, high classification accuracy is necessary because possible errors, even in areas such as livestock health monitoring, may translate to late interventions or interventions that do not hold any value. The accuracy of training data and the level of training data complexity is critical in terms of how behaviors are classified. Concrete, definite manners like walking, resting are always classified with greater accuracy than complex and partial behaviors like grooming, socializing etc. Temporal models using recurrent neural networks (RNNs) or Long Short-Term Memory (LSTM) networks can usually outperform static models where they can take full advantage of sequential identities of the behavioral patterns. Also, there are multimodal solutions that unite down estimations of the pose with other sensors. Although these developments have been made, a weakness still lies in differentiating between behaviors that present similar visible as well as those that coincide with each other. The confusion matrix analysis can oftentimes show which behaviors are typically misclassified and further model refinement or data gathering can be guided with that information. One way of doing this is by using independent data to cross-validate the classification to avoid over-fitting the classification result and the ability of such classification to be applied to other different populations or in environmental block to which they have not been trained.

A few real actual applications show how useful computer vision is in exploring animal behavior. As an example, automated camera-based protecting the wildlife with the use of models to estimate the pose has been used to track the social life and the mobility rate of endangered species by using automatic camera traps, hence, providing useful data without injuring animals. This has been reported in agriculture with farms with computer vision systems to monitor livestock finding health conditions (such as lameness or stress) earlier allowing prompt treatment and overall better welfare of the herd. High throughput screening of drugs in model organisms can be performed using automated behavior classification to achieve dramatic savings in experiment timelines and eliminating human bias, thus improving the relative merit of laboratory research. Pet monitoring commercial products incorporate activity recognition algorithms to notify the owner when the pet expresses strange activities and thus improving animal health.

These case studies bring out some successes and some of the current difficulties. As an illustration, one should remember that real-world deployment should include environmental variability, hardware constraints, user-friendliness of the software tools in use. Iterative refinement of the accuracy of models and integration of the system arises out of constant evaluation by end-users, farmers (or researchers), who subsequently provide feedback.

Key Challenges: Important ML-Related Barriers Annotation Bottlenecks, Dataset Bias, and Deployment:

Machine Learning (ML) has become increasingly important for analyzing ecological data, automating species identification, and supporting conservation decisions. However, its practical adoption is limited by several critical challenges. Among these, annotation bottlenecks, dataset bias, **and** deployment barriers some of the most significant issues that ecologists and ML practitioners must overcome. Many ecological ML tasks such as; labeling species in camera-trap images, identifying bird calls in audio recordings, or mapping vegetation types require expert knowledge. Ecologists must manually annotate thousands or even millions of samples, a process that is extremely time-consuming and labor-intensive. Experts are few, and ecological data often contain rare species, cryptic behavior, or low-quality recordings that make labeling even more difficult. This limited availability of expert-verified labels restricts the size and quality of training datasets. As a result, ML models often struggle with poor generalization, especially in tasks involving fine-grained species classification as discussed in the table 02.

Table 2. Key ML-Related Challenges in Ecological Applications

Challenge	Core Problem	Why It Happens	Impact on Ecological ML	Example
Annotation Bottlenecks	High-quality labels are slow, costly, and require expertise.	Behavioral/ species identification often needs domain experts; large video/image volumes overwhelm annotators.	Slows model development; leads to label noise or incomplete ethograms.	Thousands of hours of camera-trap footage requiring behavior classification.
Dataset Bias	Distributions of species, behaviors, environments, or sensor types are uneven.	Rare species under-sampled; sites vary in lighting, habitat, or seasons; human presence may distort data.	Models generalize poorly, inflate accuracy on common classes, and fail in real world.	Classifier over-detects deer but misses rare carnivores due to skewed data.
Deployment Challenges	Models fail when moved from controlled training environments to new field conditions.	Environmental changes, sensor mismatch, domain shift; limited on-device compute in remote areas.	Reduced accuracy, unreliable automated monitoring, or incomplete ecological insights.	Sound classifier trained on high-quality recordings fails on noisy field microphones.

Closely related the issue of dataset bias, which arises from imbalanced or uneven sampling. Ecological datasets commonly over represent certain habitats, popular sampling regions, or easily detectable species, while rare or remote species remain underrepresented. Sensor placements, seasonal variations, and human accessibility further skew data collection. Models trained on such biased data may exhibit strong performance during testing but fail when applied in new regions or under different conditions. Dataset bias can also lead to misclassification of rare or endangered species, creating serious ecological and conservation implications. Addressing this bias requires better data collection strategies, augmentation techniques, and model designs that emphasize robustness. Finally, deployment challenges pose practical barriers to applying ML in real-world ecological settings. Ecological monitoring systems often operate in remote areas with limited internet access, low-power devices, and harsh environmental conditions. Deploying heavy ML models on such hardware is difficult due to constraints in memory, energy, and processing power. Furthermore, many ecological applications require real-time or near-real-time predictions—for instance, detecting invasive species or monitoring poaching activities. Additionally, continuous monitoring introduces the problem of model drift, as environmental conditions change over time, requiring models to be updated or retrained. The annotation bottlenecks, dataset bias, and deployment challenges form core barriers that limit the full potential of ML in ecological research. Overcoming these obstacles will be critical for developing reliable, scalable, and impactful ecological ML solutions.

FUTURE WORK

Among the possible directions of future research is the creation of multispecies behavior modeling systems. Existing computer vision models may be trained to be limited to a few related species, reducing their usefulness in other animals found throughout the animal kingdom. It is possible that future models can be inbuilt in improved architectures, able to learn generalized features that are transferable between different species with varying morphology and behavior. This would entail large and varied datasets, and complex algorithm that can deal with large variations in appearance, movement patterns and environmental contexts. The robust multispecies model achievement would greatly aid the ecological studies and the conservation endeavors as it would allow an array of intense monitoring to be conducted on a variety of species within complicated ecosystems as a whole.

Real-time animal behavior monitoring is in high demand not only in the area of conservation of wildlife animals, management of livestock and pet care agencies. Creation of systems that will be able to analyze the live video streams and provide instant analysis will increase the speed of response and decision making. The real-time functionality demands low-latency and efficient- computation modeling, either based upon edge computing devices or optimized hardware usage such as GPUs or TPUs. Future research will involve the trade-off between accuracy and speed, incorporated adaptive algorithms, or those capable of updates or self-updates with incoming streams of data, as well as designing user-friendly interfaces to implement in real-life applications in a wide variety of environments. The possibility of coupling computer vision, Internet of Things (IoT) technologies, and robotics creates new opportunities with regard to automated, smart animal monitoring and care. The complementary data received by IoT sensors, which may be temperature, humidity, or physiological data may be used in combination with the visual data likely to give a fuller meaning to visual data regarding animal health and behavior. Robotics e.g. autonomous drones or ground robots equipped with cameras can actively track animals, obtain data in many different view angles or even be used to make site visits to provide selective intervention such as feeding or medical treatment. Further efforts will incorporate smooth communication between vision systems, IoT networks and robot systems to develop complete, autonomous animal monitoring systems that have started showing better precision, scalability and welfare.

Data scarcity and annotation are still the major directions of future research. Transfer learning, in which a pre-trained model is adapted to new species or new behaviours with little on-task data, is promising but remains limited when domain gaps are large. The goal of few-shot learning is allowing models to learn much on very little labeled data, and thus they are less reliant on large-scale annotation. Future work will be concentrated on advancing these methods towards increased generalization and adaptation in the future, which will include using meta-learning, data synthesis, and self-supervised learning techniques. These developments will make animal behavior computer vision democratized and available even to rare cases or research areas.

CONCLUSION

Incorporating computer vision to the phenomenon of animal behavior is a major stride to the science of animal behavior, state of wildlife, agriculture and biomedical research. Computer vision technologies to detect, track, and identify animal behaviors and movement allow researchers and practitioners to collect rich, objective, and high-throughput data related to animal behavior and movement that was not possible in the past through manual observation. The development has created new opportunities in ascertaining complex behavioral patterns, animal health and evaluation of how to better manage these animals in various situations. It has further been pointed out in this paper that data collection and annotation is critical in building out reliable computer vision models. Supervised learning can be done by designing annotated datasets, either publicly available or developed by the special experimental trial, on which the model can be further designed. The appropriate selection of powerful models structure like Convolution Neural Networks to analyze spatial information and the Recurrent Neural Networks to model the temporal dynamics has shown strength in predicting the poses of animals and the classification of a large number of behavioral events. Additional development of tools, such as DeepLabCut, has also eased pose estimation to a greater convenience such that these technologies are now within the reach of researchers who are not experts in machine learning.

The comparison of these models against major performance indicators highlights and pierces their strong and weak points. A great deal of accuracy in pose estimation and climate classification shows the possible opportunities to replace or supplement human observation. Nevertheless, there is still some shortcoming seen with challenges like occlusion, variations of light, diverse species of species, and scarcity of data. To counter these problems, you need to conduct further investigation in the fields of algorithm strength, extension of datasets, and efficiency of the annotation. There are some ethical factors that should be given priority, especially when it comes to animal welfare and information security concerning the conduct of using such technologies in a reasonable way. In the future, the influence and application of computer vision in the study of animal behavior are likely to increase due to the possibilities of future directions. Being able to generalize multispecies modeling across taxa has the potential to transform ecological monitoring and conservation. This will improve timely interventions in farming and wildlife management through real time monitoring system. The combination of computer vision, IoT, and computer-controlled robotics, produce the promise of complete autonomous and intelligent animal care solutions, which involve sensory data and the ability to respond to the data. It is hoped that future data constraints will be overcome via advances in transfer learning and few-shot learning such that it should be able to quickly train on new species and new behaviors without large amounts of labeled data.

Altogether, the combination of computer vision and a science of animal behavior will reshape the way we perceive, comprehend and communicate with the animal world. Technical and ethical barriers have yet to go away, but, due to ongoing development of algorithms, computing power, and interdisciplinary cooperation, innovation will not lack. Such tools have not only increased knowledge in science, but also helped in application which enhances the lives of animals, results in conservation, and helps in increasing agricultural output. The synergy represents the breakthrough to higher-order, more intelligent, scaleable and, humane methods of studying and managing the behavior of animals in an increasingly complex world.

REFERENCES

Arac, A., Zhao, P., Dobkin, B. H., Carmichael, S. T., & Golshani, P. (2019). DeepBehavior: A Deep Learning Toolbox for Automated Analysis of Animal and Human Behavior Imaging Data. *Frontiers in Systems Neuroscience, 13*, 20. DOI: 10.3389/fnsys.2019.00020 PMID: 31133826

Bernardes, R. C., Lima, M. A. P., Guedes, R. N. C., Da Silva, C. B., & Martins, G. F. (2021). Ethoflow: Computer Vision and Artificial Intelligence-Based Software for Automatic Behavior Analysis. *Sensors (Basel), 21*(9), 3237. DOI: 10.3390/s21093237 PMID: 34067084

Bhuiyan, M. R., & Wree, P. (2023). Animal Behavior for Chicken Identification and Monitoring the Health Condition Using Computer Vision: A Systematic Review. *IEEE Access : Practical Innovations, Open Solutions, 11*, 126601–126610. DOI: 10.1109/ACCESS.2023.3331092

Boulanger, M. D., Miyamae, J. A., Martin, T., Hish, G., & Moore, T. Y. (2025). *Computer Vision for Lesser Egyptian Jerboa (* Jaculus jaculus *) Behavioral Analysis and Animal Care Refinement. Animal Behavior and Cognition*. Advance online publication. DOI: 10.1101/2025.07.30.667603

Chen, C., Qiu, J., He, L., Wu, Q., & Lu, X. (2024). Research on the application of image recognition technology in live-stock and poultry behavior recognition. In Zhang, X., & Yin, M. (Eds.), *Fifth International Conference on Computer Vision and Data Mining (ICCVDM 2024)* (p. 60). SPIE. https://doi.org/DOI: 10.1117/12.3048224

Elbasi, E., Mostafa, N., AlArnaout, Z., Zreikat, A. I., Cina, E., Varghese, G., Shdefat, A., Topcu, A. E., Abdelbaki, W., Mathew, S., & Zaki, C. (2023). Artificial Intelligence Technology in the Agricultural Sector: A Systematic Literature Review. *IEEE Access : Practical Innovations, Open Solutions, 11*, 171–202. DOI: 10.1109/ACCESS.2022.3232485

Fernandes, A. F. A., Dórea, J. R. R., & Rosa, G. J. D. M. (2020). Image Analysis and Computer Vision Applications in Animal Sciences: An Overview. *Frontiers in Veterinary Science, 7*, 551269. DOI: 10.3389/fvets.2020.551269 PMID: 33195522

Ferrante, G. S., Rodrigues, F. M., Andrade, F. R. H., Goularte, R., & Meneguette, R. I. (2021). Understanding the state of the Art in Animal detection and classification using computer vision technologies. *2021 IEEE International Conference on Big Data (Big Data)*, 3056–3065. https://doi.org/DOI: 10.1109/BigData52589.2021.9672049

Guarnido-Lopez, P., Pi, Y., Tao, J., Mendes, E. D. M., & Tedeschi, L. O. (2024). Computer vision algorithms to help decision-making in cattle production. *Animal Frontiers : the Review Magazine of Animal Agriculture, 14*(6), 11–22. DOI: 10.1093/af/vfae028 PMID: 39764526

Gupta, S., Hasan, W., Singh, S., Kumar, D., Ansari, M. J., & Nisar, S. (2024). *Agriculture 4.0: Smart Farming with IoT and Artificial Intelligence* (1st ed.). CRC Press., DOI: 10.1201/9781003570219

Han, J., Siegford, J., Colbry, D., Lesiyon, R., Bosgraaf, A., Chen, C., Norton, T., & Steibel, J. P. (2023). Evaluation of computer vision for detecting agonistic behavior of pigs in a single-space feeding stall through blocked cross-validation strategies. *Computers and Electronics in Agriculture, 204*, 107520. DOI: 10.1016/j.compag.2022.107520

Holzinger, A., Fister, I., Fister, I., Kaul, H.-P., & Asseng, S. (2024). Human-Centered AI in Smart Farming: Toward Agriculture 5.0. *IEEE Access : Practical Innovations, Open Solutions, 12*, 62199–62214. DOI: 10.1109/ACCESS.2024.3395532

Hughey, L. F., Hein, A. M., Strandburg-Peshkin, A., & Jensen, F. H. (2018). Challenges and solutions for studying collective animal behaviour in the wild. *Philosophical Transactions of the Royal Society B: Biological Sciences, 373*(1746), 20170005. https://doi.org/DOI: 10.1098/rstb.2017.0005

Hutchison, D., Kanade, T., Kittler, J., Kleinberg, J. M., Mattern, F., Mitchell, J. C., Naor, M., Nierstrasz, O., & Rangan, P. C., Steffen, B., Sudan, M., Terzopoulos, D., Tygar, D., Vardi, M. Y., Weikum, G., Branson, S., Wah, C., Schroff, F., Babenko, B., … Belongie, S. (2010). Visual Recognition with Humans in the Loop. In K. Daniilidis, P. Maragos, & N. Paragios (Eds.), *Computer Vision – ECCV 2010* (Vol. 6314, pp. 438–451). Springer Berlin Heidelberg. https://doi.org/DOI: 10.1007/978-3-642-15561-1_32

Kabra, M., Robie, A. A., Rivera-Alba, M., Branson, S., & Branson, K. (2013). JAABA: Interactive machine learning for automatic annotation of animal behavior. *Nature Methods, 10*(1), 64–67. DOI: 10.1038/nmeth.2281 PMID: 23202433

Kling-Eveillard, F., Allain, C., Boivin, X., Courboulay, V., Créach, P., Philibert, A., Ramonet, Y., & Hostiou, N. (2020). Farmers' representations of the effects of precision livestock farming on human-animal relationships. *Livestock Science, 238*, 104057. DOI: 10.1016/j.livsci.2020.104057

Koger, B., Deshpande, A., Kerby, J. T., Graving, J. M., Costelloe, B. R., & Couzin, I. D. (2023). Quantifying the movement, behaviour and environmental context of group-living animals using drones and computer vision. *Journal of Animal Ecology*, *92*(7), 1357–1371. DOI: 10.1111/1365-2656.13904 PMID: 36945122

Li, G. (2024). Advancements, Challenges, and Future Prospects in Vision-Driven Animal Behavior Recognition Technology. In Y. Wang (Ed.), *Proceedings of the 2024 International Conference on Artificial Intelligence and Communication (ICAIC 2024)* (Vol. 185, pp. 522–532). Atlantis Press International BV. https://doi.org/DOI: 10.2991/978-94-6463-512-6_55

Menaker, T., Zamansky, A., Van Der Linden, D., Kaplun, D., Sinitica, A., Karl, S., & Huber, L. (2020). Towards a Methodology for Data-Driven Automatic Analysis of Animal Behavioral Patterns. *Proceedings of the Seventh International Conference on Animal-Computer Interaction*, 1–6. https://doi.org/DOI: 10.1145/3446002.3446126

Nazir, S., & Kaleem, M. (2021). Advances in image acquisition and processing technologies transforming animal ecological studies. *Ecological Informatics*, *61*, 101212. DOI: 10.1016/j.ecoinf.2021.101212

Nilsson, S. R., Goodwin, N. L., Choong, J. J., Hwang, S., Wright, H. R., Norville, Z. C., Tong, X., Lin, D., Bentzley, B. S., Eshel, N., McLaughlin, R. J., & Golden, S. A. (2020). Simple Behavioral Analysis (SimBA) – an open source toolkit for computer classification of complex social behaviors in experimental animals. *Animal Behavior and Cognition*. Advance online publication. DOI: 10.1101/2020.04.19.049452

Okuyama, J., Nakajima, K., Matsui, K., Nakamura, Y., Kondo, K., Koizumi, T., & Arai, N. (2015). Application of a computer vision technique to animal-borne video data: Extraction of head movement to understand sea turtles' visual assessment of surroundings. *Animal Biotelemetry*, *3*(1), 35. DOI: 10.1186/s40317-015-0079-y

Perez, M., & Toler-Franklin, C. (2023). *CNN-Based Action Recognition and Pose Estimation for Classifying Animal Behavior from Videos: A Survey* (Version 1). arXiv. https://doi.org/DOI: 10.48550/ARXIV.2301.06187

Schneider, S., Taylor, G. W., Linquist, S., & Kremer, S. C. (2019). Past, present and future approaches using computer vision for animal re-identification from camera trap data. *Methods in Ecology and Evolution*, *10*(4), 461–470. DOI: 10.1111/2041-210X.13133

Stern, U., He, R., & Yang, C.-H. (2015). Analyzing animal behavior via classifying each video frame using convolutional neural networks. *Scientific Reports*, *5*(1), 14351. DOI: 10.1038/srep14351 PMID: 26394695

Sun, J. J., Zhou, H., Zhao, L., Yuan, L., Seybold, B., Hendon, D., Schroff, F., Ross, D. A., Adam, H., Hu, B., & Liu, T. (2024). Video Foundation Models for Animal Behavior Analysis. *Animal Behavior and Cognition*. Advance online publication. DOI: 10.1101/2024.07.30.605655

Vogg, R., Lüddecke, T., Henrich, J., Dey, S., Nuske, M., Hassler, V., Murphy, D., Fischer, J., Ostner, J., Schülke, O., Kappeler, P. M., Fichtel, C., Gail, A., Treue, S., Scherberger, H., Wörgötter, F., & Ecker, A. S. (2025a). Computer vision for primate behavior analysis in the wild. *Nature Methods*, *22*(6), 1154–1166. DOI: 10.1038/s41592-025-02653-y PMID: 40211003

Vogg, R., Lüddecke, T., Henrich, J., Dey, S., Nuske, M., Hassler, V., Murphy, D., Fischer, J., Ostner, J., Schülke, O., Kappeler, P. M., Fichtel, C., Gail, A., Treue, S., Scherberger, H., Wörgötter, F., & Ecker, A. S. (2025b). Computer vision for primate behavior analysis in the wild. *Nature Methods*, *22*(6), 1154–1166. DOI: 10.1038/s41592-025-02653-y PMID: 40211003

Weinstein, B. G. (2018). A computer vision for animal ecology. *Journal of Animal Ecology*, *87*(3), 533–545. DOI: 10.1111/1365-2656.12780 PMID: 29111567

Wurtz, K., Camerlink, I., D'Eath, R. B., Fernández, A. P., Norton, T., Steibel, J., & Siegford, J. (2019a). Recording behaviour of indoor-housed farm animals automatically using machine vision technology: A systematic review. *PLoS One*, *14*(12), e0226669. DOI: 10.1371/journal.pone.0226669 PMID: 31869364

Wurtz, K., Camerlink, I., D'Eath, R. B., Fernández, A. P., Norton, T., Steibel, J., & Siegford, J. (2019b). Recording behaviour of indoor-housed farm animals automatically using machine vision technology: A systematic review. *PLoS One*, *14*(12), e0226669. DOI: 10.1371/journal.pone.0226669 PMID: 31869364

Chapter 8
Digital Transformation, Analytics, and Technological Maturity in SMEs

Rupam Hazra
https://orcid.org/0009-0003-0332-2128
Global Institute of Management and Technology, Krishnanagar, India

ABSTRACT

This chapter explores how Small and Medium Enterprises (SMEs) navigate digital transformation by developing dynamic capabilities—technological, relational, financial, and environmental—that enhance their strategic resilience in volatile markets. Recognizing the unique challenges faced by resource-limited SMEs, the volume updates current knowledge on managing digital maturity throughout the SME lifecycle. It suggests customized management frameworks and performance indicators aimed at enhancing innovation and competitiveness. The chapter also offers practical guidance for policymakers, development institutions, and entrepreneurship ecosystems focused on fostering sustainable growth and digital adoption in SMEs. Integrating perspectives from senior management, institutional economics, and sustainable entrepreneurship, this work broadens traditional views centered on large corporations, providing actionable insights for driving SME innovation and long-term economic impact.

DOI: 10.4018/979-8-3373-5483-5.ch008

1. INTRODUCTION

Sustainable entrepreneurship in SMEs emphasizes the integration of environmental, social, and economic goals into business practices, fostering long-term value creation beyond mere profit. By linking sustainability with digital innovation, SMEs can develop eco-friendly products, optimize resource use, and enhance operational efficiency, which not only reduces environmental impact but also meets growing consumer demand for responsible businesses. Social and environmental responsibility further strengthens brand reputation and stakeholder trust, opening new markets and partnerships. The long-term economic effects of digital maturity combined with sustainable entrepreneurship include increased resilience, competitiveness, and contribution to local and global economies. This holistic approach enables SMEs to balance growth with social and environmental stewardship, positioning them as key drivers of sustainable development in volatile and complex business landscapes.

2. DYNAMIC CAPABILITIES IN SMEs

Technological capabilities empower SMEs to adopt and innovate with new digital tools, driving efficiency and competitive advantage. Relational capabilities emphasize the importance of building strong networks and partnerships, enabling SMEs to access knowledge, resources, and market opportunities beyond their immediate boundaries. Additionally, it is essential to consider both ostensive and performative elements of routines that make up dynamic skills(Weaven,2021). Financial capabilities focus on the strategic management of scarce financial resources, ensuring that investments in technology and innovation yield sustainable growth. Together, these dynamic capabilities amplify the strategic resilience of SMEs, allowing them to sense opportunities and threats, adapt quickly to market volatility, and maintain agility in competitive landscapes. Developing and leveraging these interconnected capabilities is essential for SMEs aiming to thrive in today's fast-paced digital economy.

A. Technological Capabilities and Innovation

Small and medium-sized enterprises (SMEs) play a vital role in driving economic growth, and their ability to adapt to digital transformation significantly impacts their competitiveness. Technological capabilities, such as access to modern IT infrastructure, cloud computing, and data analytics, empower SMEs to streamline operations and enhance customer experiences. Innovation, often spurred by digital tools, enables these businesses to develop new products, enter new markets, and remain agile in a changing environment. This reflective idea was assessed uniquely by examining the application and utilization of technology to enhance a company's innovation(Valdez-Juárez,2021). Analytics supports informed decision-making by turning data into actionable insights, helping SMEs optimize processes and anticipate market trends. However, many SMEs face challenges such as limited budgets, skill gaps, and low technological maturity, which can hinder their transformation journey.

a) Integration of Digital Technologies

The integration of digital technologies like cloud computing, automation, and customer relationship management (CRM) systems is crucial for SMEs to compete in a digital economy. These tools enable businesses to increase efficiency, reduce operational costs, and scale services with flexibility. However, successful integration requires careful planning, change management, and training to ensure employee adoption and long-term sustainability. SMEs with limited IT resources must prioritize strategic implementation to achieve the highest return on investment.

b) Use of Data Analytics for Strategic Decisions

By analysing customer behavior, sales trends, and operational data, SMEs can tailor services and forecast future demand more accurately. Despite the benefits, many SMEs struggle with data silos or lack the tools and expertise to extract value from data. Investing in user-friendly analytics platforms and upskilling employees can help overcome these barriers.

c) Building Technological Maturity and Innovation Culture

Small and medium-sized enterprises that promote an innovative culture support experimentation, ongoing learning, and adaptability to market shifts. As SMEs mature digitally, they become more resilient, adaptable, and capable of sustained innovation in a dynamic market.

B. Relational Capabilities: Networking and Partnerships

Relational capabilities refer to an SME's ability to build, maintain, and leverage relationships with external stakeholders such as customers, suppliers, research institutions, and other businesses. In the digital era, these networks are crucial for accessing new knowledge, technology, and market opportunities that SMEs may not be able to develop internally. Strong partnerships can accelerate innovation, reduce costs through resource sharing, and provide entry into global value chains. Digital tools—such as collaboration platforms, CRM systems, and supply chain technologies—further enhance the effectiveness of these relationships. Research in strategic management aims to elucidate the differences among firms. The Resource Based View of the firm (RBV) perceives variations in assets as significant explanatory factor, however observational data1 endorses this perspective to a certain extent degree(Czakon,1999). SMEs that invest in relational capabilities tend to be more adaptive, competitive, and sustainable in fast-changing environments.

a) Strategic Alliances and Collaborations

Forming strategic alliances with technology providers, research organizations, or larger firms allows SMEs to gain access to expertise and infrastructure they might otherwise lack. These collaborations can support joint innovation efforts, co-development of products, or access to digital markets. However, the success of such alliances depends on trust, clearly defined roles, and mutual strategic alignment.

b) Customer and Supplier Engagement

Maintaining close relationships with customers and suppliers enables SMEs to co-create value, improve service delivery, and enhance supply chain responsiveness. Digital platforms allow real-time communication, feedback collection, and data sharing, which strengthen these relationships. Engaged customers often become brand advocates, while reliable suppliers contribute to operational stability. Continuous interaction fosters loyalty, innovation, and long-term collaboration.

c) Participation in Innovation Ecosystems

By joining innovation ecosystems—such as industry clusters, incubators, or digital hubs—SMEs gain exposure to cutting-edge knowledge, funding opportunities, and mentorship. These ecosystems foster collaboration among diverse actors, including startups, universities, and policymakers, facilitating rapid innovation and growth. SMEs that actively participate are more likely to adopt emerging technologies and align with market demands. This participation also builds visibility and credibility in their industry.

C. Financial Capabilities: Resource Allocation and Investment Strategies

Financial capabilities are critical for SMEs aiming to navigate digital transformation and sustain innovation. These capabilities refer to the firm's ability to effectively manage financial resources, allocate capital strategically, and invest in areas that drive long-term growth. We contend that resource allocation strategies at an early-stage impact long-term outcome. consequences for competitive edge and sustainability as they affect the company's processes of capability enhancement and external validation(Symeonidou,2022). SMEs often face funding constraints, so making informed, priority-based financial decisions is essential to avoid resource wastage and ensure impactful investments. Efficient financial planning supports the adoption of new technologies, staff training, R&D, and infrastructure development. SMEs with strong financial capabilities are more likely to scale innovations and remain resilient in competitive and volatile markets.

a) Strategic Budgeting and Prioritization

Strategic budgeting involves aligning financial planning with business goals, particularly those related to digital and innovation strategies. SMEs must prioritize investments that yield the highest value, such as automation, cloud tools, or customer-centric platforms. Limited capital means not all initiatives can be pursued, so careful evaluation of return on investment (ROI) and business impact is crucial. When financial planning is strategically aligned, it strengthens the firm's ability to sustain innovation without jeopardizing core operations.

b) Access to Funding and Financial Instruments

Access to appropriate funding sources—such as government grants, venture capital, bank loans, or crowdfunding—is vital for SMEs to finance their transformation. Each funding avenue comes with its own risks, benefits, and conditions, which SMEs must assess before committing. Knowledge of financial instruments tailored for digital innovation can significantly improve a firm's capacity to invest confidently. Building financial literacy and seeking professional advice can improve funding success and reduce reliance on internal capital alone.

c) Investment in Capacity-Building and Innovation

Allocating funds toward capacity-building—such as employee training, R&D, or upgrading infrastructure—enables SMEs to stay competitive and responsive to technological changes. These investments may not yield immediate returns, but they are crucial for long-term resilience and adaptability. Innovation funding should support both product and process improvements, as well as the development of a digital-first mindset across the organization. Financially capable SMEs treat innovation as a continuous investment, not a one-time expense.

3. ENVIRONMENTAL CAPABILITIES AND MARKET ADAPTATION

Environmental capabilities refer to an SME's ability to sense, interpret, and respond effectively to external market conditions and regulatory landscapes, which are crucial for sustaining competitiveness in volatile environments. This includes actively monitoring changes in customer preferences, technological trends, and competitor behaviour to identify new opportunities and threats. Responding to regulatory and competitive pressures requires agility and compliance strategies that enable SMEs to adapt their operations while minimizing risk. Furthermore, leveraging sustainability practices not only helps SMEs meet evolving environmental standards but also creates competitive advantages by aligning with growing consumer and stakeholder expectations for responsible business conduct. The importance and function of adaptation compared to standardization for success in overseas markets has sparked significant discussions within the scholarly community(Ciszewska-Mlinarič & Trąpczyński, 2019). By developing strong environmental capabilities, SMEs enhance their market adaptation skills, allowing them to navigate uncertainty, capitalize on emerging trends, and position themselves as resilient and forward-thinking players in their industries.

A. Sensing Market Changes and Opportunities

Sensing market changes is the ability of an SME to proactively monitor, detect, and interpret shifts in customer behavior, industry trends, emerging technologies, and competitive dynamics. In today's fast-evolving business landscape, this capability is essential for identifying new opportunities and avoiding strategic blind spots. Digital tools such as analytics, AI, and social listening platforms empower SMEs to gather real-time market intelligence. Understanding sensors and their future uses and advancements is essential for various industrial sectors and companies to seize emerging opportunities (Andersen et al., 2004). This allows businesses to adapt offerings, pivot quickly, and stay aligned with market demands. Companies that excel at sensing changes are typically more agile, innovative, and better positioned for sustainable growth.

a) Use of Digital Tools for Market Intelligence

Digital tools like customer analytics, web scraping, and business intelligence platforms allow SMEs to track market trends and consumer behavior with greater accuracy. These technologies help identify emerging preferences, product gaps, and competitor movements early on. With limited resources, SMEs must be selective and use cost-effective tools that deliver actionable insights. Leveraging such data helps them make timely and informed strategic decisions.

b) Engaging with Customers and Stakeholders

Direct engagement through surveys, feedback systems, social media, and CRM platforms allows SMEs to sense shifts in customer needs and preferences. Listening to customer feedback in real time can alert businesses to new demands, dissatisfaction, or innovation opportunities. Stakeholder engagement—such as partnerships with distributors or local community leaders—also provides ground-level insights. This two-way communication enhances responsiveness and builds stronger customer relationships.

c) Environmental Scanning and Competitor Analysis

Environmental scanning involves monitoring external factors like regulatory changes, technological advances, and macroeconomic shifts that affect the business environment. SMEs that routinely analyze their external landscape can spot both risks and innovation opportunities early. Competitor analysis, using tools like benchmarking or SWOT assessments, provides insight into market positioning and potential areas for differentiation. These practices help SMEs remain proactive rather than reactive in a dynamic market.

B. Responding to Regulatory and Competitive Pressures

SMEs operate in environments where regulatory changes and competitive pressures can rapidly affect business viability and growth. The ability to respond effectively to these external forces is crucial for sustaining operations and maintaining market relevance. Customers and authorities maintain constant pressure on companies. to create new methods that will lessen their effect on the environment surroundings(Yalabik,2011). Regulations may involve compliance with data protection, environmental standards, or industry-specific laws. while competitive pressure can stem from globalization, digital disruption, or new market entrants. SMEs that can swiftly interpret and respond to such pressures are more likely to adapt strategically, avoid penalties, and retain competitive advantage. Proactive adaptation requires agility, foresight, and the integration of compliance and market intelligence into business planning.

a) Regulatory Compliance and Adaptability

SMEs must monitor legal developments through regulatory bodies, trade associations, or advisory services. Implementing compliance-focused processes and digital tools (e.g., GDPR-ready systems or e-invoicing platforms) can ease the administrative burden. Businesses that treat regulation as a strategic component—not just an obligation—are more prepared for change and better trusted by partners and clients.

b) Competitive Benchmarking and Strategic Positioning

Competitive benchmarking allows SMEs to compare their products, services, and practices with industry leaders or emerging rivals. This practice helps identify areas for improvement, innovation, or differentiation. By understanding what competitors offer and how they operate, SMEs can position themselves strategically to attract target customers or occupy niche markets. Strategic positioning often involves enhancing value propositions, pricing models, or customer experience based on real-time market intelligence.

c) Flexibility in Business Models and Offerings

To respond to market and regulatory pressures, SMEs must be willing to modify or even transform their business models. This could include shifting from traditional sales to subscription models, adopting e-commerce, or diversifying product lines. Flexibility ensures that the business remains aligned with current market demands and regulatory conditions. Firms that embed agility into their operations can pivot quickly and sustain competitive edge in dynamic environments.

C. Leveraging Sustainability for Competitive Advantage

Sustainability is no longer just a compliance issue—it has become a strategic tool that SMEs can use to differentiate themselves and build long-term value. By integrating environmental, social, and governance (ESG) principles into their operations, SMEs can appeal to eco-conscious customers, reduce costs, and strengthen their brand reputation. Sustainable practices also help businesses future-proof against tightening regulations and supply chain risks. The enduring competitive edge is vital for the long-term durability and achievement of small and medium enterprises (Arsawan et al., 2022). When aligned with core business goals, sustainability initiatives can unlock new markets, attract investors, and foster innovation. SMEs that embed sustainability into their strategy often outperform competitors who ignore it.

a) Green Innovation and Product Differentiation

Developing eco-friendly products or using sustainable materials allows SMEs to tap into the growing demand for ethical and green offerings. Green innovation helps position the brand as responsible and forward-thinking, which can increase customer loyalty and market share. By differentiating through sustainability, SMEs can avoid price wars and instead compete on values and impact. This not only supports sales but also improves stakeholder trust and long-term business viability.

b) Operational Efficiency and Cost Savings

SMEs that monitor and manage their environmental footprint often discover inefficiencies that, once corrected, lower operational costs. These efforts also improve resilience against energy price fluctuations or supply disruptions. Thus, sustainability becomes a source of internal value creation, not just external branding.

c) Access to Sustainable Finance and Markets

Investors and funding agencies are increasingly prioritizing businesses with clear sustainability goals and reporting. SMEs that demonstrate ESG performance are more likely to attract green financing, grants, or partnerships with like-minded corporations. Many large buyers and global supply chains now require suppliers to meet sustainability standards, opening new business opportunities. Positioning as a sustainable enterprise can therefore unlock capital and expand market access.

4. STRATEGIC RESILIENCE IN HIGH-VOLATILITY SCENARIOS

Building adaptive organizational structures allows SMEs to remain flexible, enabling rapid decision-making and resource reallocation when faced with disruptions. Effective risk management and crisis response strategies are essential for minimizing the negative impact of external shocks, such as economic downturns, supply chain interruptions, or technological disruptions. Due to multiple factors like globalization, digitalization, and industry convergence, organizations face heightened volatility, uncertainty, complexity, and ambiguity (VUCA) within their business landscape(Schuh,2019). Additionally, fostering a culture of learning and continuous improvement equips SMEs to evolve from challenges, using feedback and experience to strengthen future responses. By developing strategic resilience, SMEs not only survive turbulent periods but can also seize new opportunities, ensuring long-term sustainability and competitive advantage in volatile markets.

A. Building Adaptive Organizational Structures

Adaptive organizational structures allow SMEs to stay flexible, innovative, and responsive in rapidly changing markets. Unlike rigid hierarchies, adaptive structures support faster decision-making, decentralized control, and team empowerment. Creating adaptive organizations is demonstrating to be a significant challenge, necessitating concurrent enhancements across the three aspects of organizational framework, team dynamics, and data technology (Hess et al., 2000). These organizations can integrate new technologies more effectively and pivot business strategies with minimal disruption. Building such structures is essential for SMEs undergoing digital transformation, enabling them to remain competitive and future-ready.

a) Flattening Hierarchies and Decentralizing Authority

SMEs can enhance agility by reducing management layers and giving teams more decision-making power. This decentralization speeds up responses to market changes and encourages accountability at all levels. Employees feel more involved and motivated when trusted with responsibility. As a result, innovation and responsiveness improve across the organization.

b) Encouraging Cross-Functional and Agile Teams

Creating teams that include members from different departments allows SMEs to solve problems with a broader perspective. These agile teams adapt quickly to project needs and customer feedback, delivering faster and more relevant outcomes. Cross-functional collaboration breaks down silos and fosters innovation through shared knowledge. Such teamwork is vital for navigating complex digital and competitive challenges.

c) Embedding Flexibility in Roles and Workflows

Adaptive structures rely on fluid job roles and dynamic workflows that can adjust based on evolving priorities. Employees are encouraged to develop diverse skill sets and shift roles as needed. Processes are optimized using automation and digital tools, making operations more efficient and scalable. This built-in flexibility enables SMEs to seize new opportunities without major restructuring.

B. Risk Management and Crisis Response in SMEs

Effective risk management and crisis response are vital for SMEs, especially in volatile and fast-changing business environments. SMEs often face higher vulnerability due to limited resources, making it essential to proactively identify, assess, and prepare for risks such as cyber threats, market disruptions, or supply chain issues. In recent years, the close relationship between crisis management (CM) and risk management (RM) has become especially significant for small and medium enterprises (SMEs), which have faced considerable challenges when handling crises(Crovini,2022). A strong risk management strategy enables quick recovery, protects assets, and preserves customer trust during crises. Integrating digital tools and flexible planning helps SMEs become more resilient and adaptive to unexpected challenges.

a) Developing Proactive Risk Assessment Frameworks

SMEs should regularly identify and assess operational, financial, and technological risks using structured methods like SWOT, PESTLE, or risk matrices. Proactive planning enables early detection of threats and minimizes their impact. This involves setting up monitoring systems, defining risk thresholds, and preparing response plans. Anticipating risks helps SMEs act swiftly rather than react defensively.

b) Leveraging Digital Tools for Crisis Monitoring and Communication

Digital platforms such as real-time dashboards, alert systems, and collaboration tools enhance SMEs' ability to monitor emerging risks and communicate during crises. These tools help track supply chain disruptions, market volatility, or cybersecurity incidents as they unfold. Quick, clear internal and external communication ensures transparency and coordinated action. Timely response reduces downtime and reputational damage.

c) Building Organizational Resilience and Recovery Plans

SMEs must embed resilience into their culture through flexible operations, scenario planning, and crisis simulation exercises. Recovery plans should include steps for business continuity, data backup, financial contingency, and employee safety. Training employees on crisis response protocols strengthens preparedness across all levels. Resilient SMEs not only survive crises but often emerge stronger and more innovative.

C. Learning and Continuous Improvement Processes

Learning and continuous improvement are fundamental for SMEs aiming to sustain growth and adapt to digital transformation. By fostering a culture of ongoing learning, SMEs encourage employees to acquire new skills, share knowledge, and embrace innovation. By analysing how people in organizations acquire knowledge, a learning process model is created to enhance ongoing improvement and innovation in business operations(Buckler,1996). This dynamic approach helps SMEs stay competitive, optimize resources, and respond effectively to market changes.

a) Promoting a Learning Culture

Creating an environment where employees feel encouraged to learn and experiment supports innovation and agility. SMEs can facilitate this through training programs, knowledge-sharing sessions, and open feedback loops. A learning culture reduces resistance to change and boosts morale. It empowers staff to contribute actively to business improvement.

b) Implementing Regular Performance Reviews and Feedback

Consistent evaluation of performance through metrics, customer feedback, and internal assessments helps identify areas needing improvement. SMEs should establish clear KPIs and foster transparent communication channels for constructive feedback. This process ensures alignment with business goals and timely course correction. Regular reviews drive accountability and motivate continuous progress.

c) Utilizing Data-Driven Insights for Improvement

Leveraging data analytics enables SMEs to objectively analyze operations, customer behavior, and market trends. Data-driven insights uncover bottlenecks and opportunities that may not be apparent otherwise. By integrating these insights into decision-making, SMEs can refine processes and innovate more effectively. This approach fosters evidence-based improvements rather than guesswork.

5. MANAGEMENT MODELS FOR SME DIGITAL MATURITY

Management models for SME digital maturity provide structured approaches to guide small and medium enterprises through the stages of adopting and integrating digital technologies effectively. These models recognize that SMEs progress through distinct lifecycle stages, each with unique challenges and digital needs, from initial adoption to full digital transformation. Frameworks for assessing technological maturity help organizations evaluate their current capabilities, identify gaps, and set realistic goals aligned with their strategic objectives. Integrating dynamic capabilities into management practices ensures that SMEs not only implement technology but also develop the agility to adapt to ongoing technological changes and market demands. By applying these models, SMEs can enhance their digital readiness, optimize resource use, and improve innovation capacity, positioning themselves for sustainable growth and competitive advantage in the digital economy. Considering these facts, it is evident that SMEs have an essential part of contemporary society and economy, and studies on SMEs' digital efforts can assist support the worldwide economy(Williams,2019).

A. Lifecycle Stages and Their Impact on Digital Strategy

The lifecycle stage of an SME—whether startup, growth, maturity, or renewal—significantly influences its digital strategy and transformation priorities. Early-stage firms often focus on establishing digital presence and leveraging cost-effective tools to build market entry. During growth, SMEs invest more in scalable digital infrastructure, analytics, and process automation to support expansion. The first phase of the digital product life cycle, referred to as the introduction stage, is undoubtedly among the most critical and demanding times (Elgazzar& Dawood, 2025). Mature firms emphasize optimizing digital operations, integrating advanced technologies, and enhancing customer experience for retention. Finally, in the renewal or decline stage, SMEs may pursue digital innovation or diversification to revitalize their business. Understanding lifecycle stages helps SMEs tailor digital investments and strategies to align with evolving business needs and resource capacities.

a) Startup Stage: Digital Foundations and Market Entry

At this stage, SMEs prioritize building a strong online presence, such as websites and social media, to attract customers. They often rely on affordable cloud services and digital marketing tools to minimize costs while maximizing reach. The focus is on experimenting with digital channels and gathering customer insights.

b) Growth Stage: Scaling Digital Capabilities

As SMEs grow, they need more robust digital systems to manage increased operations, sales, and customer data. Investments typically include ERP systems, automation tools, and customer analytics platforms. Digital strategy shifts towards enhancing efficiency, improving data-driven decision-making, and supporting geographic or product expansion.

c) Maturity and Renewal Stage: Digital Optimization and Innovation

In maturity, SMEs focus on refining digital processes, integrating emerging technologies like AI or IoT, and enhancing customer engagement through personalized experiences. They may explore innovation-driven digital transformation to avoid stagnation. Renewal involves reassessing business models and possibly pivoting with new digital initiatives to stay competitive. Digital strategy becomes more sophisticated and aligned with long-term sustainability.

B. Frameworks for Assessing Technological Maturity

Frameworks for assessing technological maturity help SMEs evaluate their current digital capabilities and identify areas for improvement. These models provide structured approaches to benchmark technology adoption, process integration, and innovation readiness. By using such frameworks, SMEs can prioritize investments, track progress, and align digital initiatives with business goals. A framework must be created for the advancement of emerging technologies and their capabilities can be assessed initially through an evaluation focusing on technology maturity and performance (Gove& Uzdzinski, 2013). Understanding technological maturity enables firms to develop realistic digital transformation roadmaps and reduce risks associated with technology deployment.

a) Technology Maturity Models (TMM)

Each level reflects increasing sophistication in technology use and process integration. SMEs can assess their current position and set goals for advancing to higher maturity stages. This stepwise approach supports systematic digital capability development.

b) Digital Maturity Models (DMM)

DMMs evaluate dimensions such as digital strategy, culture, customer experience, technology infrastructure, and analytics capabilities. These models help SMEs understand how well they leverage digital tools and embed innovation across the organization. DMMs often incorporate industry-specific benchmarks for more relevant insights.

c) Industry 4.0 Maturity Frameworks

Focused on advanced manufacturing and digital automation, these frameworks assess readiness for technologies like IoT, cyber-physical systems, and AI. SMEs in manufacturing sectors use these models to evaluate their integration of smart technologies and process digitization. The frameworks promote phased adoption and help avoid costly missteps. They emphasize alignment with operational goals and workforce skills development.

C. Integrating Dynamic Capabilities into SME Management Practices

Dynamic capabilities refer to an SME's ability to sense opportunities and threats, seize them through resource mobilization, and transform processes to sustain competitive advantage. Integrating these capabilities into management practices enables SMEs to navigate rapid technological changes and market uncertainties effectively.

A framework must be created for the advancement of emerging technologies and their capabilities can be assessed initially through an evaluation focusing on technology maturity and performance(Quansah,2022). This requires embedding flexibility, continuous learning, and innovation into the firm's culture and operations. SMEs with strong dynamic capabilities can adapt business models, upgrade digital technologies, and respond swiftly to environmental shifts, ensuring long-term growth and resilience.

a) Sensing Opportunities and Threats

SMEs need systems to continuously scan the external environment for changes in customer preferences, technology trends, and competitor moves. Management practices should encourage proactive market research, use of analytics, and open communication channels to capture relevant insights. Early detection allows timely strategic decisions.

b) Seizing Opportunities through Agile Resource Allocation

Once opportunities are identified, SMEs must quickly allocate financial, human, and technological resources to capitalize on them. Agile budgeting and flexible team structures support rapid project initiation and experimentation. Management should empower teams to make decisions and pivot as needed without bureaucratic delays. This responsiveness drives innovation and competitive advantage.

c) Transforming Organizational Processes and Capabilities

Dynamic capability integration requires ongoing transformation of internal processes, skills, and structures to align with changing market demands. This might involve adopting new digital tools, re-skilling employees, or redesigning workflows for efficiency.

6. PERFORMANCE METRICS AND EVALUATION

Performance metrics and evaluation are vital for SMEs to measure the effectiveness of their digital transformation efforts and guide continuous improvement. Financial and non-financial Key Performance Indicators (KPIs) help organizations assess both tangible outcomes, such as revenue growth and cost savings, and intangible benefits, including customer satisfaction and innovation capability. Measuring innovation and competitive performance involves tracking how well SMEs are adapting to market changes, introducing new products or services, and maintaining their competitive edge. Performance assessment mainly assists in recognizing weaknesses, pinpointing gaps for enhancement, and ensures that organizational activities are effectively aligned with corporate goals(Sardana,2009). Additionally, leveraging analytics for real-time decision-making enables SMEs to monitor performance continuously, identify emerging trends, and make data-driven adjustments promptly. A comprehensive evaluation framework ensures that digital initiatives align with business objectives, maximize return on investment, and contribute to long-term resilience and success.

A. Financial and Non-Financial KPIs for Digital Transformation

Tracking both financial and non-financial key performance indicators (KPIs) is crucial for SMEs to measure the success of digital transformation initiatives. Financial KPIs provide tangible evidence of return on investment, cost efficiency, and revenue growth resulting from digital efforts. Non-financial KPIs capture operational improvements, customer satisfaction, and innovation capacity, which are often leading indicators of long-term success. To run a business both effectively and efficiently, the idea of performance measurement is employed (Cakir et al., 2019). Together, these KPIs offer a balanced view to guide decision-making, optimize resource allocation, and demonstrate value to stakeholders.

a) Financial KPIs

These metrics quantify the direct economic impact of digital transformation. Monitoring cash flow changes and capital expenditure efficiency also helps assess financial sustainability. Financial KPIs are essential for justifying continued digital investments.

b) Non-Financial KPIs

Non-financial KPIs focus on process efficiency, employee digital skills development, customer engagement, and innovation metrics such as the number of new digital products launched. Metrics like customer satisfaction scores (CSAT) and Net Promoter Score (NPS) reveal how well digital initiatives improve user experience. Tracking internal adoption rates of digital tools reflects organizational readiness and culture. These KPIs highlight the operational and strategic benefits beyond immediate financial returns.

c) Balancing Both KPI Types for Holistic Assessment

A balanced dashboard combining financial and non-financial KPIs provides SMEs with comprehensive insights into digital transformation progress. This balance ensures focus on short-term gains and long-term capabilities development. Regular review of these KPIs allows timely adjustments and continuous improvement. Holistic assessment fosters sustainable digital maturity and competitive advantage.

B. Measuring Innovation and Competitive Performance

Measuring innovation and competitive performance is vital for SMEs to understand how effectively they leverage new ideas and technologies to outperform rivals. Innovation metrics assess the firm's ability to generate, implement, and commercialize novel products, services, or processes. Competitive performance indicators reveal how well the SME sustains market share, customer loyalty, and profitability compared to competitors. Together, these measurements guide strategic decisions, resource allocation, and continuous improvement to maintain relevance in fast-changing markets.

a) Innovation Metrics

Key innovation metrics include the number of new products or services launched, R&D expenditure as a percentage of sales, time-to-market, and patent filings. Tracking these indicators helps SMEs evaluate their innovation pipeline and capacity for renewal. Customer adoption rates and feedback on innovations also reflect market acceptance and success. Regular monitoring promotes a culture of experimentation and risk-taking.

b) Market Share and Growth Rates

Competitive performance can be assessed through market share analysis and sales growth relative to industry peers. Increasing or maintaining market share signals effective competitive strategies and innovation impact. Growth rates provide insight into how well SMEs capitalize on market opportunities. These metrics help identify strengths and weaknesses in positioning and execution.

c) Customer Loyalty and Brand Strength

Metrics such as customer retention rates, Net Promoter Score (NPS), and brand awareness indicate the SME's ability to build lasting relationships and differentiate itself. High customer loyalty often correlates with innovative offerings and superior value delivery. Brand strength enhances competitive positioning and pricing power. Measuring these aspects ensures SMEs focus on sustained performance, not just short-term gains.

C. Analytics for Real-Time Decision-Making

Analytics for real-time decision-making empowers SMEs to respond quickly to changing market conditions and operational challenges. By leveraging live data streams and advanced analytics tools, businesses can monitor performance, customer behavior, and external factors instantly. This immediacy enables faster, more informed decisions that reduce risks and capitalize on emerging opportunities. Real-time analytics also supports agile management practices by providing actionable insights at the right moment.

a) Implementing Real-Time Data Collection Systems

SMEs need to deploy sensors, IoT devices, and digital platforms that capture and feed data continuously into analytic systems. These sources can include sales transactions, website interactions, supply chain updates, and social media sentiment.

b) Utilizing Advanced Analytics and AI Tools

Sophisticated analytics, including machine learning and AI, help SMEs detect patterns, forecast trends, and identify anomalies instantly. These tools automate data processing, highlight key insights, and even recommend optimal actions. AI-driven analytics reduce human bias and improve predictive accuracy. Adoption of such technologies accelerates decision cycles and enhances competitiveness.

c) Enhancing Agility through Real-Time Insights

Access to real-time insights allows SMEs to adjust marketing campaigns, inventory levels, and customer service approaches dynamically. This agility minimizes losses from delays and maximizes resource utilization. Real-time feedback loops foster continuous improvement and rapid experimentation. Overall, analytics-driven decision-making increases organizational responsiveness and resilience.

7. ANALYTICS AND DATA-DRIVEN DECISION-MAKING IN SMES

Analytics and data-driven decision-making empower SMEs to leverage their data assets for strategic advantage, enhancing operational efficiency and market responsiveness. Implementing business intelligence tools allows SMEs to collect, analyze, and visualize data, facilitating informed decisions despite often limited resources and technical expertise. Overcoming data challenges such as data quality, integration, and privacy concerns is critical for effective analytics adoption. By embracing data-driven approaches, SMEs can uncover valuable insights into customer behaviour, market trends, and internal processes, enabling targeted marketing, optimized resource allocation, and improved risk management. Case studies of successful analytics implementation demonstrate how SMEs have transformed their decision-making processes, leading to increased competitiveness and innovation capacity. Overall, adopting analytics is a key driver for SME digital maturity and sustainable growth. The findings indicate that descriptive analysis is predominant in data usage among circular economy SMEs, while significant opportunities exist for enhancing the use of predictive analysis techniques(Järvenpää, 2021).

A. Implementing Business Intelligence Tools in SMEs

It suggests customized management frameworks and performance indicators aimed at enhancing innovation and competitiveness. Utilizing business intelligence (BI) tools allows SMEs to convert raw data into practical insights that aid in strategic decision-making. BI solutions help consolidate data from multiple sources, visualize trends, and monitor key performance indicators in real-time. For SMEs, adopting BI tools improves operational efficiency, uncovers new growth opportunities, and enhances customer understanding. Successful implementation requires careful selection of scalable tools aligned with business needs and staff training to maximize usage and impact.

a) Choosing the Right BI Tools

SMEs should evaluate BI platforms based on ease of use, integration capabilities, cost, and scalability. Cloud-based solutions often provide flexible, affordable options with minimal IT overhead. It's important to select tools that offer customizable dashboards and automated reporting tailored to the SME's specific goals.

b) Data Integration and Quality Management

Effective BI implementation depends on consolidating accurate and timely data from diverse sources like sales, finance, marketing, and operations. SMEs must establish data governance practices to ensure data quality and consistency. Poor data quality can lead to misleading insights and flawed decisions. Continuous monitoring and cleaning processes are essential for reliable analytics.

c) Training and Change Management

To fully benefit from BI tools, SMEs need to invest in employee training and foster a data-driven culture. Staff must understand how to interpret reports, use dashboards, and leverage insights for decision-making. Change management initiatives help overcome resistance and encourage adoption across departments. A collaborative approach ensures BI tools become integral to daily business processes.

B. Overcoming Data Challenges in Small Enterprises

In the last twenty years, businesses have experienced a significant rise in the gathering and creation of large quantities of information from various internal and external sources(Brandy,2023). Small enterprises often face significant data challenges such as limited data volume, poor data quality, and lack of expertise in data management. Overcoming these hurdles is essential for effective decision-making and leveraging digital transformation benefits. SMEs need to adopt practical strategies to improve data collection, storage, and analysis while optimizing resource use. Addressing data challenges enables small businesses to unlock insights that drive growth, efficiency, and competitive advantage.

a) Improving Data Quality and Consistency

SMEs must adopt standardized data entry methods and validation controls to minimize errors and discrepancies. Routine audits and cleansing procedures assist in keeping datasets accurate and trustworthy. Educating employees about the significance of data quality encourages responsibility. High-quality data forms the foundation for trustworthy analytics and informed decisions.

b) Leveraging Affordable and Scalable Data Solutions

Small enterprises can utilize cloud-based data storage and processing platforms that offer scalability without heavy upfront investments. These solutions provide flexible capacity to handle growing data needs and reduce IT infrastructure burdens. Open-source tools and software-as-a-service (SaaS) models are cost-effective options. Selecting the right tools ensures SMEs can manage data efficiently within budget constraints.

c) Building Data Literacy and Skills

Developing employees' data literacy through training and workshops empowers them to collect, interpret, and apply data insights effectively. SMEs may also consider hiring or partnering with data specialists to fill expertise gaps. Encouraging a data-driven mindset across the organization promotes better use of analytics in daily operations. Skilled teams are critical for turning data into actionable business value.

C. Case Studies on Successful Analytics Adoption

Examining case studies of SMEs that successfully adopted analytics highlights practical strategies, challenges overcome, and measurable benefits. These real-world examples demonstrate how analytics can transform operations, enhance customer insights, and drive competitive advantage even with limited resources. Case studies provide inspiration and actionable lessons for SMEs planning or advancing their analytics journeys. Understanding diverse industry applications showcases the versatility and impact of data-driven decision-making.

a) Retail SME Enhancing Customer Experience

A small retail company implemented customer analytics to track purchasing patterns and preferences. By analysing this data, they personalized marketing campaigns, optimized inventory, and improved customer retention. The SME saw a significant increase in repeat sales and customer satisfaction within six months. This case underscores the value of targeted analytics for revenue growth.

b) Manufacturing SME Optimizing Production Efficiency

A manufacturing SME utilized real-time analytics and IoT sensors to oversee equipment performance and anticipate maintenance requirements. The SME also leveraged data to streamline supply chain operations, achieving faster turnaround times. This case highlights how analytics drives operational excellence.

c) Service SME Driving Strategic Decision-Making

A service-based SME adopted analytics dashboards to monitor key performance indicators such as client acquisition, project timelines, and employee productivity. Enhanced visibility enabled managers to make data-driven staffing and resource allocation decisions. The SME improved profitability and client satisfaction by responding swiftly to performance trends. This example shows analytics' role in strategic management.

8. INSTITUTIONAL AND POLICY SUPPORT FOR SME INNOVATION

Institutional and policy support plays a crucial role in fostering digital transformation and innovation within SMEs by providing the necessary resources, guidance, and enabling environments. Public policies aimed at promoting digital adoption help lower barriers such as limited access to technology, funding, and skilled personnel, thereby accelerating SME competitiveness. Development institutions and entrepreneurship ecosystems offer vital support through training programs, mentorship, and funding opportunities tailored to the unique needs of smaller enterprises. Best practices for catalysing SME competitiveness emphasize collaboration among government agencies, industry bodies, and academic institutions to create integrated support networks. This multi-faceted support not only boosts SMEs' innovation capabilities but also contributes to broader economic growth and sustainability by empowering resource-limited firms to thrive in rapidly evolving markets.

A. Role of Public Policies in Promoting Digital Transformation

Public policies play a crucial role in accelerating digital transformation in SMEs by creating an enabling environment for technology adoption and innovation. Governments can provide financial incentives, such as grants, subsidies, and tax breaks, to lower the cost barriers SMEs face when investing in digital tools. Policies that support digital infrastructure development, including broadband access and cloud services, enhance connectivity and access to digital resources. Additionally, regulatory frameworks that encourage data security, privacy, and fair competition help build trust and protect SMEs during their digital transition. Together, these policy measures stimulate SME growth, competitiveness, and inclusion in the digital economy.

a) Financial Support and Incentives

Public programs offering direct funding or tax relief reduce upfront costs associated with digital technology acquisition and training. These incentives encourage SMEs to experiment with and scale digital innovations. Access to affordable financing also helps SMEs overcome liquidity constraints. Financial support accelerates digital adoption rates and lowers failure risks.

b) Infrastructure Development

Investment in digital infrastructure such as high-speed internet, cloud platforms, and data Centers is essential for SMEs to effectively implement digital solutions. Reliable infrastructure ensures SMEs can leverage advanced technologies and connect with global markets. Enhanced connectivity boosts productivity and innovation capacity.

c) Regulatory Frameworks and Standards

Clear regulations around data protection, cybersecurity, and digital transactions provide a secure environment for SMEs to operate digitally. Governments can create standards that foster interoperability and equitable competition in digital markets. These frameworks build confidence among SMEs, customers, and partners. Robust regulation minimizes risks associated with digital transformation and fosters sustainable growth.

B. Support from Development Institutions and Ecosystems

Development institutions and innovation ecosystems play a vital role in supporting SMEs through their digital transformation journeys. These entities provide access to funding, technical assistance, training programs, and networking opportunities that SMEs might otherwise lack. By fostering collaboration between government agencies, private sector players, academia, and NGOs, ecosystems create environments conducive to innovation and knowledge sharing. Such support helps SMEs overcome resource constraints, build digital capabilities, and integrate into larger value chains, enhancing their competitiveness and growth prospects.

a) Financial and Technical Assistance

Development institutions often offer grants, low-interest loans, or subsidies targeted at digital adoption and capacity building. They also provide expert consulting, technology audits, and implementation support to guide SMEs in choosing and deploying appropriate digital tools. This support lowers obstacles and boosts the chances of a successful change. SMEs benefit from tailored solutions addressing their unique challenges.

b) Training and Capacity Building

Ecosystems organize workshops, mentoring, and skill development programs to enhance SMEs' digital literacy and management capabilities. These initiatives equip entrepreneurs and employees with the knowledge needed to operate new technologies and interpret data effectively. Continuous learning fosters adaptability and innovation culture within SMEs. Improved skills translate directly into better digital outcomes.

c) Networking and Collaborative Platforms

Innovation ecosystems facilitate connections between SMEs, startups, research institutions, and large corporations. Such platforms enable knowledge exchange, partnerships, and co-creation of digital solutions. Collaboration accelerates innovation cycles and market access. Being part of a supportive network strengthens SMEs' resilience and scalability.

C. Best Practices for Catalysing SME Competitiveness

Catalysing SME competitiveness involves adopting strategies that enhance innovation, operational efficiency, and market responsiveness. Best practices include fostering a culture of continuous improvement, leveraging digital technologies for process optimization, and actively engaging customers through data-driven insights. Additionally, SMEs benefit from strategic partnerships and participation in innovation networks to access resources and knowledge beyond their internal capabilities. Implementing these practices enables SMEs to differentiate themselves, adapt to market changes, and sustain long-term growth.

a) Fostering a Culture of Innovation and Agility

Encouraging experimentation, learning from failures, and quickly adapting to feedback cultivates an environment where innovation thrives. SMEs that prioritize agility can respond faster to customer needs and competitive threats. Leadership commitment and open communication are key to embedding this culture. This mindset supports ongoing transformation and resilience.

b) Leveraging Digital Technologies

Integrating digital tools such as automation, analytics, and cloud computing streamlines operations and reduces costs. These technologies also enable personalized customer experiences and more efficient supply chain management.

c) Building Strategic Partnerships and Networks

Collaborating with other businesses, research institutions, and industry bodies expands access to expertise, innovation, and new markets. Strong partnerships accelerate growth and competitive positioning.

9. SUSTAINABLE ENTREPRENEURSHIP AND ECONOMIC IMPACT

Sustainable entrepreneurship in SMEs emphasizes the integration of environmental, social, and economic goals into business practices, fostering long-term value creation beyond mere profit. By linking sustainability with digital innovation, SMEs can develop eco-friendly products, optimize resource use, and enhance operational efficiency, which not only reduces environmental impact but also meets growing consumer demand for responsible businesses. Social and environmental responsibility further strengthens brand reputation and stakeholder trust, opening new markets and partnerships. The long-term economic effects of digital maturity combined with sustainable entrepreneurship include increased resilience, competitiveness, and contribution to local and global economies. This holistic approach enables SMEs to balance growth with social and environmental stewardship, positioning them as key drivers of sustainable development in volatile and complex business landscapes.

A. Linking Sustainability with Digital Innovation

Linking sustainability with digital innovation allows SMEs to create value while addressing environmental and social challenges. Digital technologies—such as IoT, AI, and blockchain—enable smarter resource management, energy efficiency, and transparent supply chains. By integrating sustainable practices with digital tools, SMEs can reduce waste, optimize operations, and meet growing consumer demand for responsible products. This synergy drives competitive advantage, enhances brand reputation, and opens access to new markets and funding opportunities. SMEs that strategically combine sustainability and digital innovation position themselves as leaders in the evolving green economy.

a) Enhancing Resource Efficiency through Digital Solutions

Digital tools like sensors, data analytics, and automation help SMEs monitor and reduce resource consumption in real-time. For example, IoT devices can optimize energy use or minimize water waste in manufacturing. These innovations lower operational costs and environmental footprints simultaneously. Resource efficiency is a tangible benefit of linking sustainability with technology.

b) Promoting Transparency and Traceability

Technologies such as blockchain provide immutable records of product origins, materials, and lifecycle impacts. SMEs can use this transparency to build trust with customers, comply with regulations, and differentiate themselves in the market.

c) Driving Sustainable Business Models

Digital innovation enables new business models that prioritize sustainability, such as product-as-a-service or sharing platforms. SMEs can leverage digital platforms to extend product lifecycles, encourage reuse, and reduce waste. These models align profitability with environmental stewardship and customer values. Sustainable business models supported by technology foster long-term resilience.

B. Social and Environmental Responsibility in SMEs

Social and environmental responsibility reflects an SME's commitment to ethical practices that benefit society and minimize ecological impact. SMEs adopting these responsibilities enhance their reputation, foster customer loyalty, and comply with regulatory standards. Integrating social and environmental initiatives—such as fair labour practices, community engagement, and waste reduction—can lead to operational efficiencies and innovation. Responsibly managed SMEs contribute to sustainable development while differentiating themselves in increasingly conscious markets.

a) Implementing Ethical Labor Practices

SMEs committed to social responsibility ensure fair wages, safe working conditions, and equal opportunities. Ethical labour practices also reduce risks related to legal compliance and reputational damage. Workers who feel appreciated are generally more productive and creative.

b) Minimizing Environmental Impact

Environmental responsibility involves reducing pollution, conserving resources, and managing waste effectively. SMEs can adopt energy-efficient technologies, use sustainable materials, and implement recycling programs. These efforts not only reduce costs but also align the business with global sustainability goals.

c) Engaging with Local Communities

Active participation in community development fosters goodwill and creates social capital for SMEs. This can include supporting local education, health initiatives, or environmental conservation projects. Community engagement builds stronger relationships and opens opportunities for partnerships or customer base expansion. Social responsibility rooted in local involvement enhances long-term business sustainability.

C. Long-Term Economic Effects of Digital Maturity

Digital maturity significantly influences the long-term economic performance of SMEs by enhancing efficiency, innovation, and market competitiveness. Mature digital capabilities enable SMEs to automate routine tasks, optimize resource allocation, and improve customer engagement, leading to cost savings and revenue growth. Over time, digital maturity fosters resilience by enabling rapid adaptation to market changes and reducing operational risks. Furthermore, digitally mature SMEs can access new markets and funding opportunities, contributing to sustained economic success and business scalability.

a) Increased Operational Efficiency and Cost Reduction

Digital maturity allows SMEs to streamline processes using automation, cloud computing, and data analytics, reducing errors and resource wastage. Efficient operations lower costs and increase profit margins, freeing resources for innovation and growth. Over the long term, cost savings contribute to stronger financial stability.

b) Enhanced Innovation and Product Development

Digitally mature SMEs leverage advanced technologies to accelerate innovation cycles and improve product quality. Access to real-time data and customer insights enables rapid prototyping and market testing. Continuous innovation supports differentiation and attracts new customers. Innovation driven by digital maturity fuels sustainable economic growth and market relevance.

c) Improved Market Access and Customer Engagement

With mature digital tools, SMEs can expand their reach through e-commerce, social media, and personalized marketing strategies. Enhanced customer engagement leads to higher satisfaction, loyalty, and repeat business. Digital channels also open opportunities to enter global markets without heavy physical infrastructure. This expanded market access translates into increased revenue streams and long-term economic resilience.

10. CONCLUSION

In conclusion, this chapter highlights the critical importance of digital transformation and the development of dynamic capabilities for SMEs striving to achieve resilience and competitiveness in today's rapidly changing markets. By embracing technological, relational, financial, and environmental capabilities, SMEs can effectively navigate volatility, innovate, and adapt to evolving business environments. The tailored management models and performance metrics presented provide practical frameworks to guide SMEs through their digital maturity journey, while policy recommendations underscore the vital role of institutional support in catalysing innovation. Ultimately, fostering sustainable entrepreneurship and integrating digital analytics not only enhance SME growth but also contribute significantly to broader economic development. This comprehensive approach empowers SMEs to transform challenges into opportunities, positioning them as key drivers of innovation and sustainable success in the global economy.

REFERNCES

Andersen, P. D., Jørgensen, B. H., Lading, L., & Rasmussen, B. (2004). Sensor foresight—Technology and market. *Technovation, 24*(4), 311–320.

Arsawan, I. W. E., Koval, V., Rajiani, I., Rustiarini, N. W., Supartha, W. G., & Suryantini, N. P. S. (2022). Leveraging knowledge sharing and innovation culture into SMEs sustainable competitive advantage. *International Journal of Productivity and Performance Management, 71*(2), 405–428.

Brandy, S. (2023). Overcoming challenges and unlocking the potential: Empowering Small and Medium Enterprises (SMEs) with data analytics solutions. *International Journal of Information Technology and Computer Science Applications, 1*(3), 150–160.

Buckler, B. (1996). A learning process model to achieve continuous improvement and innovation. *The Learning Organization, 3*(3), 31–39.

Cakir, G., Bezbradica, M., & Helfert, M. (2019, May). The Shift from financial to non-financial measures during transition into digital retail–a systematic literature review. In *International conference on business information systems* (pp. 189-200). Cham: Springer International Publishing.

Ciszewska-Mlinarič, M., & Trąpczyński, P. (2019). Foreign market adaptation and performance: The role of institutional distance and organizational capabilities. *Sustainability, 11*(6), 1793.

Crovini, C. (2022). Crisis Management and Risk Management in SMEs: Towards an Integrated Early Warning System. In *Crisis Management for Small and Medium-Sized Enterprises (SMEs) Strategies for External Crises* (pp. 215–233). Springer International Publishing.

Czakon, W. (1999). Relational capability of organizations–theoretical advances. *Journal of Sociology (Melbourne, Vic.), 104*(5), 48–65.

Elgazzar, M. A. G., & Dawood, M. E. T. (2025). Mastering the Phases of Design Strategy: A Comprehensive Guide to Navigating from Conceptualization to Launch in the Digital Product Life Cycle. *Journal of Art. Design and Music, 4*(2), 1.

Gove, R., & Uzdzinski, J. (2013). A performance-based system maturity assessment framework. *Procedia Computer Science, 16*, 688–697.

Hess, K. P., Entin, E. E., Hess, S. M., Hutchins, S. G., Kemple, W. G., Kleinman, D. L., ... & Serfaty, D. (2000). Building adaptive organizations: a bridge from basic research to operational exercises.

Järvenpää, A. M., Kunttu, I., Jussila, J., & Mäntyneva, M. (2021, April). Data-driven decision-making in circular economy SMEs in Finland. In *The International Research & Innovation Forum* (pp. 371–382). Springer International Publishing.

Quansah, E., Hartz, D. E., & Salipante, P. (2022). Adaptive practices in SMEs: Leveraging dynamic capabilities for strategic adaptation. *Journal of Small Business and Enterprise Development, 29*(7), 1130–1148.

Sardana, G. D. (2009). Evaluating the business performance of an SME: A conceptual framework. *International Journal of Globalisation and Small Business, 3*(2), 137–159.

Schuh, G., Patzwald, M., & Cardoso, M. C. I. (2019, August). Resilient technology strategy in volatile environments: Derivation of requirements to enable long-term strategic positioning in times of volatility, uncertainty, complexity and ambiguity. In *2019 Portland International Conference on Management of Engineering and Technology (PICMET)* (pp. 1-15). IEEE.

Symeonidou, N., Leiponen, A., Autio, E., & Bruneel, J. (2022). The origins of capabilities: Resource allocation strategies, capability development, and the performance of new firms. *Journal of Business Venturing, 37*(4), 106208.

Valdez-Juárez, L. E., & Castillo-Vergara, M. (2021). Technological capabilities, open innovation, and eco-innovation: Dynamic capabilities to increase corporate performance of SMEs. *Journal of Open Innovation, 7*(1), 8.

Weaven, S., Quach, S., Thaichon, P., Frazer, L., Billot, K., & Grace, D. (2021). Surviving an economic downturn: Dynamic capabilities of SMEs. *Journal of Business Research, 128*, 109–123.

Williams, C., Schallmo, D., Lang, K., & Boardman, L. (2019). *The International Society for Professional Innovation Management.* ISPIM.

Yalabik, B., & Fairchild, R. J. (2011). Customer, regulatory, and competitive pressure as drivers of environmental innovation. *International Journal of Production Economics, 131*(2), 519–527.

Chapter 9
Effect of IoT Devices on Animals

Nakul Sharma
https://orcid.org/0000-0003-3947-904X
Vishwakarma Institute of Technology, India

ABSTRACT

IOT devices are often in the vicinity of animals. IOT devices on animals are used for experimentation and monitoring purposes. This chapter discusses the effect of different radiations emitted from IoT devices on animals. The specific animals in focus are rats, mice and chickens. There are different short-range and long-range frequency-based radiations emitted from IoT devices. The current literature discusses the experiments conducted on these animals and the duration of exposure to radiation. The literature review provides evidence regarding their effect on different classes of animals. These effects also depend upon the type of animal and the duration of exposure to radiations. The rats showed memory, blood, brain, cell loss, learning related effects. The mouse or mice showed reduced calcium absorption, memory loss, behavioral, anxiety-related, carcinogens-related changes. The chickens showed effect on neurons and blood-brain integrity changes. There were also studies which did not point to any drastic changes in these animals.

1. INTRODUCTION

IOT technology has evolved moving from Internet-Of-Things to Internet-Of-Everything (IOE). The multidisciplinary and ubiquitous application of these devices is part of a larger landscape where these devices are deployed (Pratap & Annamalai, 2025). The animal ecosystem has also not been unaffected by the Internet of-Everything. The animal ecosystem includes creating different flora and fauna

DOI: 10.4018/979-8-3373-5483-5.ch009

which are developed by natural resources and the environment. The preservation and conservation of even one of the members of the ecosystem is needed for maintaining the food chain (Geary et al., 2020). The application of these devices is according to the specific domain in which they are deployed. In relation to an animal's ecosystem, there are the following areas where the devices are employed

1. Air pollution monitoring
2. Weather monitoring system
3. Smart agriculture system (Kamal, 2024)
4. Pharmaceutical trials (Portelange, 2020)
5. Precision Livestock Farming (OZGER, 2024)

Figure-1 shows different applications of IoT devices in relation to an animal's ecosystem.

Figure 1. Domain of application for IoT devices in relation to animals (Kamal, 2024)[4](OZGER, 2024)

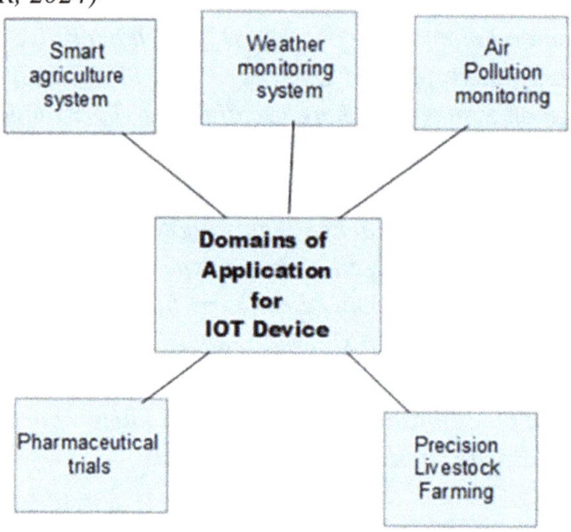

There are different types of enabling technologies that are used in the implementation of IoT-based networks. These include sensor technology, Wifi networks, Zigbee networks, IEEE 802.11, LoRa, IEEE 802.15.4, GPRS/3G/4G/5G/6G/7G, Xbee, MQTT, SigFox, RF, RFID (Portelange, 2020). These technologies are used along with different communication protocols. All these networking devices emit radiation ranging from short-range to long range frequencies (Kamal, 2024). This chapter's objectives are as follows

1. Enumerating the list of radiations emitted by the IOT devices.
2. Investigating existing literature for radiation-induced diseases in animals.

This section introduces study of IOT technology. IOT technology and its evolution along with its different applications is discussed. It also provides a summary of different enabling technology which provide support towards realization IOT devices communication.

2. IOT-DEVICE BASED RADIATIONS

There are the following different networks that operate for enabling IoT devices:

1. RFID systems
2. WiFi
3. Zigbee network
4. GPRS
5. Xbee
6. SigFox
7. LoRa (Portelange, 2020)

The list of frequencies operated by these technologies is given in Table-1.

Table 1. Name of networking technology along with their operating frequency

Sr. No.	Name of Networking Technology	Operating Frequency	Reference
1	RFID System	125 kHz to 134.2 kHz (LF: low frequency); – 13.56 MHz (HF: high frequency); – 860 to 915 MHz (UHF: ultra-high frequency); and – 2.45 GHz to 5.8 GHz (microwave frequency)	Pratap and Annamalai (2025)
2	Wifi	1-45 GHZ	Elshaekh et al. (2025)
3	Zigbee Network	915 MHz and 2.4 GHz frequency bands.	Samijayani et al. (2024)
4	GPRS/GSM	900 MHz to 2.1 GHz	Elshaekh et al. (2025)
5	5G	7 GHz	Elshaekh et al. (2025)
6	Xbee	2.4 GHz ISM band	Mahmud et al. (2024)
7	SigFox	863 MHz to 870 MHz	Garlisi et al. (2025)
8	LoRa	865.0 to 868.0 MHz	Garlisi et al. (2025)

RFID systems operate with different types of frequency. The low-level frequency operating range is 125 kHz and 134.2 kHz. The high-level frequency operating range is 13.56 MHz. The ultrahigh frequency range is from 863 to 915 MHz. The microwave frequency range is 2.45 GHz to 5.8 GHz. Wifi technology has a range from 1 to 45 GHz. The Zigbee network range is 915 MHz to 2.4 GHz. The GPRS/GSM's operating frequency is in the range of 900 MHz to 2.1 GHz. 5G technology has an operating frequency of 7 GHz. XBee is a variation of the Zigbee network having a range of 2.4 GHz. The SigFox technology has a range from 863 MHz to 870 MHz. The LoRa technology has a frequency range of 865.0 to 868.0 MHz.

This section discusses IOT-based radiations and their operating frequencies. The 5G technology is supposed to have highest operating frequency of 7GHZ while RFID technology operates at lowest at 125 Khz. This list of different enabling technologies and their operating frequencies are provided in this section.

3. EFFECT OF RADIATION ON ANIMALS

The existing literature provides the effect of radiation on animal health via radiation emitted. The animals are subjected to these radiations owing to experiments that are conducted on them. These include clinical trials conducted on animals. This section discusses the effect of RF/EMF on animals.

3.1 Rats

This sub-section discusses different literature related to the effect of radiation on rats. The main effect of these radiations also depends on the time duration of exposure to these radiations.

The short-term effect of 2.1 Ghz radiation on rats is studied by the author. These effects are said to be related to locomotor activity, recognition memory, and anxiety-related behavior. The animals were divided into three different groups for the purpose of experimentation. The animals in these different groups were subjected to different tests consisting of the Open Field (OF), novel object recognition (NOR), and marble burying (MB) tests (Gökçek-Saraç, 2020).

Shojaeifard et al. investigate the disease caused by the electromagnetic field (EMF) of mobile phone jammers. The blood factors in rats were studied in light of EMF radiation on account of increasing diseases in humans due to EMF. The authors found for mature rats the difference in several blood factors for mature rats. The authors found that for immature rats the different types of factors were affected by mobile phone jammers (Shojaeifard et al., 2018).

Marc et. al. study the effect of EMF radiation on rats across different ages. The experiment was performed to assess the effect of radiation on the neurobiological system. The rats sample population across different ages was subjected to 900 MHZ RF-EMF radiation. The effect of EMF radiation on rats were seen in their spatial learning, anxiety, and increase in hippocampal ILs cortical IL-1β. These results were however, not seen in the young rats. The authors claim that further longer-duration studies are needed for checking the effect of EMF radiation on rats (Bouji et al., 2016).

Baş et. al. study the effect of EMF radiation on newly born rats in the age bracket 13-21 days. These rats were subjected to 900 MHZ EMF applied on sample population of mothers as well as female children. A pyramidal cell loss was observed in the parts of brain for female rat pups after the radiation exposure (Migdał et al., 2025).

Tang et. al. discuss the exposure of 900 MHZ EMF radiation on the cognitive and blood-barrier changes in rats. The authors subjected a sample of 108 male rat population. Th]e duration of exposure was 3 HRS for 14-28 days. The authors performed various test to check the difference in the cognitive as well as brain-blood-barrier changes in rats before and after exposure. The authors observed ultra-structural changes in different parts of brain. The authors also observed damage caused to brain-blood-barrier damage (Tang et al., 2015).

Hao et. al. performed experimentation to study the effect of long-term EMF on different functions in rats. These included the spatial learning and memory retention in rats. A 916 MHz EMF radiation was cast on rats. The rats were subjected to radiation for 6 hours, 5 days per week for 10 weeks. The hippocampal neurons showed differences in the firing patterns, and EMF radiation can influence the learning and memory patterns. The rats also showed adaptability to EMF radiations across time (Zhao et al., 2025).

Table-2 provides the range of radiation studied in the existing literature, the effect on rats, along with the exposure time in the experimental setup.

Table 2. Radiation range its exposure time and effect on Rats

Reference No.	Radiation Range	Exposure Time	Effect on Rats	Year of Study
Gökçek-Saraç (2020)	2.1 Ghz	2 hrs, 7 days	locomotor activity, recognition memory, anxiety related behaviour	2020
Shojaeifard et al. (2018)	Varying	8 hrs, 5 days/week for 40 days	platelet count, MCHC, MVC and RDWCV changed	2018
Bouji et al. (2016)	900 MHZ RF-EMF	45 min/ day, for 30 days	Spatial learning,anxiety, increase in hippocampal Ils cortial IL-1β for aged rats	2016

continued on following page

Table 2. Continued

Reference No.	Radiation Range	Exposure Time	Effect on Rats	Year of Study
Migdał et al. (2025)	900 MHZ EMF	13-21 days	A pyramidal cell loss in female rat pups.	2013
Tang et al. (2025)	900 MHZ EMF	3 HRS for 14-28 days	blood-brain barrier damage and cognitive impairment	2015
Zhao et al. (2025)	916 MHZ EMF	6 hrs, 5 days per week for 10 weeks	Affecting overall learning and memory of rats	2013

The duration for experimentations that were carried out ranged from 70 days to 7 days by different authors. The maximum duration was 70 days for a frequency of 916 MHz for 6 hours per week. The minimum duration is 7 days with 2.1 GHz radio frequency.

3.2 Mouse

This section discusses experiments conducted on mice regarding EMF radiation exposure.

Kim et. al. performed experimentation on rats using 835 MHz frequency of EMF radiation. The experiment consisted of exposure to voltage-gated calcium channels which regulate the entry of calcium into cells. The calcium absorption in the cells was reduced as a consequence of RF-EMF radiation exposure (Kim et al., 2018).

Sienkiewicz et. al. studied the effect of a small magnetic field. The size of the magnetic field was 50 MHZ. The author claims that the spatial learning performance of mice gets 's affected by this magnetic field. This study showed that that there are negative effects on the memory and various congitive functions of mice. The duration of exposure was 45 minutes for less than 4 days (Sienkiewicz et al., 1998).

Another study by Sienkiewicz et. al. In 2001, however, they found little evidence to connect memory functions of the mice with magnetic fields of similar environments and settings. The specific task given to mice was object detection which was not affected by the 50 HZ of magnetic frequency with which the mice were exposed (Sienkiewicz et al., 2001).

A work by tamir AS and others described how 800-1900 MHz frequency mobile-based phone radiation affects the neurodevelopment and behavior of mice. The authors had conducted a controlled experiment on different groups and varying radiation exposure among the mice population (Aldad et al., 2012).

Zhang et. al. exposed newborn mice to microradiation of 9.417 GHz for a short duration. The duration of exposure ranged from 3.5 days to 18 days. Several tests related to radiation exposure were performed. The offspring male mouse demonstrated decreased learning and memory (Zhang et al., 2015).

A study in 2018 with a high-intensity EMF radiation of 1800 MHz showed several changes in mice. The age of the mice was 4-6 weeks, and the duration of exposure was 2-6 hrs. The authors observed the following effects of radiation:

1. Impact on cell cycle
2. DNA damage
3. Cell death
4. Cell signaling
5. Altered nervous system and immune system
6. Lipid metabolism
7. Carcinogenesis

Table 3. Radiation range, its exposure time and effect on Mouse / Mice

Reference No.	EMF Radiation Range	Exposure Time	Effect on Mouse	Year of Study
Kim et al. (2018)	835 MHZ	4 weeks	Reduced calcium abosrption	2018
Sienkiewicz et al. (1998)	50 HZ	45 minutes for 4 days	Memory loss and affected cognitive functions	1998
Sienkiewicz et al. (2001)	50 HZ	45 Minutes for 1 day	No effect	2001
Aldad et al. (2012)	800-1900 MHZ	16 weeks	Neuropatological changes and behaviour difference	2013
Zhang et al. (2015)	9.417-GHz	3.5-18 days	Anxiety-related behaviours for males but not seen in females	2014
Woelders et al. (2017)	1800 MHz GSM	2-6 hrs	1. Impact on cell cycle 2. DNA damage 3. Cell death 4. Cell signaling 5. Altered nervous system and immune system 6.lipid metabolism 7. carcinogenesis	2018

3.3 Chicken

Another work by Siddiqi et. al. demonstrates the effect of mobile-phone-based EMF radiation on the embryo of a chick. The normal development of the brain was severely affected by EMF radiation with exposure duration ranging from 10 to 15 days. The neurons from the exposed group had different neuron structures and swelling (Siddiqi et al., 2024).

A work by Woelders et. al. had an experiment checking the effect of GSM, UMTS,DECT, WIFI based radiation on chickens. The authors detected no significant effect of these radiations on the growing embryo. These spectrum of technologies included all of the enabling technologies for IOT devices. However, the duration of exposure per day was not clarified (Woelders et al., 2017).

Almášiová et. al. studied the effect of WiFi radiation in 2024 on the growth of chicken embryos. The authors observed after 9th day an increased collagen production, glycosaminoglycans production in the bone skeleton. Their morphological changes indicated WiFi's ability to influence histogenesis of individuals (Abdulhussain et al., 2025).

This section discusses the effect of radiations listed in section-2 on animal health. These animals include rats, chicken and mouse. There are studies suggesting effect on rats to be varying from effect in locomotor activity to affect on memory loss function. The studies carried on chickens suggested varying effect from blood-brain integrity change to effect on neurons. There also studies which suggested no impact on different types of animals.

4. RADIATION SAFETY STANDARDS

There are different radiation standards which have been proposed by different organizations, nations and union (EU). The international organizations which have created these standards are United Nation (UN), International Atomic Energy Agency (IAENG), World Health Organization (WHO), International Commission on Non-ionising Radiation Protection (ICNRP). There are also standards which are created by different nations across the globe. These laws and guidelines monitor standard duration of exposure to different radiations and also ways to mitigate long term exposure (Brady et al., 2025)[26] (Lin, 2025).

The radiation safety standards provided by UN state that IOT devices such as mobile devices do not have any cardiogenic tendencies. These findings are in contraction to previously stated findings of 2011. In the previously stated findings mobile phone and its usage had been graded ‘possibly cardiogenic’ (Lin, 2025).

5. TECHNOLOGICAL TRENDS IN IOT

The technological trends in IOT devices include change in their size, enabling technologies evolution to reflect changing needs. The technological trends also include managing different types of data generated by sensors, regulatory norms and frameworks and their evolution. The requirement of these evolving trends is moving towards higher ranges and capabilities of enabling technologies (Brady et al., 2025). The evolution of trends in technology is governed by the research and need of the specific environment in which the IOT device is deployed. This has led to miniaturization of hand held devices to micro level size. The technological environment of deployment must be in conformance to different effects on the entities present in these environment including animals.

There have been evidences in the literature pointing at no connection between radiations emitted to effect of animals. These researchers have limited duration of exposure of 50 MhZ radiation to indicate no effect on brain activity of mice (Sienkiewicz et al., 2001). A further study may hence be needed to further validate or substantiate the claims made in the research (Sienkiewicz et al., 2001).

Table 4. Radiation range, its exposure time and effect on Chicken

Reference No.	EMF Radiation Range	Exposure Time	Effect on Mouse	Year of Study
Siddiqi et al. (2024)	50 minutes	10-15 days	1. Effect on neurons seen 2. Blood-brain integrity changed	2024
Woelders et al. (2017)	Not mentioned	9 days	No effect observed	2017

6. EXISTING SURVEY PAPERS

The existing literature also consists of several surveys conducted that highlight the effect of EMF radiation on living entities.

A survey conducted in 2024 highlighted the effect of EMF radiation on the neurological network and the different cognitive behavior changes. This survey highlighted the overall effect of radiation on all the body organs as well (Abtin et al., 2024).

7. RADIATION MITIGATION STRATEGY

The existing literature also provide information on how to deal with the radiation in local environment. The most essential way of mitigating radiation is to provide education and awareness about the amount of radiation in a particular environment. These include providing various devices for measuring the amount of radiation (Monadizadeh et al., 2021). There can be different mitigation strategies which can be adopted as follows: -

1. Monitoring the amount of radiation exposed to animals.
2. Switching to lower intensity network
3. Switching to wired connection wherever possible.

These measures can be implemented at the individual level, forest level or at higher levels. The measures must align with the latest research conducted to showcase effect of IOT device radiations on animal health.

8. CONCLUSION AND FUTURE SCOPE

There is evidence in the literature that suggests that the effect of EMF radiation on animals is prominent. This study provides information about these effects on animals. The existing literature mentions the side effects of putting animals to varying degrees of radiation exposure. These experiments are primarily done on rats, mice, chickens for pre-clinical trials. In the deployment of all devices emitting radiations, special care must be taken to prevent excessive usage of these devices. The need and usage of technology can vary across 24 hours. Hence, the technology standard can be lowered to 2G/3G when not in use. The proximity of the animals can also be checked for reducing the effect of these devices. In addition, the devices can be auto-shutdown when the services of these devices are no longer needed. A more specific Systematic Mapping Study (SMS) or Systematic Literature Review (SLR) can be conducted to give more exhaustive study on this subject. A separate commissions, organizations and institutes can be setup for studying and analyzing effect of radiations on animals as well.

REFERENCES

Abdulhussain, S. H., Mahmmod, B. M., Alwhelat, A., Shehada, D., Shihab, Z. I., Mohammed, H. J., & Hussain, A. (2025). A comprehensive review of sensor technologies in IOT: Technical aspects, challenges, and future directions. *Computers*, *14*(8), 342.

Abtin, S., Seyedaghamiri, F., Aalidaeijavadi, Z., Farrokhi, A. M., Moshrefi, F., Ziveh, T., & Haghparast, A. (2024). A review on the consequences of molecular and genomic alterations following exposure to electromagnetic fields: Remodeling of neuronal network and cognitive changes. *Brain Research Bulletin*, *217*, 111090.

Aldad, T. S., Gan, G., Gao, X. B., & Taylor, H. S. (2012). Fetal radiofrequency radiation exposure from 800-1900 mhz-rated cellular telephones affects neurodevelopment and behavior in mice. *Scientific Reports*, *2*, 312.

Bouji, M., Lecomte, A., Gamez, C., Blazy, K., & Villégier, A. S. (2016). Neurobiological effects of repeated radiofrequency exposures in male senescent rats. *Biogerontology*, *17*(5), 841–857.

Brady, A. P., Brkljačić, B., McNulty, J., Loewe, C., Coffey, M., & Visvikis, D. & European Society of Radiology, in Conjunction with ESTRO, EFRS, EFOMP and EANM. (2025). A project to define standards for radiology staffing, education and training across the European Union. *Insights Into Imaging*, *16*(1), 56.

Elshaekh, D. N., Mohamed, H. A., Menam, L. Y. A. E., Sharshar, K. A., & Kayed, S. I. (2025). Multiband printed rectenna for radio frequency energy harvesting (RF-EH). *Discover Electronics*, *2*(1), 39.

Garlisi, D., Pagano, A., Giuliano, F., Croce, D., & Tinnirello, I. (2025). Interference Analysis of LoRaWAN and Sigfox in Large-Scale Urban IoT Networks. *IEEE Access : Practical Innovations, Open Solutions.*

Geary, W. L., Bode, M., Doherty, T. S., Fulton, E. A., Nimmo, D. G., Tulloch, A. I., & Ritchie, E. G. (2020). A guide to ecosystem models and their environmental applications. *Nature Ecology & Evolution*, *4*(11), 1459–1471.

Gökçek-Saraç, Ç. (2020). Effects of 2.1 GHz Electromagnetic Radiation on Locomotor Activity, Recognition Memory, and Anxiety-Related Behavior in Rats. *Neurophysiology*, *52*, 261–266. DOI: 10.1007/s11062-021-09881-w

Kamal, R. (2024). *Internet of Things Architecture and Design Principles* (Fifth). MacGraw Hill Education (India) PVT LTD.

Kim, J. H., Sohn, U. D., Kim, H. G., & Kim, H. R. (2018). Exposure to 835 MHz RF-EMF decreases the expression of calcium channels, inhibits apoptosis, but induces autophagy in the mouse hippocampus. *The Korean Journal of Physiology & Pharmacology; Official Journal of the Korean Physiological Society and the Korean Society of Pharmacology, 22*(3), 277.

Lin, J. C. (2025). Health and safety practices and policies concerning human exposure to RF/microwave radiation. *Frontiers in Public Health, 13*, 1619781.

Mahmud, S. A., Ji, M., Nasim, I., Agarwal, V., Mishra, A., & Kasera, S. (2024). *Co-existence of Heterogeneous Wireless Networks in 2.4 GHz and 5 GHz Spectrum* (No. INL/RPT-24-80712-Rev000). Idaho National Laboratory (INL), Idaho Falls, ID (United States).

Melnick, R. L., Moskowitz, J. M., Héroux, P., Mallery-Blythe, E., McCredden, J. E., & Herbert, M., & International Commission on the Biological Effects of Electromagnetic Fields (ICBE-EMF). (2025). The WHO-commissioned systematic reviews on health effects of radiofrequency radiation provide no assurance of safety. *Environmental Health : A Global Access Science Source, 24*(1), 70.

Migdał, P., Plotnik, M., Bieńkowski, P., Berbeć, E., Latarowski, K., Białecka, N., & Murawska, A. (2025). The Influence of an Electromagnetic Field at a Radiofrequency of 900 MHz on the Behavior of a Honey Bee. *Agriculture, 15*(12), 1266.

Monadizadeh, S., Kibert, C. J., Li, J., Woo, J., Asutosh, A., Roostaie, S., & Kouhirostami, M. (2021). A review of protocols and guidelines addressing the exposure of occupants to electromagnetic field radiation (EMFr) in buildings. *Journal of Green Building, 16*(2), 55–81.

Ozger, Z. B., Cihan, P., & Gokce, E. (2024). A Systematic Review of IoT Technology and Applications in Animals. *Kafkas Üniversitesi Veteriner Fakültesi Dergisi, 30*, 411–431.

Portelange, N. (2020). Implementing an IoT-based information system for the data recording in a phase I clinical center.

Pratap, A. R., & Annamalai, S. (2025). Sensor Technologies for Environmental Data Collection. *Environmental Monitoring Using Artificial Intelligence*, 133-166.

Samijayani, O. N., Sabrina, A., Thamrin, R. Z., Astharini, D., & Syahriar, A. (2024). Performance of Interconnected Hybrid ZigBee-Optic for Extended Wireless Sensor Networks. *J. ICT Res. Appl, 18*, 69–80.

Shojaeifard, M. B., Jarideh, S., Owjfard, M., Nematollahii, S., Talaei-Khozani, T., & Malekzadeh, M. (2018). Electromagnetic Fields of Mobile Phone Jammer Exposure on Blood Factors in Rats. *Journal of Biomedical Physics & Engineering*, *8*(4), 403–408.

Siddiqi, N., Moin, F., & Al-Kindi, M. (2024). The Impact of Mobile Phone Electromagnetic Waves on the Neurons and Blood Brain Barrier Integrity in the Chick Embryo. *Biomedical & Pharmacology Journal*, *17*(3).

Sienkiewicz, Z. J., Bartram, R., Haylock, R. G., & Saunders, R. D. (2001, January). Single, brief exposure to a 50 Hz magnetic field does not affect the performance of an object recognition task in adult mice. *Bioelectromagnetics*, *22*(1), 19–26. DOI: 10.1002/1521-186x(200101)22:1<19:aid-bem3>3.0.co;2-0

Sienkiewicz, Z. J., Haylock, R. G., Bartrum, R., & Saunders, R. D. (1998). 50 Hz magnetic field effects on the performance of a spatial learning task by mice. *Bioelectromagnetics*, *19*(8), 486–493. DOI: 10.1002/(sici)1521-186x(1998)19:8<486:aid-bem6>3.0.co;2-k

Tang, J., Zhang, Y., Yang, L., Chen, Q., Tan, L., Zuo, S., & Zhu, G. (2015). Exposure to 900 MHz electromagnetic fields activates the mkp-1/ERK pathway and causes blood-brain barrier damage and cognitive impairment in rats. *Brain Research*, *1601*, 92–101.

Woelders, H., de Wit, A., Lourens, A., Stockhofe, N., Engel, B., Hulsegge, I., & Zwamborn, P. (2017). Study of potential health effects of electromagnetic fields of telephony and Wi-Fi, using chicken embryo development as animal model. *Bioelectromagnetics*, *38*(3), 186–203.

Zhang, Y., Li, Z., Gao, Y., & Zhang, C. (2015). Effects of fetal microwave radiation exposure on offspring behavior in mice. *Journal of Radiation Research*, *56*(2), 261–268.

Zhao, J., Yue, H., Wang, K., Sun, Y., Zhao, L., Wang, H., & Peng, R. (2025). Disruption of the linear relationship between recognition memory and local functional connectivity of left perirhinal cortex in rat after acute 2.856 GHz microwave exposure. *Brain Research*, 149816.

Chapter 10

Ethical and Societal Implications of AI in Human-Animal Communication:

Towards Responsible Interspecies Technology

Chika Lilian Onyagu
https://orcid.org/0009-0004-1845-2432
Margaret Lawrence University, Ika North East, Nigeria

Etaoghene Paul Polo
https://orcid.org/0009-0001-1606-0273
University of Delta, Agbor, Nigeria

ABSTRACT

This chapter explores the implications of AI technologies used in decoding animal vocalisations, emotion recognition, behavioural prediction, and robotic interaction. Employing a qualitative, interdisciplinary methodology that draws on philosophical analysis, ethical theory, including animal ethics and AI ethics, and empirical case studies, the work critically examines the impact of AI design choices on animal agency, consent, and welfare. Attention is also paid to broader societal issues, such as anthropomorphism, data bias, and the unequal effects of AI adoption in contexts like conservation and industrial farming. Furthermore, the chapter situates these interspecies technologies within the wider discourse on AI ethics, addressing concerns around fairness, transparency, accountability, data privacy, and digital

DOI: 10.4018/979-8-3373-5483-5.ch010

inequality. It advocates for an ethics-by-design approach and participatory governance frameworks that incorporate the perspectives of ethicists, technologists, affected communities, and policymakers.

1. INTRODUCTION

Background to Human-Animal Communication

Human-animal communication has long fascinated philosophers, linguists, ethologists, and cognitive scientists, reflecting deep questions about what it means to be human and how humans relate to other living beings. Historically, communication with animals was largely symbolic, ritualistic, or based on intuitive interpretation of behaviour, such as shepherds understanding the calls of their herds or hunters interpreting the tracks and sounds of wildlife (Bradshaw, 2021).

Over time, domestication led to a more structured interaction between species, where certain animals, such as dogs and horses, were trained to respond to human signals, creating a rudimentary form of two-way communication (Miklósi, 2015). Advances in animal cognition research demonstrated that many non-human animals possess sophisticated communicative abilities. Primates such as chimpanzees and bonobos have been taught to use sign language or lexigrams, while dolphins and parrots have shown capacity for symbolic understanding and mimicry (Pepperberg, 2019; Savage-Rumbaugh *et al.*, 1993). These studies disrupted anthropocentric assumptions about the uniqueness of human language and opened debates on interspecies understanding. However, such communication remains limited, requiring human mediation and training.

The advent of Artificial Intelligence (AI) introduces a transformative shift, as AI offers the possibility of "decoding" animal vocalisations, gestures, and emotional states without intensive training or conditioning. This development holds profound implications for human-animal relations, ranging from conservation and agriculture to companionship and even philosophical conceptions of animal personhood (Crist, 2019).

The Rise of AI in Interspecies Interaction

Artificial Intelligence is increasingly employed to analyse vast datasets of animal sounds, behaviours, and biological signals. Machine learning and natural language processing (NLP) techniques are applied to detect patterns in whale songs, elephant rumbles, and bird calls, while computer vision is used to monitor behavioural cues in livestock and wildlife (Kershenbaum *et al.*, 2016; Stowell, 2022). Robotic technologies have also been developed to interact with animals in controlled settings, such as drones guiding herds or robotic fish influencing schools of live fish (Kovac *et al.*, 2021).

While such applications promise to enhance conservation, farming efficiency, and companionship, they also raise questions of ethics and governance. Should animals' "voices" be interpreted through human-designed AI systems? Can animals give consent to being surveilled or having their communications decoded? These questions strike at the intersection of AI ethics and animal ethics.

Statement of the Problem

Despite the promise of AI in improving interspecies understanding, its integration into human-animal communication generates critical ethical and societal concerns. On the one hand, these technologies can promote animal welfare, aid conservation, and deepen empathy. On the other hand, they risk reinforcing anthropocentrism, perpetuating data biases, and reducing animals to instrumental objects for human use (Fitzpatrick *et al.*, 2022). Moreover, AI systems may impose technological determinism, where the algorithmic interpretation of animal communication is taken as objective truth, even though these systems are shaped by human assumptions and limitations (Cave & Dihal, 2020).

Purpose and Significance of the Chapter

The aim of this chapter is to critically examine the ethical and societal implications of AI in human-animal communication, emphasising the need for responsible design, participatory governance, and sensitivity to animal agency. The chapter situates these debates within broader discourses in philosophy of technology, animal ethics, and AI governance. It seeks to contribute to emerging interspecies ethics by highlighting the risks, opportunities, and responsibilities of integrating AI into contexts that directly involve non-human lives.

Methodology and Theoretical Orientation

The chapter employs a qualitative, interdisciplinary methodology, drawing on philosophical analysis, ethical theory, and empirical case studies. Philosophical inquiry provides the normative framework for assessing issues of animal agency, consent, and dignity, while empirical studies supply real-world grounding for technological developments in conservation, farming, and research. The theoretical orientation is informed by animal ethics (Singer, 1975; Regan, 2004), AI ethics (Floridi *et al.*, 2018; Jobin *et al.*, 2019), and the philosophy of technology (Ihde, 1990; Verbeek,

2011). These perspectives allow a layered exploration of both ethical concepts and technological realities.

Chapter Structure Overview

The chapter contains 8 sections, with the first section being introduction. Section 2 clarifies the core concepts underpinning the debate, such as AI, human-animal communication, interspecies technology, and ethics-by-design. Section 3 outlines the technological landscape, surveying current AI applications in animal vocalisation decoding, behavioural prediction, and robotic interaction. Section 4 delves into ethical implications, focusing on questions of agency, anthropomorphism, data bias, and transparency. Section 5 examines societal implications, such as socioeconomic impact in conservation and farming, digital inequality, and technological access. Section 6 discusses governance, policy, and ethical frameworks. Section 7 establishes future directions towards responsible interspecies technology. Section 8 concludes the chapter with a call for responsible stewardship of interspecies AI.

2. CONCEPTUAL CLARIFICATIONS

Artificial Intelligence (AI): Meaning and Scope

Artificial intelligence (AI) refers to the ability of machines, particularly computer systems and computer-controlled robots, to simulate human intelligence by performing tasks that typically require human-like thinking, reasoning, and behaviour (Polo, Emmanuel & Obeka, 2024). This means that AI is not merely about machines carrying out instructions, but about their capacity to process information in ways that demonstrate a form of rationality comparable to human cognitive processes. For example, when a machine solves a mathematical problem, interprets a sentence, or navigates through a city using sensors, it is engaging in activities that mimic aspects of human reasoning and adaptation. In practical terms, AI technologies are designed to bridge the gap between human mental capabilities and machine efficiency, thereby making it possible for machines to undertake complex functions that would otherwise require significant human involvement.

In other words, artificial intelligence refers to computational systems capable of performing tasks that typically require human intelligence, such as perception, reasoning, learning, and decision-making (Russell & Norvig, 2020). Perception in this sense involves a machine's ability to interpret sensory data, such as recognising faces in images or detecting objects in a video stream. Reasoning refers to the logical processes through which AI systems can evaluate alternatives and make rational choices. Learning, particularly through machine learning and deep learning, allows AI systems to improve performance over time without explicit reprogramming, by identifying patterns and adjusting behaviour accordingly. Decision-making, perhaps one of the most critical aspects, enables machines to weigh multiple factors and select actions that are most likely to lead to desired outcomes. Hence, AI can be understood as a multidisciplinary field of study and application, integrating computer science, mathematics, cognitive science, and even philosophy, to build systems that extend human problem-solving capacities (Russell & Norvig, 2020).

From the foregoing, we can surmise that artificial intelligence is intelligence displayed by machines, devices, or apparatuses, usually imitating natural human intelligence. It performs tasks or exhibits attributes (traits) ordinarily typical of humans (Olojede & Polo, 2025). Such traits include creativity, adaptability, and problem-solving. For instance, an AI-powered chess engine demonstrates strategic planning comparable to that of a human grandmaster, while AI-driven translation systems can interpret and reproduce linguistic nuances with increasing accuracy. Beyond technical definitions, AI also carries a symbolic dimension: it reflects humanity's attempt to recreate its own intelligence in external forms. This imitation is not absolute, machines do not "think" in the human sense, but the resemblance is sufficient for them to replicate results traditionally achieved by humans.

In the context of interspecies communication, AI encompasses machine learning algorithms that analyse animal vocalisations, neural networks that model behavioural patterns, and robotics that enable physical interaction with animals. These applications illustrate AI's capacity to move beyond human-centred problem-solving into areas where communication and interaction across species becomes possible. For example, researchers use AI to decode the complex calls of whales, enabling a deeper understanding of their social structures and behaviours. Similarly, neural networks have been employed to model the migratory and feeding patterns of birds, which helps ecologists in both conservation and the prediction of environmental changes. Robotics powered by AI allow for safe and non-intrusive interaction with animals in laboratory or field settings, making it possible to study behaviour in ways that would be impractical or harmful if humans were directly involved. Thus, AI broadens the horizons of communication, research, and understanding not only among humans but also across the natural world.

Importantly, AI is not neutral, it reflects the assumptions of its designers, the biases in its training data, and the socio-economic contexts in which it is deployed (Crawford, 2021). This means that AI systems are shaped by the cultural, political, and historical conditions under which they are created. For instance, an AI model trained predominantly on data from Western societies may fail to adequately recognise or interpret information from non-Western contexts, thereby reproducing global inequalities. Moreover, the developers' perspectives, whether consciously or unconsciously, are embedded in the algorithms, influencing how fairness, accuracy, and efficiency are prioritised. On a socio-economic level, AI can reinforce existing disparities, as powerful corporations and governments often control access to the most advanced AI systems, potentially marginalising those without the same resources. Therefore, while AI appears as a neutral, technical tool on the surface, it is in fact deeply value-laden, and its outcomes cannot be divorced from the human contexts that shape its creation and application (Crawford, 2021).

Human-Animal Communication: Definitions and Modalities

Human-animal communication refers to the processes by which humans and non-human animals exchange signals, information, or affective states. At its core, this communication enables two-way interaction, where humans attempt to interpret animal expressions, sounds, or behaviours, and animals, in turn, respond to human cues. This process is not restricted to vocal interaction but encompasses a wide spectrum of modalities that allow cross-species understanding. For humans, decoding these communicative acts is fundamental for strengthening the human-animal bond, enhancing domestication practices, and advancing scientific research on animal cognition and behaviour (Miklósi, 2015).

Modalities include vocalisation (e.g., bird calls, dolphin whistles), body language (e.g., tail wagging in dogs, trunk gestures in elephants), chemical signals (pheromones), and technologically mediated channels (such as collars with AI sensors) (Miklósi, 2015). Vocalisation is perhaps the most immediately recognisable form of animal communication, where sounds serve diverse functions such as warning others of danger, attracting mates, or coordinating group activities. Body language is equally important; for instance, a dog's wagging tail may express excitement, submission, or agitation depending on the context, while elephants' trunk gestures can communicate comfort, threat, or invitation. Chemical signals like pheromones demonstrate a deeper layer of communication, particularly among insects and mammals, by conveying messages related to reproduction, territory, or social hierarchy. In recent years, technologically mediated channels have added a new dimension, with innovations such as smart collars embedded with AI sensors that track physiological signals, interpret stress levels, and translate them into forms comprehensible to human caretakers. These modalities illustrate that communication is not bound by speech alone but can manifest through multiple channels, each uniquely adapted to the ecological and social needs of species (Miklósi, 2015).

Unlike human language, animal communication systems are often specific to survival functions, mating, alarm, food discovery, but research increasingly shows cognitive and emotional depth across species (Sebeok, 1990; Bradshaw, 2021). For example, alarm calls among primates are not mere instinctive reactions but can carry specific information about the type of predator, prompting different responses depending on the threat. Similarly, mating calls in birds often involve complex songs that not only signal reproductive readiness but also display learning and creativity. Food discovery signals, such as the "waggle dance" of honeybees, demonstrate remarkable precision in conveying spatial information to hive members. Despite these survival-oriented functions, scholars argue that animal communication is richer than once assumed. Studies reveal that elephants exhibit mourning behaviours, dolphins recognise themselves in mirrors (an indication of self-awareness), and dogs respond to human emotional states, suggesting that animal communication extends beyond immediate utility to encompass affective and cognitive dimensions (Sebeok, 1990; Bradshaw, 2021).

Interspecies Technology: Emerging Trends

"Interspecies technology" refers to technological systems explicitly designed to mediate or facilitate communication and interaction between humans and non-human animals. This concept highlights the ways in which digital innovations and engineering are being harnessed not simply to observe animals but to actively engage with them. At its core, interspecies technology recognises that humans and animals share communicative capacities, albeit through different channels, and therefore attempts to bridge these channels through artificial means. It is rooted in both scientific curiosity, understanding how animals think, behave, and interact, and practical applications, such as conservation, veterinary care, and companion animal welfare (Kershenbaum *et al.*, 2016).

These range from AI collars that interpret dog emotions, to drones that monitor elephant movements, to machine learning models attempting to decode whale songs (Kershenbaum *et al.*, 2016). AI collars are designed with sensors that can track a dog's heart rate, stress levels, and vocalisations, and then translate these biological and behavioural signals into human-readable data. This allows pet owners not only to understand their animals' needs more accurately but also to respond to signs of discomfort or illness earlier than they otherwise might. Drones used in conservation settings play a different role: they provide a bird's-eye view of elephant herds, monitoring their movements across vast landscapes, which helps prevent human-elephant conflict and allows conservationists to respond swiftly to threats such as poaching. Machine learning models designed to decode whale songs push the boundaries even further, as they attempt to penetrate the complexities of cetacean communication, potentially uncovering social structures, mating rituals, and even cultural behaviours within pods. These examples demonstrate that interspecies technology does not simply observe but actively translates, mediates, and, in some cases, transforms the communicative possibilities between species (Kershenbaum *et al.*, 2016).

Interspecies technology also extends to robotic interfaces, such as robotic bees assisting pollination, or robotic fish guiding marine populations (Kovac *et al.*, 2021). Robotic bees have been developed as a response to the global decline in natural bee populations, aiming to sustain pollination processes that are vital for agricultural productivity. By mimicking the flight patterns and pollinating behaviours of bees, these robots interact with plants in a way that substitutes or supplements the role of living pollinators. Similarly, robotic fish are designed to integrate into marine environments, either to guide schools of fish away from hazardous areas such as oil spills or to encourage sustainable migration patterns in changing climates. Both examples illustrate that robotic interfaces are not merely imitations of animals but active agents within ecosystems, interacting with and influencing animal behaviour in real time. These technologies therefore raise profound questions about the boundaries between natural and artificial systems, and the extent to which machines can act as ecological participants (Kovac *et al.*, 2021).

The rise of such technologies challenges existing categories of "tool use" and raises questions about co-agency between humans, animals, and machines. Traditionally, tools have been regarded as objects wielded by humans to manipulate their environment, with animals occasionally noted as tool users in exceptional cases, such as primates using sticks to extract termites. However, interspecies technology does not fit neatly into this paradigm. Rather than being a static object of use, these technologies often operate autonomously, gather their own data, and respond dynamically to both human and animal inputs. This complicates the notion of agency: if a robotic fish alters the movement of a school, is it acting on behalf of humans, or has it become a co-agent within the ecological system? Similarly, if an AI collar allows a dog to express emotions in ways that shape human behaviour, does the animal, through the technology, gain a new form of communicative power? Such questions push the boundaries of philosophy, ethics, and technology studies, suggesting that interspecies technologies are not neutral tools but mediators that create hybrid forms of interaction in which humans, animals, and machines share influence and agency.

Ethics-by-Design and Participatory Governance: Key Concepts

"Ethics-by-design" refers to embedding ethical principles, such as transparency, fairness, and accountability, into technological systems from the outset, rather than retrofitting them after deployment (Floridi *et al.*, 2018). The essence of this approach is that ethical reflection is not treated as an afterthought once potential harms have already emerged, but rather as a guiding principle at every stage of technological innovation. Transparency ensures that both developers and users understand how systems operate, what data they process, and the limitations of their outputs. Fairness demands that systems are designed in ways that avoid bias, whether against particular human communities or against the interests of non-human animals. Accountability requires that individuals and organisations remain answerable for the consequences of the technologies they create, ensuring that responsibility cannot be deferred to "the machine" itself. In the interspecies context, this requires accounting for animal welfare and dignity alongside human values. It means recognising animals not merely as passive subjects of technology but as beings with needs, preferences, and vulnerabilities that deserve to be respected. For example, the design of AI collars for dogs should not only prioritise human convenience but also consider whether continuous monitoring causes stress or whether feedback mechanisms align with the animal's natural behavioural patterns (Floridi *et al.*, 2018).

"Participatory governance" refers to involving a wide range of stakeholders, including ethicists, technologists, animal welfare organisations, indigenous communities, and policymakers, in shaping how technologies are designed and regulated (Stahl *et al.*, 2022). This approach emphasises inclusivity and democratic decision-making, recognising that technology has wide-reaching effects and therefore should not be controlled solely by corporations or small groups of experts. Ethicists contribute philosophical and moral perspectives, ensuring that questions of justice and dignity are foregrounded. Technologists provide practical knowledge about what systems are capable of achieving, as well as limitations that need to be acknowledged. Animal welfare organisations bring insights into the lived realities of animals, ensuring that welfare standards are not compromised in the pursuit of technological novelty. Indigenous communities often contribute traditional ecological knowledge, which can guide responsible and sustainable interaction with non-human species. Policymakers play a vital role in establishing regulations and frameworks that formalise ethical practices into enforceable standards. By bringing these diverse groups together, participatory governance fosters a collective sense of responsibility, balancing competing interests, and enabling more socially and ecologically sustainable outcomes (Stahl *et al.*, 2022).

Both concepts are crucial for ensuring that interspecies technologies are developed responsibly. Without ethics-by-design, technologies risk perpetuating hidden biases or causing unintentional harm to both humans and animals. Without participatory governance, decisions about such technologies may be driven by narrow commercial interests, sidelining ethical, ecological, and cultural considerations. Together, they establish a framework that integrates moral foresight with democratic inclusivity, ensuring that technological progress does not occur at the expense of justice, sustainability, or respect for non-human life. By embedding ethical principles early and ensuring a plurality of voices in governance, societies can cultivate innovations that genuinely benefit humans, animals, and the ecosystems they share.

Distinguishing Anthropocentrism From Interspecies Ethics

Anthropocentrism privileges human interests and perspectives, often at the expense of non-human life. It assumes that human needs, values, and modes of understanding are central and superior, relegating animals and other species to secondary or instrumental roles. Within this orientation, animals are frequently viewed primarily in terms of their usefulness to humans, whether as sources of food, labour, companionship, or research subjects. This approach tends to obscure the intrinsic worth of non-human life, reducing animals to objects of utility or scientific curiosity rather than recognising them as beings with their own inherent value. As a result, anthropocentrism reinforces hierarchical distinctions between humans and animals, making it difficult to conceptualise genuine moral obligations toward non-human beings (Donaldson & Kymlicka, 2011).

In contrast, interspecies ethics seeks to recognise the moral considerability of animals as sentient beings with their own forms of agency and value (Donaldson & Kymlicka, 2011). The central premise is that animals are not mere background entities in human-centred worlds, but participants in shared ecological and social systems. By acknowledging their sentience, their capacity to feel, perceive, and experience, this ethical perspective accords animals a moral status that obliges humans to treat them with respect and care. Agency, in this context, refers to animals' ability to make choices, exhibit preferences, and influence their environments in ways that are meaningful to their lives. For example, migratory birds demonstrate agency in selecting routes that respond to climate shifts, while domesticated animals express preferences for social interaction or solitude. Interspecies ethics therefore encourages humans to move beyond anthropocentric assumptions and cultivate relationships based on mutual respect, recognising animals as co-inhabitants of a shared moral community rather than as resources to be managed or exploited (Donaldson & Kymlicka, 2011).

While anthropocentrism risks interpreting animal communication solely in human terms, interspecies ethics advocates for humility, recognising the limits of human understanding and the need to avoid projecting human categories onto non-human communicative systems (Crist, 2019). Anthropocentric interpretations often seek to translate animal communication into human linguistic or behavioural frameworks, which can distort or oversimplify the complexity of non-human interactions. For instance, describing dolphin whistles purely as "language" risks neglecting unique acoustic structures that may not conform to human grammar or semantics but nonetheless carry profound social meaning within dolphin communities. Interspecies ethics, by contrast, calls for caution and humility: it urges humans to approach animal communication with an openness that acknowledges both the richness of non-human forms of expression and the epistemic limits of human interpretation.

This humility prevents overconfident projections of human categories, such as assuming that all communication must resemble speech, and instead supports a more careful and respectful study of the ways animals interact with each other and with humans (Crist, 2019).

Distinguishing these orientations is critical to developing ethical frameworks for AI in interspecies communication. If AI technologies are designed from an anthropocentric standpoint, they risk instrumentalising animals further, treating communicative signals as data points to be translated solely for human benefit. Such an approach could perpetuate the reduction of animals to tools in human projects rather than recognising them as communicative partners. On the other hand, grounding AI design in interspecies ethics would prioritise technologies that respect animal welfare, preserve ecological balance, and foster genuine cross-species understanding. This distinction matters because it shapes not only how technologies are built but also how humans conceive of their relationships with non-human life. By moving away from anthropocentrism and embracing interspecies ethics, researchers and designers can create frameworks that encourage co-agency, respect, and moral responsibility across species boundaries.

3. TECHNOLOGICAL LANDSCAPE OF AI IN HUMAN-ANIMAL COMMUNICATION

Overview of Current AI Tools and Applications

AI technologies are increasingly used to decode animal communication, predict behaviour, and mediate human-animal interactions. These developments represent a growing convergence between computer science, ethology, and conservation studies, where advanced computational models are applied to questions that were once confined to traditional zoology or behavioural science. By processing large amounts of data far beyond human cognitive capacity, AI allows researchers to identify patterns and correlations that shed light on how animals communicate, interact, and respond to their environments. This not only broadens scientific understanding but also opens up possibilities for improving human-animal relationships, enhancing conservation strategies, and supporting animal welfare initiatives. In this sense, AI serves as a powerful mediator between human and non-human systems, transforming the scope of interspecies communication research (Stowell, 2022).

These include deep learning models that classify animal sounds, wearable devices that track physiological states, and robotic platforms that interact with animals in controlled ways. Deep learning models, which rely on artificial neural networks, are particularly well suited to the analysis of bioacoustic data because they can detect subtle variations in pitch, frequency, and rhythm that are difficult for humans to discern. This enables the classification of vocalisations not only by species but also by context, for example, distinguishing between alarm calls, mating calls, and social calls within the same species. Wearable devices, on the other hand, extend the study of animal communication into the physiological domain. Equipped with sensors that monitor heart rate, body temperature, and movement, these devices provide real-time data that help researchers and caretakers to understand how animals experience stress, excitement, or relaxation. Robotic platforms introduce yet another dimension by enabling controlled interaction; robots designed to mimic animal movement or behaviour can be used to study group dynamics, predator-prey interactions, or the social responses of animals to new stimuli. Collectively, these tools demonstrate the versatility of AI in bridging the gap between observation and interaction, allowing for a more nuanced appreciation of animal lives.

For example, the Earth Species Project uses machine learning to analyse bioacoustic data in an effort to uncover the structures of animal communication systems (Stowell, 2022). This initiative exemplifies how AI can be harnessed for large-scale, non-invasive research. By compiling vast libraries of sound recordings from diverse species, ranging from birds and whales to primates, the project applies computational models to search for recurring structures and syntactic rules, much like linguists would in the study of human languages. The ambition is not merely to catalogue sounds but to uncover whether they represent structured communicative systems that convey complex meanings. If successful, such work could revolutionise our understanding of non-human intelligence and open the possibility of deeper cross-species dialogue. Beyond pure research, these findings also have practical applications, including informing conservation strategies by providing insights into animal social organisation, reproductive behaviour, and responses to ecological pressures (Stowell, 2022).

Similarly, AI-enabled collars for pets claim to translate barks into categories of emotions such as hunger, playfulness, or distress. While these devices are sometimes marketed with a degree of exaggeration, they nonetheless represent a broader trend of integrating AI into everyday human-animal relationships. The underlying principle is that by analysing the acoustic qualities of a dog's bark alongside physiological and behavioural indicators, AI systems can generate interpretations that help owners respond more effectively to their pets' needs. For instance, distinguishing between a bark signalling distress and one indicating excitement could improve welfare by allowing timelier interventions. Such tools also raise intriguing questions about the extent to which technology can make animal inner states more accessible to humans, and whether this represents a genuine enhancement of empathy or a simplified projection of human categories onto animal behaviour. Either way, these consumer-oriented applications demonstrate that AI's role in interspecies communication is not limited to the research lab but is increasingly entering the realm of everyday life.

Animal Vocalisation Decoding Technologies

One of the most promising applications is the decoding of animal vocalisations. AI models are used to identify patterns in whale songs, bird calls, and elephant rumbles that may correspond to specific meanings or functions (Kershenbaum et al., 2016). Bioacoustics combined with machine learning allows the classification of calls at a scale and precision not possible through traditional human observation. For example, researchers have used convolutional neural networks to automatically detect bat echolocation calls, aiding in biodiversity monitoring (Stowell, 2022). While promising, such decoding remains probabilistic and may risk anthropomorphic over-interpretation.

Behavioural Prediction and Emotion Recognition Systems

AI systems are also being developed to predict animal behaviour and recognise emotional states. This area of innovation is grounded in the understanding that animals, like humans, display behavioural patterns and physiological cues that can be systematically observed, measured, and analysed. By capturing these signals through digital sensors, video analysis, and algorithmic modelling, AI offers the possibility of identifying not only what animals are doing at a given moment but also what they are likely to do next. Such predictive and interpretative capacities can be applied across multiple domains, from farming and veterinary medicine to wildlife conservation and companion animal care, making them a rapidly expanding field of research and application. Importantly, these systems are not limited to observation; they actively process large datasets, learn from repeated patterns, and generate insights that humans alone might overlook, thereby transforming the study of animal behaviour into a more data-driven and proactive enterprise.

Computer vision can monitor livestock for signs of stress, illness, or discomfort, enabling early intervention in farming (Tullo *et al.*, 2019). Cameras installed in barns, fields, or enclosures continuously record the movements and postures of animals, while AI models process the footage to identify subtle deviations from normal behaviour. For instance, reduced feeding, altered gait, or increased restlessness can be early indicators of illness or injury. By detecting such changes at an early stage, farmers and veterinarians are able to intervene more swiftly, potentially reducing animal suffering and improving productivity in agricultural systems. This technology also assists in monitoring environmental factors, such as overcrowding or poor ventilation, that may contribute to stress. In this way, AI not only benefits animal health but also supports more sustainable farming practices by minimising losses and reducing reliance on routine antibiotics. Thus, behavioural prediction through computer vision is becoming an essential tool for advancing both welfare and efficiency in animal husbandry (Tullo *et al.*, 2019).

Similarly, algorithms trained on video data of dogs or horses can detect tail and ear positions associated with mood. These algorithms rely on machine learning techniques that map specific body postures to particular emotional states, such as contentment, agitation, fear, or excitement. In dogs, for example, the speed and angle of a tail wag may signal different intentions, while in horses, the movement of ears can indicate attentiveness, irritation, or anxiety. By automating the detection of these signals, AI systems provide caretakers with objective feedback that supplements human interpretation, which can sometimes be inconsistent or influenced by anthropocentric bias. Such technologies are increasingly being integrated into smart collars, wearable sensors, or stable-monitoring platforms, thereby providing continuous, real-time assessments of animals' emotional wellbeing. This can be particularly valuable in contexts such as equestrian sports or therapy animals, where monitoring stress and comfort levels is essential for ethical practice.

These tools have potential to improve animal welfare, but they also raise ethical questions about surveillance and the reduction of complex behaviours into simplistic categories. Continuous monitoring may enhance safety and wellbeing, but it simultaneously introduces a level of surveillance that could alter the naturalness of animal lives, particularly in captive or farmed contexts. There is also the risk that the rich and multifaceted nature of animal behaviour might be oversimplified when reduced to a set of algorithmic labels. For example, categorising a dog's bark or a horse's ear flick exclusively as "stress" could overlook contextual factors such as playfulness, curiosity, or environmental disturbances. Moreover, reliance on automated systems could encourage humans to defer to technological interpretations, potentially sidelining empathetic engagement and nuanced observation. Ethical concerns therefore centre on striking a balance: harnessing the benefits of AI for welfare without allowing technology to dominate or distort our understanding of animal lives. In this regard, behavioural prediction and emotion recognition systems exemplify both the promise and the challenge of interspecies AI, ffering opportunities for care and insight while demanding careful reflection on their broader implications.

Robotic Interfaces and Animal-Responsive Machines

Robotics offers another frontier for interspecies communication, one that pushes the boundaries of how technology can mediate and shape interactions between humans, animals, and machines. Unlike traditional tools that merely extend human control or observational power, robotic systems are increasingly designed to interact with animals directly, responding to their behaviours and even influencing their choices in ways that bypass human mediation. This makes robotics a particularly fascinating domain, as it does not simply involve interpreting or predicting animal signals but actively creating new forms of embodied, dynamic interaction between species. Such developments open up innovative possibilities for improving animal management, welfare, and companionship, while simultaneously generating profound philosophical and ethical questions about what it means to engage with non-human beings through artificial agents.

Robotic fish have been designed to integrate into schools of live fish, influencing group movements without direct human intervention (Kovac *et al.*, 2021). These biomimetic machines are programmed to replicate the movement patterns, shapes, and even colouring of real fish, thereby gaining acceptance within a shoal. By adjusting their swimming behaviour, they can steer or reorganise groups of live fish, subtly altering patterns of migration, feeding, or avoidance of predators. This demonstrates that animals may perceive and respond to artificial agents as if they were members of their own species, illustrating the capacity of robotics to operate not just as external observers but as active participants in animal societies. Such experiments reveal both the technical sophistication of modern robotics and the complexity of animal perceptual worlds, while also highlighting the possibility of unintended manipulation, since the fish, unaware of the machine's artificial nature, may be influenced in ways that serve human objectives rather than their own natural inclinations (Kovac *et al.*, 2021).

In agriculture, robotic herders and drones are used to guide livestock, providing a practical alternative to traditional labour-intensive methods. These machines can be programmed to maintain optimal distances, generate specific sounds, or use movement cues that encourage animals to move towards grazing areas or enclosures. The advantage lies in efficiency and safety: robotic herders can reduce the need for constant human presence in potentially dangerous situations, such as moving large cattle herds, while drones offer aerial perspectives that allow farmers to oversee wide pastures. By integrating with animal behavioural tendencies, these systems reduce stress compared to harsher manual herding techniques, as they can employ gentler, more consistent signals that animals come to recognise over time. At the same time, their adoption reflects a broader trend in agriculture towards automation, where the management of animal life becomes increasingly mediated by machines rather than by direct human-animal relationships. This shift raises questions about whether such technologically mediated interactions diminish opportunities for empathy and attentiveness in human caretaking.

Robotic pets offer companionship for elderly people, further expanding the scope of animal-responsive machines beyond agriculture and scientific research into the domain of social and emotional care. These artificial companions are designed to mimic the behaviours of real animals, such as purring, wagging, or responding to touch, thereby providing comfort, reducing loneliness, and even improving mental health in individuals who may be unable to care for live animals. For instance, robotic cats or dogs can offer a sense of presence and routine without the practical challenges of feeding, grooming, or veterinary care. While these machines cannot replicate the full richness of genuine animal companionship, they demonstrate how robotics can be designed to engage with human needs by simulating interspecies interaction. Their use highlights both the therapeutic potential of robotics and the philosophical dilemma of authenticity: does simulated companionship diminish or replace the meaningfulness of real relationships with living beings, or does it represent an acceptable substitute in contexts where real pets are impractical?

Such technologies blur the boundaries between biological and artificial agents, raising questions about authenticity, manipulation, and co-agency. On one hand, the ability of animals to treat machines as fellow agents suggests that robotics can create new forms of interspecies communication, enriching our capacity to understand and interact with non-human life. On the other hand, it also reveals the risk of manipulation, where animals are influenced by artificial entities without the possibility of recognising them as such, thereby undermining the autonomy and natural agency of non-human beings. The issue of authenticity further complicates matters: if relationships with robotic pets or animal-responsive machines come to substitute for relationships with living beings, we may inadvertently cultivate a culture where simulations are valued over genuine interspecies bonds. Finally, the question of co-agency arises, if animals respond meaningfully to robotic agents, can these interactions be considered genuine forms of collaboration, or are they always asymmetrical, with machines ultimately programmed to serve human goals? These questions show that while robotic interfaces expand the horizons of interspecies communication, they also demand critical reflection on the ethical and philosophical implications of merging life with artificiality in increasingly intimate ways.

Case Studies: Wildlife Monitoring, Smart Farming, and Wearable AI

Several case studies illustrate the diversity of applications of artificial intelligence in human-animal relations, each showing how technological innovation can be harnessed to address pressing practical needs while simultaneously raising deeper philosophical and ethical concerns about the nature of our engagement with non-human life. These examples help to demonstrate not only the potential of AI to improve welfare and conservation outcomes, but also the risks of overreliance on technological mediation in contexts that traditionally required closer, more organic human-animal relationships.

- **Wildlife monitoring:** AI-powered drones are used to track elephant populations in Africa, assisting in anti-poaching efforts (Puri *et al.*, 2019). These drones are equipped with advanced cameras and machine learning algorithms that can identify elephants from the air, often in vast and difficult terrains where traditional human patrols would be less effective. By continuously gathering visual data, the drones can detect unusual patterns such as rapid herd movements, signs of distress, or the presence of human intruders. This information is transmitted in real time to conservationists and law enforcement, enabling faster and more targeted interventions against poaching activities. Beyond surveillance, these systems also provide valuable long-term ecological data on migration routes, feeding patterns, and reproductive cycles, which can inform sustainable conservation strategies. At the same time, the use of AI in this domain raises critical questions about technological dependency: will conservation efforts become overly reliant on machines, potentially reducing the importance of fostering community-based stewardship of wildlife? Furthermore, there is an underlying tension between the protective function of drones and their intrusive nature, as continuous aerial monitoring could be seen as a form of surveillance that intrudes upon the natural lives of elephants, whose movements are now constantly under watch (Puri *et al.*, 2019).
- **Smart farming:** AI sensors in barns monitor cows' movements and health, optimising feeding and reducing disease spread (Tullo *et al.*, 2019). These systems typically use accelerometers, thermal cameras, and behavioural tracking tools to monitor subtle changes in posture, feeding habits, or levels of activity that may signal illness, discomfort, or reproductive readiness. Farmers benefit from receiving automated alerts that allow them to intervene before diseases spread or before productivity losses occur. For instance, early detection of lameness or mastitis can significantly reduce suffering for the animals while also ensuring economic efficiency for the farm. By tailoring feed distribution and identifying specific nutritional needs, AI can also help to improve overall herd health, reduce waste, and make farming practices more environmentally sustainable. Yet, while such systems can be justified on welfare grounds, they also risk reframing animals primarily as units of productivity to be optimised. The very act of constantly measuring, categorising, and predicting their behaviour may reduce the richness of their lived experience into quantifiable data points, thereby reinforcing a view of livestock as biological machines rather than sentient beings with intrinsic value. This dual role of AI—as both a tool for welfare and a tool for commodification—reveals the ambivalence at the heart of smart farming technologies (Tullo *et al.*, 2019).

- **Wearable AI:** Collars fitted with accelerometers and machine learning models can track pets' activity levels and emotional states (Brugarolas *et al.*, 2020). These devices are marketed to pet owners as tools for monitoring health, ensuring adequate exercise, and even gaining insight into the emotional wellbeing of their animals. For example, sudden reductions in activity can alert owners to potential illness, while patterns of movement or tail positioning may be interpreted as indicators of stress or excitement. In this way, wearable AI promises to deepen the bond between humans and their companion animals by providing continuous feedback about their condition. Such technologies, however, come with their own set of ethical complexities. On the one hand, they may help to prevent neglect by giving owners more information to act responsibly. On the other hand, they risk creating a culture in which pet care is reduced to data interpretation, potentially displacing the attentiveness and empathy that arise from lived, embodied interaction with the animal. Moreover, questions arise about accuracy and interpretation: can an algorithm truly capture the emotional states of pets, or does it simplify complex behaviours into categories that may mislead owners and generate misplaced responses (Brugarolas *et al.*, 2020)?

Model Interpretability and Transparency in Animal-Centred AI

One emerging challenge in deploying AI for decoding animal vocalisations or predicting behaviour is model interpretability. Deep learning systems, especially those trained on acoustic spectrograms or multimodal behavioural datasets, often operate as "black boxes." This presents two concerns.

First, opaque models risk misclassification due to biases in training data (e.g., overrepresentation of a single species, environment, or context). Second, lack of interpretability makes it difficult for researchers, conservationists, and animal welfare experts to validate whether model outputs align with species-specific ethological knowledge.

Explainable AI (XAI) tools, such as feature attribution maps, interpretable decision trees, or rule-based classifiers, are increasingly recommended when working with animal datasets. These methods help uncover what aspects of a vocalisation, gesture, or physiological pattern a model is using to make predictions, enabling more responsible, scientifically grounded analysis.

4. ETHICAL IMPLICATIONS

The integration of Artificial Intelligence (AI) into human-animal communication systems is unfolding with unprecedented speed, ushering in a series of profound ethical challenges that extend far beyond conventional animal welfare concerns. These challenges delve into complex questions of animal agency, their representation within technologically mediated interfaces, and the very essence of interspecies relationships in a rapidly digitising world. Understanding and proactively addressing these ethical dimensions is crucial for ensuring a responsible and sustainable future for human-animal interactions.

Animal Agency and Consent in AI-Mediated Interactions

A fundamental ethical concern revolves around the concept of animal agency and the intricate question of consent in contexts where AI is deployed. When AI systems are utilised to predict, influence, or even manipulate animal behavior, for instance, in optimised smart farming operations or sophisticated wildlife conservation strategies, it fundamentally raises questions about whether animals are genuinely treated as autonomous beings capable of expressing preferences, or merely as data-generating entities whose behaviors are optimised for human benefit (Kim *et al.*, 2024). This instrumental view risks reducing sentient life to programmable inputs and outputs, diminishing their intrinsic value in the process. While AI can undoubtedly decode complex animal emotional states or vocalisation patterns, offering unprecedented insights into their physiological needs and psychological well-being, this technological capability does not readily translate into a straightforward system of true consent, which remains a deeply complex, often human-centric, construct (Cowie *et al.*, 2023).

The inherent asymmetrical power dynamic between human designers and animal subjects underscores the urgent need for a thorough re-evaluation of existing ethical frameworks. These frameworks must be robust enough to genuinely account for animal interests, a critical topic frequently explored within the burgeoning and increasingly vital field of AI ethics that explicitly recognizes the profound impact of technology on non-human entities (Potluri *et al.*, 2025).

Emerging research, however, indicates a potential silver lining: thoughtfully designed animal-computer interactions have the capacity to enhance animal agency by providing them with greater control over their immediate environment (Martínez-Vega & García-Martínez, 2025). Yet, this potential can only be realised if the design considerations are meticulously aligned with authentic animal preferences and ethological requirements, rather than solely serving human operational goals. The ethical imperative here is to move beyond mere efficiency and towards a symbiotic partnership where AI tools empower animals, rather than just manage them.

The Perils of Anthropomorphism and Ethical Misrepresentation

The increasing sophistication of AI in interpreting animal signals also presents a significant ethical dilemma related to anthropomorphism and the subsequent risk of ethical misrepresentation. While AI systems are engineered to bridge the communication gap between species, they frequently achieve this by inadvertently projecting human-like cognitive abilities, emotional states, or social paradigms onto animals. This projective process, while seemingly benign, can inadvertently create a false sense of understanding, leading to interventions that are not genuinely in the animal's best interest (Miranda-de la Lama & Mota-Rojas, 2020). For instance, an AI interpreting a specific animal vocalisation as "sadness" might trigger a response based on a human emotional paradigm, such as attempting to comfort, rather than a more nuanced, species-specific behavioral or physiological context. The animal's distress might stem from a completely different, non-human-like cause, rendering the AI's "empathetic" response ineffective or even detrimental. This ethical pitfall highlights the critical need for a critical, interdisciplinary approach that synergistically combines cutting-edge technological expertise with profound ethological insights into animal behavior, psychology, and species-specific communication, complemented by robust philosophical grounding in animal ethics. Without such a holistic perspective, AI runs the risk of creating "communication" systems that are more reflective of human preconceptions than of genuine animal realities, potentially leading to misdirected efforts in welfare, conservation, and management.

Data Bias, Technological Determinism, and Accountability

The deployment of AI in interspecies contexts is inextricably linked with the significant risks of data bias and technological determinism. AI models, by their very nature, are trained on vast datasets, and if these foundational datasets are inherently biased, for example, by overrepresenting certain species, geographical environments, or specific human-animal interaction types, the resultant technologies will inevitably reflect, and consequently perpetuate, these ingrained biases (Bazarganu et al., 2024). This phenomenon can lead to a form of technological determinism, where the AI's pre-programmed logic and biased operational parameters dictate outcomes, effectively overriding more context-sensitive, ethically informed, and humane approaches.

Consider an AI designed to maximise agricultural output: if its training data disproportionately emphasizes production metrics over animal well-being indicators, it might recommend practices that, while deemed "efficient" by its biased algorithms, are demonstrably detrimental to animal welfare. This underscores why the issues of transparency, fairness, and accountability are paramount in the development of interspecies AI. When an AI system makes a critical decision, such as recommending the culling of a herd based on predictive health analytics or altering the habitat of a wildlife population, the lack of transparency inherent in its decision-making process, often referred to as the "black box" problem, makes it exceedingly difficult to ascertain accountability or ensure that the decision was fair and justifiable from the animals' perspective (Cambraia & Pyrrho, 2025). This "black box" problem is a central, long-standing concern in broader AI ethics and applies with equal, if not greater, force to interspecies technologies. The ethical imperative demands the development and implementation of Explainable AI (XAI), systems capable of providing clear, human-understandable justifications for their decisions, thereby allowing for critical ethical oversight and the assignment of responsibility when adverse outcomes occur.

Surveillance, Privacy, and the Instrumental Value Dilemma

Finally, the pervasive use of surveillance and the erosion of privacy in animal monitoring systems represent a significant and rapidly expanding ethical frontier. While AI-driven monitoring can offer profoundly beneficial applications for crucial conservation efforts, such as tracking endangered species or detecting poaching activities, and for optimising animal health management in controlled environments, it simultaneously raises profound questions about the animals' fundamental "right to be left alone" and the potential for ubiquitous, intrusive, and continuous surveillance.

The constant collection of detailed data, even when motivated by benevolent intentions, can inadvertently infringe upon the natural behaviors, social structures, and private lives of animals, potentially altering their fundamental existence without their "consent" or even human awareness of the full impact. This continuous digital gaze transforms natural environments into data farms, raising questions about whether such pervasive monitoring is merely a tool for observation or a subtle form of control. The core of these intricate ethical dilemmas lies in the ongoing philosophical debate between instrumentalisation versus the intrinsic value of animals.

AI systems, by their very design, are frequently developed and deployed for instrumental purposes: to increase efficiency, enhance human productivity, or simplify human tasks. This instrumental lens can inadvertently devalue animals, systematically reducing them to mere objects of data collection, algorithmic analysis, and human manipulation, rather than respecting their inherent worth as sentient beings with lives, experiences, and purposes of their own (Potluri *et al.*, 2025). A truly responsible approach mandates that the relentless pursuit of technological advancement does not overshadow, but rather actively reinforces, the fundamental ethical imperative to recognize, protect, and uphold animal dignity in all its forms.

5. SOCIETAL IMPLICATIONS

The integration of artificial intelligence (AI) into human-animal communication systems heralds a transformative era, bringing with it a complex tapestry of societal ramifications. These consequences span from profound economic restructuring within key industries to subtle yet significant shifts in cultural perspectives and established power dynamics. The economic implications, particularly within the interconnected realms of conservation and agriculture, warrant meticulous examination.

Economic Transformations and the Widening Digital Divide

In contemporary agricultural practices, AI-driven automation is poised to significantly enhance productivity. This is achieved through sophisticated systems that optimise feeding regimens, implement continuous animal health monitoring, and streamline various operational processes, thereby promising a substantial elevation in profitability for large-scale operations (Potluri *et al.*, 2025). For instance, AI algorithms can analyse real-time data from sensors attached to livestock, enabling individualized feed adjustments based on an animal's age, weight, activity, and even genetic predisposition, minimising waste and maximising growth efficiency.

Furthermore, computer vision and acoustic monitoring systems can detect early signs of illness or distress in animals, allowing for timely intervention and reducing losses due to disease outbreaks. Precision farming techniques, facilitated by AI, can also optimize resource allocation for pasture management, ensuring sustainable land use and improved animal nutrition. This technological surge offers unprecedented levels of efficiency and control, presenting a compelling economic incentive for adoption.

However, this rapid technological advancement simultaneously carries the inherent risk of exacerbating existing disparities within the agricultural sector. The substantial capital outlay required for AI-driven systems, including specialised sensors, robotics, advanced software subscriptions, and the necessary digital infrastructure, can be prohibitively expensive for small-scale farmers (Martínez-Vega & García-Martínez, 2025). This creates a burgeoning digital divide, where larger, well-capitalised enterprises can leverage AI to gain a competitive edge, while smaller farms, often operating on tighter margins and with limited access to credit or technical expertise, risk economic marginalisation. The consequence is a potential consolidation of power and wealth within the agricultural landscape, threatening the viability of traditional family farms and the diversity of food production systems.

Similarly, in conservation efforts, AI-powered surveillance tools offer transformative potential for safeguarding endangered species and monitoring biodiversity. Technologies such as drone surveillance, automated acoustic monitoring for detecting poacher activity or specific animal calls, and even AI-driven facial recognition for individual animals, can provide unprecedented data for protection and population tracking (Damián-Rodríguez & Gómez-Gómez, 2024). These capabilities can significantly bolster anti-poaching initiatives and provide crucial data for habitat management.

Yet, the considerable expenses associated with acquiring and maintaining these sophisticated systems, coupled with the demand for highly specialized technical expertise (e.g., data scientists, AI engineers, drone pilots), often lead to an increased reliance on affluent nations, large international non-governmental organisations, or well-resourced academic institutions. This dependence can inadvertently sideline local communities and indigenous populations who traditionally coexist with these animal populations and often possess invaluable traditional ecological knowledge (TEK). Their deep-rooted understanding of local ecosystems and species behavior, accrued over generations, risks being devalued or overlooked in favor of technologically driven, external solutions, fostering a new form of conservation inequality.

Digital Inequality and the Specter of Techno-Colonialism

The emerging scenario profoundly underscores pervasive concerns regarding digital inequality and the equitable access to advanced AI technology. The benefits generated by AI in interspecies communication are demonstrably not distributed uniformly across the global landscape. Instead, global AI adoption frequently mirrors and reinforces existing power asymmetries, where industrially developed nations and dominant multinational corporations largely dictate the trajectory of technological development, deployment strategies, and the very ethical frameworks governing these technologies (Cohen & Galily, 2024). This often transpires with insufficient meaningful input from marginalised communities or indigenous populations, many of whom possess profound, intergenerational relationships with animals and unique cultural practices concerning interspecies interactions.

Such dynamics risk ushering in a contemporary manifestation of techno-colonialism, where technologically advanced solutions, frequently imbued with Western-centric epistemologies and values, are imposed upon diverse local contexts (Cambraia & Pyrrho, 2025). This imposition can lead to the erosion of traditional ecological knowledge systems, which are often holistic, adaptive, and deeply integrated with local socio-cultural structures. For example, AI-driven systems designed for wildlife management, while effective in their technical scope, might overlook or actively disrupt indigenous hunting practices, sacred land-use patterns, or traditional methods of animal husbandry that are intrinsically linked to cultural identity and sustainable coexistence. This is not merely a matter of technology transfer; it is a process by which specific technological solutions and their underlying rationales become dominant, potentially leading to new forms of dependence on foreign expertise and proprietary systems, thereby undermining local self-determination and cultural sovereignty in environmental and animal management. The extraction of vast datasets from biodiversity hotspots by external entities, without equitable benefit-sharing or genuine community participation in data governance, exemplifies a modern form of resource extraction, reminiscent of historical colonial practices.

Labor Displacement and the Reshaping of the Human-Animal Bond

A critical concern emanating from the widespread adoption of interspecies AI revolves around the potential for labor displacement and pervasive automation within industries traditionally reliant on animal interaction. As AI and robotics progressively assume a wider array of tasks, such as automated milking, precision feeding, robotic cleaning of enclosures, sophisticated health diagnostics, and even complex herding operations, the human workforce in these sectors confronts a future demanding extensive reskilling or facing significant redundancy (Bazarganu et al., 2024).

The economic consequences for these workers, often in rural and agricultural communities, can be severe, leading to job losses, increased economic insecurity, and potentially forced migration. The challenges of reskilling an often-traditional workforce, potentially with limited access to advanced education or training programs, are substantial and require proactive policy interventions.

Beyond the purely economic shifts, this transformative process fundamentally reshapes the traditional human-animal bond. The direct, hands-on care, observation, and intuitive understanding that historically characterized human interaction with animals are increasingly substituted with mediated, data-centric management protocols (Kim *et al.*, 2024). When an AI system monitors an animal's health through sensors and algorithms, the human caregiver's role shifts from direct physical interaction and empathetic observation to interpreting data dashboards and responding to algorithmic alerts. This qualitative alteration in human engagement means that the nuanced, often unspoken communication and the empathetic connection forged through physical presence and shared labor may diminish.

The psychological and ethical implications for both humans and animals are profound: humans may experience a sense of detachment, while animals might be managed as mere data points rather than sentient beings with individual needs and personalities. This reductionist approach risks eroding the very foundation of reciprocal relationships that have defined human-animal coexistence for millennia.

Cultural Attitudes and the Dilemma of Perception

Finally, the widespread integration of interspecies AI is poised to significantly influence cultural attitudes toward animals, presenting a dual-edged sword of potential enlightenment and potential dehumanisation. On one hand, the capacity to "converse" with animals via sophisticated AI interfaces, for example, through AI-driven translation of animal vocalisations, interpretation of subtle behavioral cues, or the generation of insights into their cognitive states and emotional experiences, could foster a heightened sense of empathy and compel a profound re-evaluation of their cognitive capabilities (Cowie *et al.*, 2023).

By providing what appears to be a window into the inner lives of animals, AI might cultivate more respectful and compassionate interspecies relationships, challenging anthropocentric biases and fostering a deeper appreciation for animal sentience and subjective experience. This could lead to a societal shift where animals are increasingly seen as subjects rather than mere objects.

Conversely, there is an inherent risk that this very capacity could lead to a more clinical, utilitarian perception of animals. An over-reliance on algorithmic outputs might inadvertently reduce animals to quantifiable data points, de-emphasising their individual subjectivity, unique personalities, and complex emotional lives (Miranda-de la Lama & Mota-Rojas, 2020). When an animal's "well-being" is reduced to a set of sensor readings and algorithmic scores, the direct, empathetic understanding gained through lived experience and shared interaction can be diminished. This could reinforce an instrumental view, where animals are primarily valued for their utility, productivity, or the data they generate, rather than for their intrinsic worth. Such a perspective could inadvertently normalise or even optimise their instrumental use within human systems, potentially entrenching existing exploitative practices by making them more efficient and less visible, rather than challenging the fundamental ethical questions surrounding animal use.

Consequently, the societal reception and integration of these transformative technologies will play a pivotal role in determining whether they ultimately serve to elevate the status of animals within human moral consideration or primarily function to render their exploitation more scientifically efficient and culturally palatable. Public discourse, ethical education, and media representation will be crucial in shaping how these powerful technologies are perceived and integrated into our evolving societal norms regarding our non-human companions on this planet.

6. GOVERNANCE, POLICY, AND ETHICAL FRAMEWORKS

Establishing robust governance for artificial intelligence (AI) applications in human-animal interactions is paramount to fostering a responsible and ethically sound future. A thorough review of existing policies and identified regulatory gaps underscores that the current frameworks are largely inadequate to address the unique and intricate challenges presented by interspecies AI (Martínez-Vega & García-Martínez, 2025). Predominantly, AI regulations have been human-centric, primarily concentrating on aspects like data privacy, algorithmic bias, and accountability as they pertain to human subjects. This narrow focus leaves a significant void in the provision of explicit protections for animals within AI systems. Consequently, there is an urgent need for a novel methodological approach that effectively integrates comparative ethical frameworks. This necessitates drawing comprehensively from well-established animal ethics,which typically foregrounds concepts such as animal sentience, intrinsic value, and welfare, and contemporary AI ethics, with its emphasis on principles of fairness, transparency, and accountability in algorithmic decision-making.

A particularly promising avenue for responsible development lies in the adoption of an ethics-by-design approach within interspecies AI contexts (Bazarganu et al., 2024). This philosophical stance advocates for the proactive embedding of ethical considerations into the foundational core of AI system development, from the initial conceptualisation phase through to deployment and ongoing operation. Practically, this implies that animal welfare, the preservation of animal autonomy where applicable, and the systematic minimisation of potential harm must be prioritized as central design constraints. These considerations should be integrated at the outset, rather than being treated as reactive afterthoughts or mere compliance-based add-ons. For example, an AI system engineered for precision livestock management ought to be programmed with sophisticated welfare metrics that are afforded equitable weighting alongside traditional productivity metrics, ensuring that technological advancements do not inadvertently compromise animal well-being (Kim *et al.*, 2024). Such an approach moves beyond simply avoiding harm to actively promoting positive animal outcomes.

The inherent complexity of these ethical and governance issues necessitates a robust model of participatory governance, which deliberately includes a diverse array of stakeholders beyond the conventional spheres of technologists and policymakers. An effective governance structure must actively incorporate the perspectives and voices of ethicists, conservationists, dedicated animal welfare advocates, indigenous communities (who often possess deep traditional knowledge of human-animal relationships), and farmers themselves. This inclusive approach ensures that emerging technologies are developed with a profound, on-the-ground understanding of their practical impact, cultural resonance, and local ecological contexts (Miranda-de la Lama & Mota-Rojas, 2020). By fostering this multi-stakeholder engagement, the governance model can help to mitigate existing power asymmetries and work towards a more equitable distribution of the benefits derived from interspecies AI technologies, preventing the concentration of technological advantage in a few hands.

Finally, the development of comprehensive global regulatory considerations and best practices is not merely advisable but essential. Given the intrinsically global nature of both AI development and the transboundary realities of many animal populations (e.g., migratory species, global agricultural supply chains), a fragmented patchwork of national regulations will inevitably prove insufficient. International collaboration is therefore imperative to establish shared ethical principles and universally applicable regulatory guidelines. These guidelines must be sufficiently flexible to be adapted effectively to diverse local contexts, while simultaneously upholding a universal and unwavering commitment to animal dignity and welfare (Potluri *et al.*, 2025). Such international frameworks could encompass shared data protocols designed to protect both animal privacy and the integrity of scientific research, alongside standardized requirements for algorithmic transparency that seamlessly traverse national borders, fostering a unified approach to responsible interspecies AI.

7. TOWARDS RESPONSIBLE INTERSPECIES TECHNOLOGY

The trajectory toward the development and deployment of truly responsible interspecies technology necessitates a unified and deliberate effort to translate abstract ethical principles into concrete, actionable practices that span the entire technological development lifecycle. At the core of this imperative lies the establishment of robust ethical design principles specifically tailored for AI applications within human-animal contexts. These foundational principles must be deeply rooted in a profound respect for animal agency, an unwavering commitment to minimising all forms of harm, and an active pursuit of promoting positive welfare outcomes for animals (Kim *et al.*, 2024). Moving beyond a reactive stance of merely preventing harm, ethical design in this domain should proactively strive to enhance the lives of animals, for instance, by conceptualising and creating technologies that significantly enrich their natural or captive environments, facilitate more nuanced and positive interspecies interactions, or even empower animals with greater choice and control over their immediate surroundings. This proactive approach demands a fundamental shift from human-centric utility to animal-centric well-being, acknowledging that technological advancement should not solely serve human interests but also contribute meaningfully to the flourishing of non-human life. Designing for animal agency, for example, could involve developing AI systems that detect and respond to animal preferences, allowing them to choose feeding times, activity levels, or social groupings within defined parameters. Such systems would move beyond simple automation to a more interactive and respectful coexistence, fostering a sense of control for the animals themselves. The integration of complex biometric and behavioral data, processed by advanced AI models, allows for unprecedented insights into animal states, yet these insights must be ethically managed to ensure they empower, rather than merely surveil or control.

Building upon these critical ethical design principles, a comprehensive series of recommendations for policy and practice can be meticulously formulated to guide stakeholders across various levels. At the macro policy level, this involves actively advocating for and implementing progressive legislation that specifically addresses the nuanced complexities of AI utilisation within animal-based industries, such as intensive agriculture and aquaculture, as well as crucial conservation initiatives. A central tenet of such legislation should be the mandated requirement for transparency and stringent accountability mechanisms for all AI systems impacting animals (Bazarganu et al., 2024). For example, regulatory bodies could establish explicit requirements for any AI system slated for deployment in an agricultural setting to undergo a rigorous, independent ethical impact assessment prior to its implementation. This assessment would scrutinise potential negative impacts on animal welfare, behavioral patterns, and environmental sustainability, ensuring a holistic evaluation. Furthermore, international cooperation is paramount, given the transboundary nature of many animal populations (e.g., migratory species) and the globalized supply chains of animal products. Harmonised international policies could prevent regulatory arbitrage and ensure a consistent standard of animal protection regardless of geographical location. This necessitates frameworks for shared data protocols that respect both animal privacy, by anonymising individual data where possible and aggregating for population-level insights,and the integrity of research, alongside international standards for algorithmic transparency that transcend national borders, fostering a unified and responsible approach (Potluri *et al.*, 2025). At the micro practice level, there is an urgent need to develop clear, unambiguous, and implementable guidelines for developers, researchers, and institutions actively engaged in building and deploying these transformative technologies. These guidelines should precisely specify robust requirements for data collection,ensuring data is ethically sourced and representative,rigorous algorithm training that minimises bias against certain animal characteristics or breeds, and comprehensive system validation protocols where animal welfare is not merely an afterthought but a central, quantifiable metric of success (Martínez-Vega & García-Martínez, 2025). Moreover, these practical guidelines should mandate thorough and accessible documentation of all AI systems, enabling independent external audits, ethical reviews, and public scrutiny to ensure ongoing adherence to established ethical standards and best practices.

It is unequivocally clear that addressing the intricate challenges of interspecies AI is not a task for a single discipline; rather, it necessitates fostering profound cross-disciplinary collaboration. The future trajectory of interspecies AI cannot, and must not, be solely entrusted to the purview of computer scientists and engineers. It demands a sustained, dynamic, and iterative dialogue among a diverse consortium of experts: technologists who can articulate the capabilities and limitations of AI; ethicists who can provide normative guidance and identify potential moral pitfalls; animal behaviorists and ethologists who possess deep understanding of species-specific needs and cognitive capacities; sociologists and anthropologists who can shed light on the societal impacts and cultural contexts of human-animal interactions; and policymakers and legal experts who can translate ethical aspirations into enforceable legal frameworks. This institutionalised collaboration should manifest in the form of interdisciplinary research grants, dedicated academic centers focused on interspecies AI ethics, and, crucially, ethical review boards for AI projects that actively include a diverse range of expertise. These boards would serve as critical junctures for scrutinizing design choices, data practices, and deployment strategies from multiple ethical, scientific, and societal vantage points. The challenges in such collaborations, such as differing terminologies and methodologies, can be overcome through concerted efforts to establish a common language, shared objectives, and a mutual appreciation for each discipline's invaluable contribution. This integrated approach ensures that technological innovation is consistently guided by a holistic understanding of its multifaceted implications, moving beyond purely technical efficacy to encompass moral rectitude and societal benefit.

Ultimately, the overarching goal of these concerted efforts is to establish a comprehensive framework that extends beyond the mere mitigation of risks. Instead, it aims to embed animal dignity and welfare as foundational tenets within every stage of AI development and application. This pivotal shift entails moving decisively beyond a purely utilitarian calculus, which traditionally assesses value based on usefulness or economic output. It demands a paradigm where animals are recognised not merely as resources to be managed or instruments to be optimized, but as inherently sentient beings deserving of profound respect and moral consideration (Miranda-de la Lama & Mota-Rojas, 2020). As AI continues to dramatically deepen our scientific and public understanding of non-human life, revealing new facets of animal cognition, emotion, and social complexity, we bear an profound ethical responsibility to ensure that these powerful technologies are not only designed to be smart and efficient but also to be inherently kind, just, and respectful of the incredibly diverse forms of life with whom we share this planet. This future vision of interspecies technology is one where ethical consciousness is as integral to its architecture as its algorithms, fostering a profound re-evaluation of humanity's role as stewards of the animal kingdom and pioneering new frontiers of respectful coexistence.

Table 1. Ethical Challenges and Mitigation Strategies in Interspecies AI

Ethical Issue	Description	Recommended Mitigation Strategy
Anthropomorphism	Risk of projecting human categories onto animal signals	Use species-specific ethological baselines; incorporate expert review
Data Bias	Training datasets overrepresent certain species/contexts	Diversify data collection; apply bias-detection tools
Loss of Agency	Animals affected by AI-guided behavioural interventions	Design for optionality, preference-testing, low intrusiveness
Surveillance	Excessive monitoring of wildlife or livestock	Establish monitoring thresholds; require welfare justification
Lack of Transparency	Users cannot interpret how models make decisions	Incorporate XAI; document model decision pathways

Practical Design Frameworks for Responsible Interspecies AI

Building on ethics-by-design, practical frameworks can guide developers and practitioners:

a. **Species-Sensitive Data Standards:** Data collection must follow ethological evidence, ensuring recordings and behavioural observations reflect natural rather than stressed or artificial states.
b. **Tiered Transparency Protocols:** Models should include documented assumptions, training conditions, and known limitations.
c. **Animal Preference Testing:** Interfaces (e.g., robotic agents, collars) should be validated through voluntary interaction patterns rather than enforced exposure.
d. **Context-Adaptive Deployment:** Technologies should respond to ecological, cultural, and cross-species differences rather than applying one-size-fits-all standards.
e. **Welfare-Centred Key Performance Indicators (KPIs):** Success metrics must include reduced stress, increased autonomy, and behavioural normalcy.

Cross-Cultural Ethical Considerations

Human–animal relations vary widely across cultures. Technologies designed in Western contexts may misalign with the animal ethics frameworks of indigenous, pastoral, or agrarian communities. Integrating cross-cultural ethical analysis ensures that interspecies AI does not universalise narrow values or impose external priorities.

This includes recognising:

- Indigenous ecological knowledge and traditional modes of reading animal signals.
- Non-Western relational ontologies, which may conceptualise animals as kin, spiritual entities, or cohabitants.
- Context-specific welfare concepts (e.g., communal grazing norms, ritual significance).

This cross-cultural lens ensures culturally responsive and globally responsible interspecies technologies.

8. CONCLUSION

Recap of Key Insights

This chapter has traced the emergence of AI as a transformative force in human-animal communication. Computational bioacoustics, multimodal machine learning, passive acoustic monitoring, and robotic interfaces together form a dynamic technological ecosystem that can potentially revolutionise how we perceive and interact with non-human minds.

Yet, these technological advancements cannot be considered value-neutral. Issues of animal agency, anonymity, anthropomorphism, misinterpretation and unequal access permeate the field. Without deliberate ethical frameworks, AI may reproduce or worsen existing asymmetries, even as it promises new forms of interspecies connection.

Emphasis on the Need for Caution and Responsibility

Progress in interspecies AI must proceed with caution. Key ethical concerns include:

- **Agency and Consent:** Animals cannot grant informed consent, but designers must strive to minimise intrusiveness and maximise welfare.
- **Anthropomorphism:** Interpreting animal signals through a human lens risks distortion. Critical anthropomorphism, grounded in species-specific understanding, is essential.

- **Transparency and Accountability:** AI systems should be explainable; users should understand model limitations and potential biases.
- **Bias and Misrepresentation:** Data imbalances, e.g., over-representation of certain species or contexts, can produce skewed outputs. Bias mitigation must be proactive.
- **Surveillance and Instrumentalisation:** Passive data gathering can morph into control or exploitation. Governance must safeguard against misuse.

Final Reflections on the Future of Interspecies AI

If pursued ethically, interspecies AI may reshape our relationship with animals, from viewing them as data points to recognising them as communicative agents. Legal scholars speculate that AI-mediated insight into animal communication could catalyse transformations in animal rights law, potentially altering animals' legal status (Rodríguez-Garavito, 2025).

Conservation strategies may become more responsive and nuanced-guided by real-time communication cues. Agriculture could shift toward welfare-centric systems. Public empathy may grow through direct, AI-facilitated glimpses into animal experience.

However, the path forward demands collaboration across disciplines: ethicists, technologists, biologists, animal care professionals, policy makers and communities must co-design ethical and effective systems.

Call to Action for Ethical Stewardship and Inclusive Innovation

To ensure responsible interspecies AI, stakeholders must commit to:

- **Ethics-by-Design:** Embed welfare, fairness, transparency and minimal intrusiveness from the earliest design stages.
- **Participatory Governance:** Include diverse voices, especially animal advocates, ethicists and local communities, in policy formation and oversight.
- **Regulatory Frameworks:** Develop legal standards to govern data use, surveillance thresholds, interpretative limits and the rights of non-human communicators.
- **Cross-disciplinary Collaboration:** Leverage insights from AI, animal ethics, behavioural ecology, law and public policy to shape humane innovation.
- **Public Advocacy and Education:** Promote critical anthropomorphism and informed discourse, avoiding sensationalism while fostering empathy and awareness.

In embracing these principles, interspecies AI can evolve responsibly, fostering deeper understanding across species without undermining the dignity of those we seek to listen to.

REFERENCES

American Bar Association. (2025). Bridging species: Legal and ethical implications of AI- assisted animal communication. *Tort Trial & Insurance Practice Law Journal*. https://www.americanbar.org/groups/tort_trial_insurance_practice/resources/brief/202 5-winter/legal-ethical-implications-ai-assisted-animal-communication/

Bradshaw, G. A. (2021). *Elephants on the edge: What animals teach us about humanity*. Yale University Press.

Brugarolas, R., Lee, J., & Martins, N. R. B. (2020). Wearable devices for animal health monitoring. *Frontiers in Veterinary Science*, *7*(63), 1–11. DOI: 10.3389/fvets.2020.00063

Cambraia, A., & Pyrrho, S. (2025). Generative artificial intelligence and the risk of technodigital colonialism. *Frontiers in Political Science*, *7*, 1628139. DOI: 10.3389/fpos.2025.1628139

Cave, S., & Dihal, K. (2020). *AI narratives: A history of imaginative thinking about intelligent machines*. Oxford University Press.

Cohen, A., & Galily, Y. (2024). Behavioral and Psychosocial Dynamics of Engagement: The Digital Divide in Artificial Intelligence [AI]-Driven Sports Podcasts. *Technologies*, *14*(10), 911. DOI: 10.3390/technologies14100911

Cowie, J., Cowie, S., & Gentry, K. (2023). Artificial intelligence technologies and compassion in healthcare: A systematic scoping review. *Journal of Medical Systems*, *47*(1), 1–17. DOI: 10.1007/s10916-023-01931-1

Crawford, K. (2021). *Atlas of AI: Power, politics, and the planetary costs of artificial intelligence*. Yale University Press.

Crist, E. (2019). *Abundant earth: Toward an ecological civilization*. University of Chicago Press.

Damián-Rodríguez, M., & Gómez-Gómez, A. (2024). Artificial Intelligence Alone Will Not Democratise Education: On Educational Inequality, Techno-Solutionism and Inclusive Tools. *Sustainability*, *16*(2), 781. DOI: 10.3390/su16020781

Donaldson, S., & Kymlicka, W. (2011). *Zoopolis: A political theory of animal rights*. Oxford University Press.

Fitzpatrick, J., Browning, H., & Brakel, S. (2022). Ethical challenges in precision livestock farming. *Journal of Agricultural & Environmental Ethics*, *35*(2), 1–19. DOI: 10.1007/s10806-022-09898-3

Floridi, L., Cowls, J., Beltrametti, M., Chatila, R., Chazerand, P., Dignum, V., & Vayena, E. (2018). AI4People—An ethical framework for a good AI society: Opportunities, risks, principles, and recommendations. *Minds and Machines*, *28*(4), 689–707. DOI: 10.1007/s11023-018-9482-5

Ihde, D. (1990). *Technology and the lifeworld: From garden to earth.* Indiana University Press.

Jobin, A., Ienca, M., & Vayena, E. (2019). The global landscape of AI ethics guidelines. *Nature Machine Intelligence*, *1*(9), 389–399. DOI: 10.1038/s42256-019-0088-2

Kershenbaum, A., Blumstein, D. T., Roch, M. A., Akçay, Ç., Backus, G., Bee, M. A., & Yuan, C. (2016). Acoustic sequences in non-human animals: A tutorial review and prospectus. *Biological Reviews of the Cambridge Philosophical Society*, *91*(1), 13–52. DOI: 10.1111/brv.12160

Kim, S., Park, J., & Lee, J. (2024). AI-Based Smart Monitoring Framework for Livestock Farms. *Applied Sciences (Basel, Switzerland)*, *15*(10), 5638. DOI: 10.3390/app151005638

Kovac, M., Bonnet, F., & Floreano, D. (2021). Robotic animals for scientific research and conservation. *Annual Review of Control, Robotics, and Autonomous Systems*, *4*(1), 185–209. DOI: 10.1146/annurev-control-062220-092241

Martínez-Vega, J., & García-Martínez, A. (2025). AI Algorithms in the Agrifood Industry: Application Potential in the Spanish Agrifood Context. *Applied Sciences (Basel, Switzerland)*, *15*(4), 2096. DOI: 10.3390/app15042096

Miklósi, Á. (2015). *Dog behaviour, evolution, and cognition.* Oxford University Press.

Miranda-de la Lama, G. C., & Mota-Rojas, D. (2020). Association between Attitude and Empathy with the Quality of Human-Livestock Interactions. *Animals (Basel)*, *10*(8), 1304. DOI: 10.3390/ani10081304

Olojede, H. T., & Polo, E. P. (2025). "In Praise of Normative Science: Arts and Humanities in the Age of Artificial Intelligence," pages 1-9. In *International Journal ofSocial Sciences and Humanities: Africa Research Corps Network (ARCN) Journals.* ISSN: 2360-9877. Volume 11, Issue 2, (January, 2025). DOI: . Available at https://arcnjournals.org/images/27265774111101121-1.pdfDOI: 27265774111101121-1

Pepperberg, I. M. (2019). *The Alex studies: Cognitive and communicative abilities of grey parrots.* Harvard University Press.

Polo, E. P., Emmanuel, H., & Obeka, O. B. (2024). Examining the epistemological status of AI-aided research in the information age: Research integrity of Margaret Lawrence University in Delta State. *International Journal of Social Sciences and Humanities: Africa Research Corps Network (ARCN). Journals, 11*(1), 197–207. https://www.arcnjournals.org/images/27265774111101125.pdf

Potluri, S., Satpathy, S., & Basa, S. S. (Eds.). (2025). *AI in Agriculture for Sustainable and Economic Management*. Taylor & Francis., DOI: 10.1201/9781003451648

Puri, V., Nayyar, A., & Raja, L. (2019). Agriculture drones in India: A review. *Computers and Electronics in Agriculture, 162*, 164–177. DOI: 10.1016/j.compag.2019.04.001

Regan, T. (2004). *The case for animal rights*. University of California Press.

Russell, S., & Norvig, P. (2020). *Artificial intelligence: A modern approach* (4th ed.). Pearson.

Savage-Rumbaugh, S., Shanker, S. G., & Taylor, T. J. (1993). *Apes, language, and the human mind.* Oxford University Press.

Sebeok, T. A. (1990). *Essays in zoosemiotics*. Springer.

Singer, P. (1975). *Animal liberation*. HarperCollins.

Stahl, B. C., Timmermans, J., & Mittelstadt, B. (2022). The ethics of computing: A survey of professional bodies and academic institutions. *Computer, 55*(6), 65–75. DOI: 10.1109/MC.2022.3164232

Stowell, D. (2022). *Computational bioacoustics with deep learning.*

ADDITIONAL READING

Anderson, D. J., & Adolphs, R. (2014). A framework for studying emotions across species. *Cell, 157*(1), 187–200. DOI: 10.1016/j.cell.2014.03.003

Beer, J. M., & Takayama, L. (2011). Mobile remote presence systems for older adults: Acceptance, benefits, and concerns. In Proceedings of the 6th ACM/IEEE International Conference on Human-Robot Interaction (pp. 19–26). *Association for Computing Machinery*. https://doi.org/DOI: 10.1145/1957656.1957665

Bensky, M. K., Gosling, S. D., & Sinn, D. L. (2013). The world from a dog's point of view: A review and synthesis of dog cognition research. In H. J. Brockmann, T. J. Roper, M. Naguib, J. C. Mitani, L. W. Simmons, & L. Barrett (Eds.), Advances in the study of behavior (Vol. 45, pp. 209–406). Academic Press., 407186-5.00005-7.DOI: 10.1016/B978-0-12-

Coeckelbergh, M. (2021). *AI ethics*. MIT Press.

Dell, A. I., Bender, J. A., Branson, K., Couzin, I. D., de Polavieja, G. G., Noldus, L. P., & Straw, A. D. (2014). Automated image-based tracking and its application in ecology. *Trends in Ecology & Evolution*, *29*(7), 417–428. DOI: 10.1016/j.tree.2014.05.004

Despret, V. (2016). *What would animals say if we asked the right questions?* University of Minnesota Press.

Haraway, D. J. (2008). *When species meet*. University of Minnesota Press.

Mancini, C. (2011). Animal–computer interaction: A manifesto. *Interaction*, *18*(4), 69–73. DOI: 10.1145/1978822.1978836

Marconi, F. (2020). *Newsmakers: Artificial intelligence and the future of journalism*. Columbia University Press.

Meijer, E. (2019). *When animals speak: Toward an interspecies democracy*. New York University Press.

KEY TERMS AND DEFINITIONS

Anthropocentrism: An ethical perspective that prioritises human interests and interpretations, often at the expense of recognising animals' intrinsic moral status.

Artificial Intelligence (AI): Computational systems capable of performing tasks that typically require human intelligence, such as pattern recognition, prediction, language processing, applied here to decoding animal communication.

Bioacoustics: The study of sound production, dispersion, and reception in animals, often used with AI to analyse communication signals.

Critical Anthropomorphism: A nuanced interpretive approach that uses informed empathy to understand animal behaviour, while avoiding inaccurate projection of human emotions or motivations.

Ethics-by-Design: The principle of embedding ethical reflection, such as welfare considerations, transparency, accountability and fairness, throughout the life cycle of a technological tool.

Human–Animal Communication: The exchange of signals, vocal, gestural, behavioural or physiological, between humans and animals, interpreted or facilitated through AI systems.

Interspecies Ethics: An ethical stance aiming to acknowledge animals' capacity for experience and value, promoting respect for their autonomy, welfare and modes of existence.

Interspecies Technology: Tools and systems developed to enable or enhance interaction between humans and other species, such as bioacoustic decoders, robotic interfaces, emotion-recognition systems, and animal wearables.

Participatory Governance: A policy-making approach that includes diverse stakeholders, scientists, ethicists, technologists, welfare advocates, local communities, in designing, supervising, and regulating AI applications.

Robotic Interface (Interspecies): A technological system capable of engaging with animals through responsive behaviour, such as playing sounds or enabling the animal to trigger responses (e.g., CHAT for dolphins).

Chapter 11
Robotic Systems for Animal Interaction

Shamim Ahmad Khan
Shibli National College, MSD University, Azamgarh, India

Kalim Ahmad
Shibli National College, MSD University, Azamgarh, India

ABSTRACT

Robotic systems for animal interaction form a rapidly advancing interdisciplinary field uniting robotics, artificial intelligence, ethology, and welfare science. These technologies monitor, engage with, and assist animals across settings such as precision livestock farming, wildlife conservation, and laboratory research. Applications include automating routine care, enriching environments, supporting behavioral studies, and enabling safe human–animal–robot collaboration. Advances in sensing, machine learning, and adaptive control now allow robots to interpret animal behavior in real time and respond appropriately, enhancing efficiency, safety, and welfare outcomes. Yet challenges remain, including ethical concerns, species-specific modeling, cost, and reliability in varied conditions. This chapter reviews system types, design principles, and applications, presents case studies from agriculture, research, and conservation, and examines future directions such as bio-inspired robotics, multimodal perception, and fully autonomous welfare monitoring platforms.

1. INTRODUCTION

Animal-robot interaction has become a vibrant and cross-disciplinary area of concern in the field of robotics. It draws on the knowledge of animal behavior, artificial intelligence, engineering and welfare science(Romano et al., 2024). The

DOI: 10.4018/979-8-3373-5483-5.ch011

first moves in this area were mostly trials, where they featured the creation of simple automated devices that could watch or change animal behavior in a closed environment. Gradually, the improvements in sensing technologies, machine learning and autonomous navigation have changed the first models to advanced robotic systems that are able to make decisions on the spot, adapt to the situation and interact physically in a safe way with different animal species (Afzal et al., 2025). As a result, robots are now functioning in farms, research centers, homes and natural habitats, which is opening up new ways to better animal management, increase monitoring and implement welfare oriented (Ghavi Hossein-Zadeh, 2025).

Robotic systems are becoming increasingly important for animal welfare, husbandry and scientific research. Animal farming is made more efficient and less stressful for the animals thanks to intelligent milking robots, automated feeders and environmental sensors(Mahato & Neethirajan, 2025). These technologies also allow for continuous health monitoring. Bio-inspired robots in research and behavioural studies resemble real animals to get genuine social reactions, allowing scientists to study communication, social hierarchy and group dynamics without interfering with the natural habitats. Interactive robotic toys and assistive devices, which are among the trendiest innovations in pet care, provide domestic animals with enrichment, behavioural stimulation and companionship, especially in situations where human presence is limited(Horsevad et al., 2022). The use of drones and autonomous ground vehicles in wildlife conservation is, a very effective way to gather data, monitor habitats and support the fight against poaching. Together, these examples demonstrate how robotics is transforming the traditional human–animal relationship by integrating automation with welfare centered (Afzal et al., 2025).

Essentially, this chapter wants to communicate a detailed comprehension of robot systems that are made to interact with animals, pointing out their technological aspects, uses, impact on welfare and the upcoming developments. The chapter consolidates the existing knowledge from different disciplines to demonstrate how robotics can be a tool for the welfare of animals, becoming a source of both scientific and practical possibilities. Besides that, the chapter also intends to explain the ways that lead to the interaction between animals and robots, which involve the use of sensors, feedback loops and the robot's ability to adapt. Through the examples of the use of such systems in the real world, the chapter helps the reader to understand the scientific theories by showing the results which can be seen in animal husbandry, research labs and nature.

2. TECHNOLOGICAL FOUNDATIONS OF ANIMAL–ROBOT INTERACTION

2.1 Robotics in Animal Environments

Animal-interacting robots have to be prepared for situations that are not only changing and uncertain, but also biologically complex. In contrast to human-centered robotics, which usually involves predictable behaviours and controlled spaces, animal environments are characterized by spontaneous movement, group dynamics and unpredictable responses. Hence, robots intended for animals have to follow the principles of adaptability, robustness and behavioural sensitivity. They have to be capable of identifying even the smallest indications of the posture, vocalizations or stress-related movements and changing their behaviour accordingly(Zhao et al., 2024). Non-invasiveness is another main idea among the principles robots should not frighten or upset the animals; rather, they should become a part of the animal's physical and social space in a natural way. Most of the time this means mimicking nature in terms of shape or movement, being quiet and having behaviour patterns that are predictable that the animals can understand and feel safe. Moreover, robots for farms, wildlife habitats or laboratories should be equipped to deal with the local weather and environmental conditions like rough ground, very hot or cold weather, humid conditions and the presence of dust or biological waste. All these principles together are a guarantee that such robotic systems will be able to interact not only efficiently but also morally with different animal (Mansoor et al., 2025).

2.2 Sensors and Actuators for Behavioural/ Physiological Monitoring

Without sensor technology, it would be impossible for robots to perceive and interpret animal behaviour and physiological states. Some of the most common sensors are cameras for visual tracking, microphones for vocalization analysis, accelerometers and gyroscopes for movement detection and infrared or thermal sensors for monitoring body temperature and heat signatures. Advanced biosensors can pick up on heart rate, respiration, stress hormones or gait abnormalities, giving more profound insights into animal welfare(Lamanna et al., 2025). Environmental sensors in livestock settings keep track of variables such as air quality, humidity and ammonia levels to make living conditions better. Actuators, however, help robots to have a physical response through locomotion, gesture or tactile feedback. Among these are motors, pneumatic systems, robotic arms, feeding dispensers and vibration units that simulate natural cues. The coordination of sensors and actuators enables the facilitation of interactive behaviours, for instance, the guidance of animals, the provision of enrichment or the performance of automated husbandry tasks(Li et al., 2023). The sensing–acting synergy is what makes up the core of the most advanced, real-time and responsive interaction phenomena between robots and animals in the natural.

2.3 AI and ML Models for Interpreting Animal Signals

Artificial intelligence and machine learning have changed the field of animal-robot interactions in a way that robots are now able to understand complicated animal signals more accurately. Machine learning models have the capability to analyse huge datasets obtained from sensors in order to identify behavioural patterns, detect anomalies or even determine emotional states(Rudroff, 2024). As an example, convolutional neural networks examine the visual data to detect posture changes, lameness, grooming behaviour or aggression. Natural language processing algorithms and acoustic models are used to get the meaning of animal vocalizations, which can indicate distress, social calls and mating signals. Reinforcement learning gives a robot the ability to change its behaviour as a result of the feedback, the interactions become more personalized and aware of the context. Predictive algorithms in livestock systems can, for example, foresee health problems or animal behavioural needs intervention at the early stage is possible(Burnham, 2023). Also, in research environments, AI models are instrumental in providing behavioural analysis free from bias and also in issuing the replication of certain cues through biomimetic robots (**Figure 1**). In essence, AI and ML are a link between the raw sensory data and the robotic responses that can be acted upon, thus robots are allowed to be intelligent partners rather than just mechanical (Jiang et al., 2024).

Figure 1. Representing the AI in animal-inspired robots, Adapted with permission from

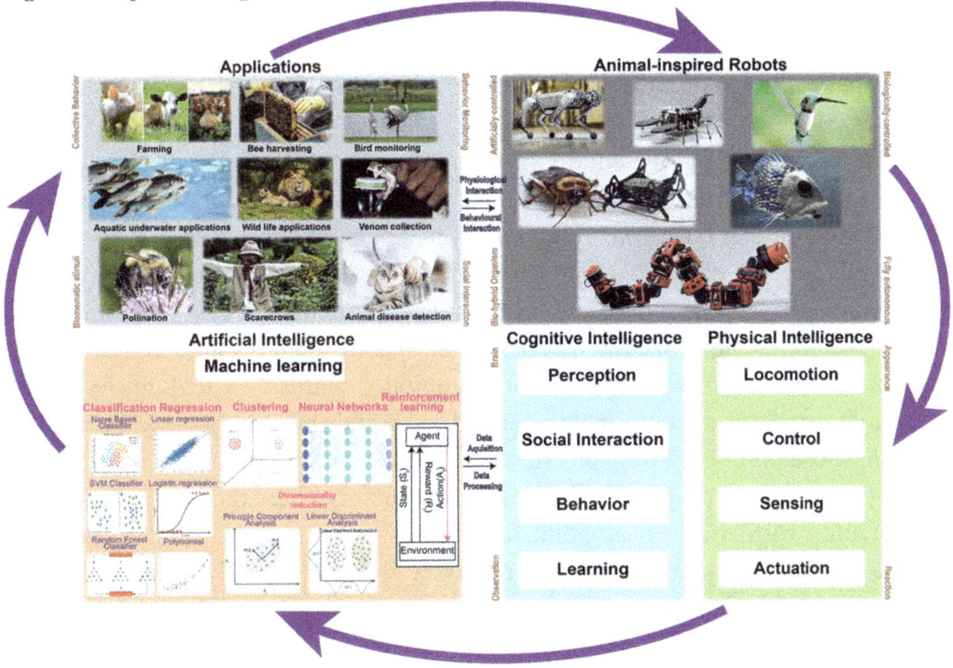

Source: Afzal et al. (2025)

2.4 Autonomous Navigation and Environmental Perception

Autonomous navigation is a must if robots are to work in farms, living habitats or homes where there might not be much human supervision. Robots use an advanced perception system like LiDAR, depth cameras, stereo vision, ultrasonic sensors and GPS modules to map their surroundings and be free of obstacles. Simultaneous localisation and mapping algorithms enable robots to create up-to-the-minute maps of complicated and changing environments, and they can move safely around animals without scaring them or colliding with them(Yerebakan & Hu, 2024). Different kinds of locomotion mechanisms, such as wheeled, tracked, legged and aerial platforms, enable robots to work in any environment ranging from barn floors to forested areas as they are terrain-adaptive. The environment perception systems also locate animals, their movement patterns, and proximity and direct robots to keep a suitable distance and behave accordingly. It is very important because if the animals' movements are unexpected or erratic, they can get scared very easily. By tying navigation to behavioural analysis, robots can come closer, follow or move away as the case might be, thereby interactions being safe, non-intrusive and behaviourally (Aoi et al., 2017; Kotha et al., 2024).

2.5 Safety, Durability and Bio-compatibility Considerations

Safety and durability are vitally important to robots that interact directly with animals. In this regard, structurally, robots must be strong enough to endure an accidental kick, bite or collision, at the same time, be light enough so as not to cause any harm(Guiochet et al., 2017). Their surfaces should be smooth, non-toxic and resistant to corrosion, moisture and biological contaminants. Using bio-compatible materials helps to minimize the risk of allergic reactions and behavioural aversion. To make the situation safer, robots can be equipped with safety features like force limitation, soft robotics components, padded edges and emergency stop mechanisms that can also help in case of accidents. Furthermore, robots should be dependable during prolonged use, especially in the farming sector, where the weather might be unfavorable. From a moral standpoint, it is necessary to ensure that the robots remain durable in the long run and are designed in a way that is friendly to the welfare of those being (Kóczi & Sárosi, 2025).

3. TYPES OF ROBOTIC SYSTEMS USED IN ANIMAL INTERACTION

3.1 Companion and Social Robots

Companion and social robots are an increasingly significant class of animal-interaction technologies, particularly in domestic and therapeutic settings. To household pets, such robots are a means of enrichment, stimulation and behavioural engagement. As an example, there may be a robotic interactive toy that is responsive to movement, vocalization or closeness, thereby exercising and rousing a dog, cat and even a small mammal(Miles et al., 2024). Most of these gadgets also have AI-powered behavior patterns that are designed to be more natural and socially engaging by simulating the movement of prey or social-verbal cues. Robotic therapists expand such features to the clinical or transition-to-life stages. The primary purpose of the robots is to provide emotional regulation and to be a source of relief for both animals and their human caregivers. For instance, a robot that is made to interact with a therapy dog or a horse can assist in the organization of the training exercises, keep the behavioural stimuli at a constant level and offer controlled social interactions to animals that are undergoing rehabilitation(Voysey et al., 2025). Adoption can be accelerated if behavioural robots can identify and help the emotionally deprived animals who are confined in places like animal shelters and hence their socialization can be enriched too through these technologies.

The emotional communication systems constitute another rising group that includes the use of sensors and adaptive AI in mood or arousal level detection in animals. Robots, by changing their gestures, lights, sounds and movements, can soothe an animal that is anxious, excite one that is lethargic and even encourage the repetition of positive behaviours. As the sophistication of social robot designs increases, they become indispensable in ensuring the welfare of animals, complementing human presence, and leading to better animal–environment (Tulsulkar et al., 2021).

3.2 Assistive and Monitoring Robots

Assistive and monitoring robots represent the main precision welfare management tools that can be used across farms, laboratories, shelters and homes. The robots collect continuous behavioural and physiological data, which makes the early distress, illness and environmental imbalance detection possible. Animal welfare surveillance robots are usually mobile or aerial and they observe animal movement patterns, social interactions, feeding behaviour or resting times without the animals' habituation to natural routines being disturbed(Neethirajan et al., 2024). The robots' autonomous operation allows for silent, 24/7 observation, which is a great source of welfare assessment and scientific studies. Tracking robots employ GPS, RFID, UWB (ultra-wideband) and computer vision technologies to single out animals in a group. They facilitate measurement of welfare indicators, location, activity level, social bonding and isolation. Precision monitoring systems not only collect environmental sensor data (temperature, humidity, air quality) but also behavioural signals to rescue a complete welfare picture of the animals. There are some robots that are equipped with the capabilities of accomplishing the targeted tasks, such as medication delivery, movement guidance or enrichment devices activation using real-time monitoring (Waworundeng, 2024).

These tools have been instrumental in significantly reducing labour demands and increasing accuracy when compared with manual observation. For example, drones can serve as remote-monitoring platforms that can reach parts that are inaccessible or dangerous and thus are ideal for wildlife and free-ranging livestock. In general, assistive and monitoring robots are the main pillars of present-day animal surveillance that lead to the enhancement of welfare through continuous, data-driven (Liang et al., 2023).

3.3 Farm and Livestock Robotics

Farm and livestock robotics represent one of the most sophisticated and deeply penetrated types of animal-interaction technology. Among the innovations in dairy farming, robotic milking systems are perhaps the most revolutionary. In these systems, the use of automated arms, a teat-cleaning mechanism and computer vision allows distributing a milking unit independently, hence, animals do not get stressed and milking routines become flexible. Also, the data recorded during milking, e.g., yield, flow rate and udder health, and facilitates early detection of mastitis, metabolic disorders and reproductive status. On top of that, the robotic milking systems through which cows are free to select their milking times promote independence and welfare gets better (Holloway et al., 2014).

An automated feeding robot will guarantee that the feed delivery for the cattle, poultry or other livestock is not only accurate and timely but also nutritionally optimized. The robots have the capability to change portions as per individual dietary requirements, thus helping in controlling growth rates, health conditions and productivity. By implementing a feeding robot, the problem of rivalry for food supplies is taken care of since animals are provided with equal access to feed (Zhang et al., 2022).

Cleaning robots play an essential role in maintaining the environment's hygiene by taking care of waste removal, bedding and ammonia level control. These actions that have been done directly for comfort purposes lower the risks of respiratory diseases, hoof infections and parasitic infestations. The fact that these tasks have been automated ensures the cleaning of the environment all the time, which is hardly achievable even manually (Megalingam et al., 2025).

Welfare sensors scattered around the barns and mostly installed in robots watch over the temperature, humidity, light intensity and air quality. Along with the help of cameras or thermal sensors fixed on a mobile robot, lameness, injury or even abnormal behaviour can be detected. Some highly developed mechanisms even utilize robotic herders for moving animals calmly from one place to another, thus, not only sorting or transportation but also stress alleviation is achieved (Özentürk et al., 2024).

On the whole, these robotic systems are geared towards a comprehensive method of precision livestock farming which is aimed at augmenting animal welfare, productivity and environmental sustainability, along with the labour easements and decision-making improvements for.

3.4 Research Robots in Behavioural Studies

In research, robots act as controllable and repeatable instruments to help scientists understand animal behaviour, cognition and communication. Biomimetic robots which are made to look like fish, birds, insects and mammals, help scientists to give the same stimuli to animals without the influence of human bias. For instance, robotic fish can swim with real fish to observe shoaling behaviour, whereas robotic birds can simulate courtship or show territorial behaviour (Jung et al., 2025).

These robots provide the means to precisely alter movement patterns, colours and signals, opening the way to understanding social interaction, sensory perception, decision-making and leadership in animal groups. Experimental control becomes better as robots act in the same way in all trials, unlike live stimulus animals, whose behaviour may change. Robots also help in experiments that would be very difficult or morally wrong if done with live animals, for example, testing predator-prey scenarios or introducing new social cues. In this way, research robots are a major factor in the progress of behavioural science, while they also lessen the sources of variation in (Tan & Zheng, 2013).

3.5 Wildlife Conservation Robots

Mitigating the effects of climate change on animals and plants and the whole ecosystem, wildlife conservation robots act as non-invasive tools that facilitate monitoring, protection and ecological research in harsh and often remote environments. For instance, anti-poaching robotic systems such as airborne drones and autonomous ground vehicles are equipped with thermal imaging, motion sensors and AI-driven pattern recognition to detect illegal activities (Pedrazzi et al., 2025). They can also cover vast terrains and the dependence on human patrols is reduced while the speed of the response is enhanced.

Drones are also used for population surveys, nest monitoring, habitat mapping and tracking migratory species. They minimize disturbance while at the same time, they provide high-resolution data, which would be very difficult or even dangerous for humans to obtain. Autonomous field sensors, sometimes embedded in robotic platforms, are always ready to collect acoustic, visual and environmental data of elusive or nocturnal species. Besides that, these devices are able to monitor habitat changes, human–wildlife conflict zones and breeding sites (Aliane, 2025).

By facilitating uninterrupted, real-time observation as well as rapid intervention, robots used in wildlife conservation are among the main contributors to the survival of biodiversity and the implementation of global conservation.

4. MECHANISMS OF INTERACTION: THE AI–SENSOR–ANIMAL FEEDBACK LOOP

4.1 Data Collection (Behaviour, Vocalization, Physiology)

Effective animal-robot collaboration is the result of thorough data gathering that reflects the behavioural, vocal and physiological aspects of the animal. To observe posture, movement patterns, feeding behaviour and social interactions, robots employ a variety of sensors like cameras, microphones, accelerometers, thermal imagers and biosensors. The analysis of vocalization offers a variety of clues to the identification of distress, joy, communication or reaction to the environment (Ding et al., 2025). Physiological monitoring, such as heart rate, temperature, respiration and stress markers, is aimed at welfare assessment at a deeper level. Environmental sensors can also help to explain animal behaviour by providing information about temperature, humidity and air quality. With the integration of these multimodal datasets, robots have a real-time understanding of an animal's condition and environment (Essien & Neethirajan, 2025). Such a data-rich base is indispensable for accurate interpretation of needs as well as for the implementation of welfare-conscious robotic behaviour that is.

4.2 Real-Time Processing and Decision-Making

After sensory data is collected, onboard processors and AI algorithms analyse it in real time to find patterns or changes in normal behaviour. Machine learning models identify different behaviours such as feeding, resting, aggression or play and an anomaly-detection system signals signs of illness, stress or environmental discomfort. Decision-making frameworks usually use reinforcement learning or rule-based logic to figure out the next robotic action, e.g., if it is to guide an animal, deliver enrichment, adjust environmental conditions and notify caregivers (Nazat et al., 2025). Temporal analysis allows the robot to follow the trends over time, thus increasing its predictive abilities. Real-time processing acts to make the robotic responses compatible with the animal's immediate context and supports safer, more informed interaction. Such quick understanding is very important in the case of dynamic environments where animals' behaviours are changing (Castro et al., 2021).

4.3 Adaptive Robotic Response and Behaviour Modulation

Robots utilize information that is processed to provide adaptive and emotionally intelligent responses, which either support or change the behaviour of the animals. The change may include the movement, sounds or even the temperature of the surroundings (Maroto-Gómez et al., 2025). For example, robots in companion settings can behave playfully or in a calming manner depending on the mood or emotional state detected. In environments with livestock or research, robots can lead animals, keep the right distances or even prevent the occurrence of some accidents (Fogelson et al., 2022). Adaptive control makes sure that the robot does not frighten the animals or shock them and is still predictable enough to gain trust. Eventually, learning algorithms get better and better at making the robot respond correctly for each individual based on their behaviour, thus creating personalized communication patterns. Such flexibility is the core through which human–robot–animal ecosystems can be effective and oriented towards the welfare of all the parties (Fernández-Rodicio et al., 2024).

4.4 Conceptual Diagram Explanation

The AI–Sensor–Animal feedback loop may be imagined as a never-ending chain: sensing → interpretation → robotic action → animal response → updated sensing. Sensors initially capture behavioural and physiological signals, which AI models interpret to determine needs or intentions. The robot then acts either physically or by changing the environment, thus eliciting an animal response. This new response is again sensed, thereby enabling the robot to adjust its actions. The cycle is an assurance that the interactions are fluid, lifelike and based on the latest animal feedback in time (Lin et al., 2023).

4.5 Challenges in Interpretation and Accuracy

Despite the progress made, understanding animal signals correctly is still a difficult task. Differences between species, breeds and individuals can cause signals to be misidentified or responses to be delayed. In noisy places, sounds or movements can be picked up incorrectly. Without support from the context, physiological signs may be hard to interpret. If too much trust is put in AI predictions, there is a danger of treating complex behaviours in a too simple way (Saarela & Podgorelec, 2024). Resolving these issues also involves having models for each species, using different kinds of sensors and always being.

5. CASE STUDIES

5.1 Robotic Milking: Europe vs. Asia

Robotic milking systems (RMS) have revolutionized dairy farms by automating milking tasks, raising welfare standards and continuously monitoring cow health in real-time. Nevertheless, their usage is quite different in Europe and Asia due to variations in farm sizes, labor costs and technological readiness (Gaworski & Kic, 2024; Hogenboom et al., 2019).

For instance, in Europe, countries such as the Netherlands, Denmark, and Germany extensively utilize RMS. European dairy farms are typically of medium size, well-managed and heavily technology-oriented; they provide the perfect setting for robotic systems. Farmers get rid of labor dependency and gain more freedom of work as cows come to the milking station on their own. This self-control decreases the stress level, increases comfort and facilitates natural behaviors (Barkema et al., 2015). European research consistently shows that the health of udders becomes better, somatic cell counts decrease and lameness or mastitis can be detected early through continuous sensor data. Also, there are some gains in productivity: RMS makes it possible to milk more frequently, which leads to an increased milk yield per cow and a stable lactation performance (Martin et al., 2022).

Meanwhile, the shift to RMS in Asia is progressing gradually. For instance, Japanese and South Korean farms have invested in robotic milking, which is prevalent in technologically advanced farms with a scarcity of labor (Lee et al., 2024). At the same time, large dairy-producing countries like India and China struggle with issues such as high costs, fragmented farm structures and diverse management practices (Yu et al., 2024). RMS capital costs hinder the huge number of smallholder farms that dominate most parts of Asia. Nevertheless, some Asian farmers where it is implemented say that the same welfare benefits appear less animal stress, better hygiene and more stable milking routines (Devendra & Thomas, 2002).

There is a great difference in production results between Europe and Asia, which is mainly due to management factors. Typically, European farms are able to exploit robotic systems to the fullest because they have well-designed facilities and animals are properly trained on the other hand, in Asia, integration may imply that barns need to be adapted, herds have to be homogenized and data management has to be reliable. RMS is progressively acknowledged in both areas as a very important instrument for dairy farming of the future that is welfare-friendly, precise in monitoring and sustainable even though there are these differences in welfare and productivity (Waddington et al., 2010).

5.2 Wildlife Drones for Conservation

Drones are increasingly being used in wildlife conservation as they provide an efficient and non-intrusive way of monitoring large areas or areas that are difficult to access. Drones are fitted with high-resolution cameras, thermal imaging and AI-based recognition systems to carry out the tasks of population surveys, anti-poaching patrols and habitat assessment (Pedrazzi et al., 2025). They are able to monitor endangered animals, keep an eye on nests, map migration routes and detect illegal encroachment more accurately than if done manually in the field. In areas where poaching is heavily practiced, drones are able to send up-to-the-minute information to enforcement teams, thus greatly cutting down their response time. Also, the fact that they are able to operate quietly and at a high altitude means that they do not cause any disturbance to wildlife. Drones, through their combination of aerial mobility and advanced analytics, offer conservationists cheap and scalable means of protecting the environment and managing wildlife in a data-driven (Aliane, 2025).

5.3 Biomimetic Fish/Bird Robots in Research

Biomimetic robots, which are built to imitate the look and movement of real animals, are a popular choice in behavioural research as they offer consistent, controllable stimuli. In water-related studies, fish-like robots are utilized to examine shoaling, leadership and collective decision-making. Such robots can imitate tail-beat frequencies, colour patterns and swimming trajectories, allowing researchers to find out how real fish respond to slight changes (Gao et al., 2019). In a similar way, bird-based robots, such as robotic decoys or wing-flapping models, are employed to investigate courtship behaviour, territoriality and social signalling in bird species. As these robots can be accurately programmed, they do away with the variability of live stimulus animals, enhancing experimental control (Cui et al., 2023). Biomimetic robots also make it possible to carry out ethically difficult experiments, for instance, predator-prey simulations, and at the same time, they lower the stress level of animals. Their involvement has been a major factor in the advancement of our comprehension of social cognition, communication and adaptive (Tamborini, 2025).

5.4 AI-Driven Domestic Pet Robots

AI-powered pet robots are becoming more popular as a means of making the existence of companion animals more delightful, particularly in homes where there is less human interaction. These robots utilize machine learning to identify behavioural patterns, vocal cues and activity levels of dogs, cats and small animals(Salma et al., 2025). According to the mood or energy that is sensed, they change the play patterns, movement speed or reward giving. A few are merely interactive toys that, through play, exercise and the alleviation of anxiety, become a living part of the pet's world, whereas others facilitate training by providing the same cues consistently. The most sophisticated models, in fact, combine cameras and sensors to give updates to the owners, allowing them to monitor their pets from a distance and also the first-hand recognition of behavioral and health changes. These robots are very effective in improving animal welfare as they deliver the necessary care and their engagement is always (Díaz-Boladeras, 2023).

6. ETHICAL, SOCIAL AND PRACTICAL CONSIDERATIONS

6.1 Animal Welfare and Behavioural Ethics

As robots become more and more a part of animal care, the main ethical concern has been how these technologies affect the behaviour, stress and general welfare of the animals. Robots should be bound to respect the needs that are specific to each species, the natural rhythms and the social structures. For instance, cows in the milking process have certain patterns for feeding, resting and social interaction; a robot that interferes with these cycles can make the animal stressed, socially displaced and less productive (d'Isa, 2025). Designing ethically calls for the implementation of animal care technologies that animals adapt to, not the other way around. Their motion, noise levels, the way of physical contact and distance must be considered to make the animal's fear response as low as possible. Just as important is the transparency of the intention to operate: robots should be facilitators of positive human–animal relationships, not replacers. Behavioural ethics also necessitate the assessment of the effects that will happen further down the road. A robot that is considered dependable can make the animal comfortable and less anxious, whereas one that acts unpredictably may cause the animals to develop avoidance or aggression as a response (Tzafestas, 2018). Therefore, welfare-centered design is the basis of the conditions under which the robots can be responsibly.

6.2 Data Privacy and Farm-Level Confidentiality

AI-powered robotic systems produce enormous datasets about animal health, productivity, environmental conditions and farm management practices. These data streams, while they allow for precision care, also bring up concerns about ownership, access and unintentional economic exposure. If sensitive information, such as disease prevalence, reproductive performance or feed efficiency, is misused, it can affect market value (Anastasiou et al., 2025). Farmers are concerned that manufacturers or third-party analytics companies may use the insights to implement pricing strategies, gain competitive intelligence and be scrutinized by regulatory authorities. Thus, it is necessary to ensure data sovereignty.

Secure storage, encryption, and transparent consent policies should be in place along with any robotic deployment. Moreover, anonymisation protocols and establishing trust through data sharing can ensure that trust is maintained. Proper data governance should give farmers the power to make decisions and ensure that the use of technology for their benefit does not lead to a breach of their (Ghafoor et al., 2025).

6.3 Human–Robot–Animal Acceptance and Trust

Successful integration of robot systems requires the faith of all those involved, the animals, the handlers and the owners in the technology. As animals are guided mainly by their senses, robots need to be characters that the animals already know, that do not scare them and are socially consistent. The acceptance of humans depends on factors like usefulness, ease of operation, and emotional comfort around automated devices. A robot that makes the workflow more efficient while being visibly safe and respectful towards animals will be trusted by the whole system (Pinto et al., 2025). Apart from that, training programs, interface simplicity and transparency in operation also contribute to the trust level. At the end of the day, acceptance comes about when robots help, instead of interrupting, the existing routines and thus, facilitate positive human–animal.

6.4 Risks of Dependency and Over-Automation

Heavy dependence on robotic systems can deteriorate human observational skills as well as the hands-on knowledge of animal behavior. Besides, over-automation may expose the system to various risks, like technical failures, software errors or power disruptions, which can stop the most important operations. Animals that are habituated to robotic routines may become confused when the systems break down and are taken for maintenance (Horvat et al., 2024). Additionally, a farm can be exposed to economic risks if increasing operational costs or limiting the flexibility become the consequences of long-term dependence. The use of robots in a balanced manner, where they supplement the work of a skilled human caretaker, helps to keep the management safe and oriented (Brandao, 2025).

6.5 Regulatory Frameworks

Rules for animals interacting with robots are still inconsistently spread out, with different sets of rules depending on the sector and the region. There is a great need for uniform rules addressing safety, welfare impact, data governance and operator responsibility. Well-defined systems will enable the implementation of ethical principles, guarantee the presence of a responsible body and provide safety not only to animals but also to users as these systems will be increasingly used in the two domains of the agricultural and the companion-animal (Bonnet et al., 2019).

7. LIMITATIONS AND CHALLENGES

Robotic animal interaction systems have the potential to revolutionize animal agriculture, conservation and companion-animal sectors. However, their development is still in the early stages and they face numerous challenges (**Table 1**). These challenges are of a technological nature, relate to the environment, interpretation and infrastructure. They define the extent to which AI-driven systems can work under real-world conditions.

Table 1. Applications vs. Benefits vs. Challenges of Robotic Systems in Animal Interaction

S. No.	Applications	Benefits	Challenges
1	Livestock Farming	Improved productivity, precision feeding and early disease detection	High cost, technical maintenance and farmer resistance
2	Behavioral Research	Continuous observation, reduced human disturbance and reliable long-term data	AI misinterpretation, dataset limitations and ethical considerations
3	Health Monitoring (All Settings)	Real-time vitals tracking, early disease prediction and automated alerts	Sensor calibration, data overload and cybersecurity risks
4	Companion Animal Care	Enrichment, training assistance, remote interaction and rehabilitation support	Adaptation difficulties, data privacy and limited emotional understanding
5	Wildlife Conservation	Non-invasive monitoring, anti-poaching surveillance and habitat mapping	Weather limitations, battery life and risk of disturbing wildlife

Technical costs are still among the major obstacles. The robots with multiple sensory arrays, navigation units and high-processing AI units require sizeable upfront and upkeep costs. Medium and small-scale farms, especially those in the developing world, are likely to find these expenses unmanageable. Besides the purchasing cost, there are also hidden costs such as software subscription, parts, calibration service and operator training that add up further. Economic viability frequently falls behind technological promises, notably in the areas where there are tight profit margins(Fascista, 2022).

Interpretation mistakes also affect performance. Models of AI-driven systems are extremely dependent on the quality, amount and uniformity of training data. The problem is that animal behavior varies across different species, breeds, habitats or even individual temperaments. Inappropriately handling the robotic response due to wrong interpretation of behaviour (like mistaking stress, curiosity, or aggression) can lead to a decrease of animals' welfare and operational efficiency. The presence of noise in the sensor data, obstructions in vision systems or ambiguous behaviour patterns makes it difficult to distinguish between false positives and negatives(-Nessler, 2025). Despite the progress made in the field of machine learning, getting accurate results in all environments is still a problem.

The scope of operations is limited due to environmental reasons. Exposure to dust, mud, temperature extremes, uneven terrain and unpredictable behavior patterns of animals and humans in the open agricultural and wildlife settings may adversely affect the performance of robots. Extreme weather can also affect battery life, sensor performance and movement stability(Anderson & Donath, 1990). On the other hand, for underwater or aerial systems, factors such as wind, turbidity and vegetation density may influence navigation or data capture. Building tough, weatherproof platforms adds to the cost and complications of engineering.

Data interoperability issues present another major obstacle. Different manufacturers supply these systems, which use incompatible formats, proprietary software and closed ecosystems. The fragmentation of this data prevents the integration of sensors, robots and farm-management platforms data in a seamless manner. Consequently, users are unable to access integrated dashboards or perform long-term analytics, which is essential for the use of multi-robot or multi-sensor systems(Dewberry et al., 2025). Although there are some standardization initiatives in place, they are far from completion.

Together, these restrictions highlight the necessity of more advanced engineering, global data standards, cost-efficient models and situationally aware AI before robot systems can be implemented in a large scale in the various fields related to.

8. FUTURE DIRECTIONS

The next generation of robotic systems for animal interaction is expected to be more intelligent, responsive and sensitive to the context, ultimately leading to welfare, sustainability and accessibility improvement in different sectors. Through these, animals and robots will not only live together but also work together in a harmonious way.

One of the most exciting areas is emotion-aware robotics. The main components of this technology are affective computing, multimodal sensing, and behavioural analytics. These components help robots to figure out the minute emotional changes of animals. In the future, robots may use a combination of thermal imaging, vocalization analysis, micro-expressions and movement signatures to create more empathetic, welfare-oriented interactions. Robots, in such cases, may change their voice, distance, speed or action pattern instantly depending on the animal's emotional state.

Biological and robotic hybrids will open up new opportunities. Such robots could have cells resembling those of an organism or components that move like an organ in a body. The scientists could use bio-inspired materials for building robots in such a way that they interact with living tissues safely. Besides, a system of small robots working with animals to do things like pollination, navigation or behaviour research is another option that could be possible in the future. With the help of soft robotics technology and biological signals, the interactions will be more natural and safer.

Edge AI and affordable solutions are going to be the main contributors in the process of making robot technology accessible to everyone. Upgrading processing on the device, using energy-efficient chips and promoting open-source software are some of the ways in which the costs can be lowered and the need for cloud connection can be removed. By these, it will be possible to do precision animal monitoring in small farms, shelters and areas for conservation even if they are far away.

In the end, sustainability and worldwide usage will be the main factors determining the locations of the upcoming robotic systems for animal interaction. The focus will be on recyclable materials, systems consuming little energy and designs adaptable to different climates and economic conditions. Thus, the integration of robot systems worldwide will be made possible through a network of partners, including governments, NGOs and industry, who will provide the services to all, regardless of class.

9. CONCLUSION

Animal-interactive robotic systems have become intelligent, adaptive and multisensory tools that are transforming animal care, research and conservation. The chapter explored their technological base, usage scenarios, methods of interaction, contactless research and the burning ethical questions concerning the use of such technology. To date, robots are used on farms, in laboratories, homes and the wild to increase precision, lower labour, upgrade monitoring and make possible more humane, data-driven management. Major findings indicate that robotics is a powerful instrument for animal welfare as they enables continuous behaviour tracking, early distress detection, individualized care and less invasive research. Emotion-aware technologies, sophisticated sensors and AI-driven interpretation provide more profound insight into the animal's emotional and mental states and enable more interactive responses. On the way ahead, moral fairness, data security, cost-effectiveness and environmental-friendliness will be the main factors for the responsible integration of robotics. The use of robotics should be a human–animal relationship complement, not a substitute. Robotic technologies, if well-designed and widely accepted, may facilitate the establishment of a more compassionate, efficient and scientifically informed future of global animal.

ACKNOWLEDGMENT

I very much appreciate the support of my workmates for the stimulating discussions and the researchers whose contributions to robotics, animal welfare and conservation have been a source of my inspiration in writing this chapter. I am full of thanks to my family and friends for their long-suffering, cheerful support and unshakable faith, which have been instrumental in the realization of this.

REFERENCES

Afzal, N., & Rehman, M. ur, Seneviratne, L., & Hussain, I. (2025). The Convergence of AI and animal-inspired robots for ecological conservation. *Ecological Informatics*, *85*, 102950. DOI: 10.1016/j.ecoinf.2024.102950

Aliane, N. (2025). Drones and AI-Driven Solutions for Wildlife Monitoring. *Drones (Basel)*, *9*(7), 455. DOI: 10.3390/drones9070455

Anastasiou, E., Kasimati, A., Papadopoulos, G., Vatsanidou, A., Gemtou, M., Schwierz, F., Meyer-Aurich, A., & Fountas, S. (2025). Assessing the agroecological impact of digital tools in livestock production: A systematic review. *Smart Agricultural Technology*, *12*, 101480. DOI: 10.1016/j.atech.2025.101480

Anderson, T. L., & Donath, M. (1990). Animal behavior as a paradigm for developing robot autonomy. *Robotics and Autonomous Systems*, *6*(1), 145–168. DOI: 10.1016/S0921-8890(05)80033-8

Aoi, S., Manoonpong, P., Ambe, Y., Matsuno, F., & Wörgötter, F. (2017). Adaptive Control Strategies for Interlimb Coordination in Legged Robots: A Review. *Frontiers in Neurorobotics*, *11*. Advance online publication. DOI: 10.3389/fnbot.2017.00039

Barkema, H. W., von Keyserlingk, M. A. G., Kastelic, J. P., Lam, T. J. G. M., Luby, C., Roy, J.-P., LeBlanc, S. J., Keefe, G. P., & Kelton, D. F. (2015). *Invited review:* Changes in the dairy industry affecting dairy cattle health and welfare. *Journal of Dairy Science*, *98*(11), 7426–7445. DOI: 10.3168/jds.2015-9377

Bonnet, F., Mills, R., Szopek, M., Schönwetter-Fuchs, S., Halloy, J., Bogdan, S., Correia, L., Mondada, F., & Schmickl, T. (2019). Robots mediating interactions between animals for interspecies collective behaviors. *Science Robotics*, *4*(28), eaau7897. Advance online publication. DOI: 10.1126/scirobotics.aau7897

Brandao, P. R. (2025). The Impact of Artificial Intelligence on Modern Society. *AI*, *6*(8), 190. DOI: 10.3390/ai6080190

Burnham, R. (2023). Animal Calling Behaviours and What This Can Tell Us about the Effects of Changing Soundscapes. *Acoustics (Basel, Switzerland)*, *5*(3), 631–652. DOI: 10.3390/acoustics5030039

Castro, A., Silva, F., & Santos, V. (2021). Trends of Human-Robot Collaboration in Industry Contexts: Handover, Learning, and Metrics. *Sensors (Basel)*, *21*(12), 4113. DOI: 10.3390/s21124113

Cui, Z., Li, L., Wang, Y., Zhong, Z., & Li, J. (2023). Review of research and control technology of underwater bionic robots. *Intelligent Marine Technology and Systems*, *1*(1), 7. DOI: 10.1007/s44295-023-00010-3

d'Isa, R. (2025). Robotic animals as new tools in rodent neuroscience research: Proposed applications of zooinspired robots for mouse behavioral testing. *Frontiers in Behavioral Neuroscience*, *19*, 1545352. DOI: 10.3389/fnbeh.2025.1545352

Devendra, C., & Thomas, D. (2002). Smallholder farming systems in Asia. *Agricultural Systems*, *71*(1), 17–25. DOI: 10.1016/S0308-521X(01)00033-6

Dewberry, N. K., AlHmoud, I., Benton, K., Suarez, D., Chen, Y.-P., Karkaria, V., Tsai, Y.-K., Brock, M., Alazzawi, N., Chowdhury, S., Chen, W., Cao, J., & Gokaraju, B. (2025). A real-time VR-enabled digital twin framework for multi-user interaction in Industry 4.0. *Manufacturing Letters*, *44*, 1486–1497. DOI: 10.1016/j.mfglet.2025.06.168

Díaz-Boladeras, M. (2023). Bond formation with pet-robots: An integrative approach. *Current Psychology (New Brunswick, N.J.)*, *42*(4), 2591–2608. DOI: 10.1007/s12144-022-03792-7

Ding, L., Zhang, C., Yue, Y., Yao, C., Li, Z., Hu, Y., Yang, B., Ma, W., Yu, L., Gao, R., & Li, Q. (2025). Wearable Sensors-Based Intelligent Sensing and Application of Animal Behaviors: A Comprehensive Review. *Sensors (Basel)*, *25*(14), 4515. DOI: 10.3390/s25144515

Essien, D., & Neethirajan, S. (2025). Multimodal AI systems for enhanced laying hen welfare assessment and productivity optimization. *Smart Agricultural Technology*, *12*, 101564. DOI: 10.1016/j.atech.2025.101564

Fascista, A. (2022). Toward Integrated Large-Scale Environmental Monitoring Using WSN/UAV/Crowdsensing: A Review of Applications, Signal Processing, and Future Perspectives. *Sensors (Basel)*, *22*(5), 1824. DOI: 10.3390/s22051824

Fernández-Rodicio, E., Castro-González, Á., Gamboa-Montero, J. J., Carrasco-Martínez, S., & Salichs, M. A. (2024). Creating Expressive Social Robots That Convey Symbolic and Spontaneous Communication. *Sensors (Basel)*, *24*(11), 3671. DOI: 10.3390/s24113671

Fogelson, D. M., Rutledge, C., & Zimbro, K. S. (2022). The Impact of Robotic Companion Pets on Depression and Loneliness for Older Adults with Dementia During the COVID-19 Pandemic. *Journal of Holistic Nursing*, *40*(4), 397–409. DOI: 10.1177/08980101211064605

Gao, Z., Shi, Q., Fukuda, T., Li, C., & Huang, Q. (2019). An overview of biomimetic robots with animal behaviors. *Neurocomputing*, *332*, 339–350. DOI: 10.1016/j.neucom.2018.12.071

Gaworski, M., & Kic, P. (2024). Assessment of Production Technologies on Dairy Farms in Terms of Animal Welfare. *Applied Sciences (Basel, Switzerland)*, *14*(14), 6086. DOI: 10.3390/app14146086

Ghafoor, A., Symeonidis, I., Rydberg, A., Lindahl, C., & Abbasi, A. Q. (2025). Towards Empowering Stakeholders Through Decentralized Trust and Secure Livestock Data Sharing. *Cryptography*, *9*(3), 52. DOI: 10.3390/cryptography9030052

Ghavi Hossein-Zadeh, N. (2025). Artificial intelligence in veterinary and animal science: Applications, challenges, and future prospects. *Computers and Electronics in Agriculture*, *235*, 110395. DOI: 10.1016/j.compag.2025.110395

Guiochet, J., Machin, M., & Waeselynck, H. (2017). Safety-critical advanced robots: A survey. *Robotics and Autonomous Systems*, *94*, 43–52. DOI: 10.1016/j.robot.2017.04.004

Hogenboom, J. A., Pellegrino, L., Sandrucci, A., Rosi, V., & D'Incecco, P. (2019). *Invited review:* Hygienic quality, composition, and technological performance of raw milk obtained by robotic milking of cows. *Journal of Dairy Science*, *102*(9), 7640–7654. DOI: 10.3168/jds.2018-16013

Holloway, L., Bear, C., & Wilkinson, K. (2014). Re-capturing bovine life: Robot–cow relationships, freedom and control in dairy farming. *Journal of Rural Studies*, *33*, 131–140. DOI: 10.1016/j.jrurstud.2013.01.006

Horsevad, N., Kwa, H. L., & Bouffanais, R. (2022). Beyond Bio-Inspired Robotics: How Multi-Robot Systems Can Support Research on Collective Animal Behavior. *Frontiers in Robotics and AI*, *9*. Advance online publication. DOI: 10.3389/frobt.2022.865414

Horvat, D., Ivanišević, R., & Gluščević, L. (2024). Risks associated with robotic process automation. *Journal of Process Management and New Technologies*, *12*(1–2), 72–82. DOI: 10.5937/jpmnt12-50617

Jiang, T., Sun, Z., Fu, S., & Lv, Y. (2024). Human-AI interaction research agenda: A user-centered perspective. *Data and Information Management*, *8*(4), 100078. DOI: 10.1016/j.dim.2024.100078

Jung, H., Park, S., Joe, S., Woo, S., Choi, W., & Bae, W. (2025). AI-Driven Control Strategies for Biomimetic Robotics: Trends, Challenges, and Future Directions. *Biomimetics*, *10*(7), 460. DOI: 10.3390/biomimetics10070460

Kóczi, D., & Sárosi, J. (2025). Safety Engineering for Humanoid Robots in Everyday Life—Scoping Review. *Electronics (Basel)*, *14*(23), 4734. DOI: 10.3390/electronics14234734

Kotha, S. S., Akter, N., Abhi, S. H., Das, S. K., & Islam, Md. R., Ali, Md. F., Ahamed, Md. H., Islam, Md. M., Sarker, S. K., Badal, Md. F. R., Das, P., Tasneem, Z., & Hasan, Md. M. (2024). Next generation legged robot locomotion: A review on control techniques. *Heliyon*, *10*(18), e37237. DOI: 10.1016/j.heliyon.2024.e37237

Lamanna, M., Bovo, M., & Cavallini, D. (2025). Wearable Collar Technologies for Dairy Cows: A Systematized Review of the Current Applications and Future Innovations in Precision Livestock Farming. *Animals (Basel)*, *15*(3), 458. DOI: 10.3390/ani15030458

Lee, Y.-G., Han, K., Chung, C., & Ji, I. (2024). Effects of Smart Farming on the Productivity of Korean Dairy Farms: A Case Study of Robotic Milking Systems. *Sustainability*, *16*(22), 9991. DOI: 10.3390/su16229991

Li, W., Hu, D., & Yang, L. (2023). Actuation Mechanisms and Applications for Soft Robots: A Comprehensive Review. *Applied Sciences (Basel, Switzerland)*, *13*(16), 9255. DOI: 10.3390/app13169255

Liang, H., Lee, S.-C., Bae, W., Kim, J., & Seo, S. (2023). Towards UAVs in Construction: Advancements, Challenges, and Future Directions for Monitoring and Inspection. *Drones (Basel)*, *7*(3), 202. DOI: 10.3390/drones7030202

Lin, C., Wang, J., & Ji, L. (2023). An AI-based Wild Animal Detection System and Its Application. *Biodiversity Information Science and Standards*, *7*, e112456. DOI: 10.3897/biss.7.112456

Mahato, S., & Neethirajan, S. (2025). Integrating Artificial Intelligence in dairy farm management – biometric facial recognition for cows. *Information Processing in Agriculture*, *12*(3), 312–325. DOI: 10.1016/j.inpa.2024.10.001

Mansoor, S., Iqbal, S., Popescu, S. M., Kim, S. L., Chung, Y. S., & Baek, J.-H. (2025). Integration of smart sensors and IOT in precision agriculture: Trends, challenges and future prospectives. *Frontiers in Plant Science*, *16*, 1587869. DOI: 10.3389/fpls.2025.1587869

Maroto-Gómez, M., Álvarez-Arias, S., Rodríguez-Huelves, J., Segura-Bencomo, A., & Malfaz, M. (2025). A Robot Companion with Adaptive Object Preferences and Emotional Responses Enhances Naturalness in Human–Robot Interaction. *Electronics (Basel)*, *14*(18), 3711. DOI: 10.3390/electronics14183711

Martin, T., Gasselin, P., Hostiou, N., Feron, G., Laurens, L., Purseigle, F., & Ollivier, G. (2022). Robots and transformations of work in farm: A systematic review of the literature and a research agenda. *Agronomy for Sustainable Development*, *42*(4), 66. DOI: 10.1007/s13593-022-00796-2

Megalingam, R. K., Vadivel, S. R. R., Kotaprolu, S. S., Nithul, B., Kumar, D. V., & Rudravaram, G. (2025). Cleaning Robots: A Review of Sensor Technologies and Intelligent Control Strategies for Cleaning. *Journal of Field Robotics*, *42*(5), 2234–2259. DOI: 10.1002/rob.22515

Miles, A., Fields, N. L., Bennett, M., Xu, L., Magruder, K., Stringfellow, M. K., Sesay, B. J., & Handique, S. (2024). Robotic Animal Use among Older Adults Enrolled in Palliative or Hospice Care: A Scoping Review and Framework for Future Research. *Robotics (Basel, Switzerland)*, *13*(6), 92. DOI: 10.3390/robotics13060092

Nazat, S., Alayed, W., Li, L., & Abdallah, M. (2025). Ensemble Learning Framework for Anomaly Detection in Autonomous Driving Systems. *Sensors (Basel)*, *25*(16), 5105. DOI: 10.3390/s25165105

Neethirajan, S., Scott, S., Mancini, C., Boivin, X., & Strand, E. (2024). Human-computer interactions with farm animals—Enhancing welfare through precision livestock farming and artificial intelligence. *Frontiers in Veterinary Science*, *11*, 1490851. DOI: 10.3389/fvets.2024.1490851

Nessler, J. N. (2025). Editorial: Artificial intelligence in animal behaviour, veterinary behaviour and neurology. *Frontiers in Veterinary Science*, *12*, 1607623. DOI: 10.3389/fvets.2025.1607623

Özentürk, U., Chen, Z., Jamone, L., & Versace, E. (2024). Robotics for poultry farming: Challenges and opportunities. *Computers and Electronics in Agriculture*, *226*, 109411. DOI: 10.1016/j.compag.2024.109411

Pedrazzi, L., Naik, H., Sandbrook, C., Lurgi, M., Fürtbauer, I., & King, A. J. (2025). Advancing animal behaviour research using drone technology. *Animal Behaviour, 222*, 123147. DOI: 10.1016/j.anbehav.2025.123147

Pinto, A., Solovov, A., Simões, A. C., & Menezes, P. (2025). User Acceptance in Human–Robot Interaction: Exploring the Role of Anthropomorphic Mechanisms in Manufacturing Environments—A Systematic Literature Review. *Journal of Intelligent & Robotic Systems, 111*(2), 46. DOI: 10.1007/s10846-025-02248-x

Romano, D., Porfiri, M., Zahadat, P., & Schmickl, T. (2024). Animal–robot interaction—An emerging field at the intersection of biology and robotics. *Bioinspiration & Biomimetics, 19*(2), 020201. DOI: 10.1088/1748-3190/ad2086

Rudroff, T. (2024). Artificial Intelligence as a Replacement for Animal Experiments in Neurology: Potential, Progress, and Challenges. *Neurology International, 16*(4), 805–820. DOI: 10.3390/neurolint16040060

Saarela, M., & Podgorelec, V. (2024). Recent Applications of Explainable AI (XAI): A Systematic Literature Review. *Applied Sciences (Basel, Switzerland), 14*(19), 8884. DOI: 10.3390/app14198884

Salma, Z., Hijón-Neira, R., Pizarro, C., & Abdul Moqeet, A. (2025). Effectiveness of Robot-Mediated Learning in Fostering Children's Social and Cognitive Development. *Applied Sciences (Basel, Switzerland), 15*(7), 3567. DOI: 10.3390/app15073567

Tamborini, M. (2025). The ethics of bioinspired animal-robot interaction: A relational meta-ethical approach. *Journal of Responsible Technology, 22*, 100116. DOI: 10.1016/j.jrt.2025.100116

Tan, Y., & Zheng, Z. (2013). Research Advance in Swarm Robotics. *Defence Technology, 9*(1), 18–39. DOI: 10.1016/j.dt.2013.03.001

Tulsulkar, G., Mishra, N., Thalmann, N. M., Lim, H. E., Lee, M. P., & Cheng, S. K. (2021). Can a humanoid social robot stimulate the interactivity of cognitively impaired elderly? A thorough study based on computer vision methods. *The Visual Computer, 37*(12), 3019–3038. DOI: 10.1007/s00371-021-02242-y

Tzafestas, S. G. (2018). Roboethics: Fundamental Concepts and Future Prospects. *Information (Basel), 9*(6), 148. DOI: 10.3390/info9060148

Voysey, I., Baillie, L., Williams, J., & Herrmann, M. (2025). Co-Designing Educational Robotic Pet Concepts with Animal Welfare Educators and Children. *International Journal of Social Robotics*. Advance online publication. DOI: 10.1007/s12369-025-01328-0

Waddington, S. R., Li, X., Dixon, J., Hyman, G., & de Vicente, M. C. (2010). Getting the focus right: Production constraints for six major food crops in Asian and African farming systems. *Food Security*, *2*(1), 27–48. DOI: 10.1007/s12571-010-0053-8

Waworundeng, J. (2024). IoT-based Environmental Monitoring with Data Analysis of Temperature, Humidity, and Air Quality. *CogITo Smart Journal*, *10*(1), 271–284. DOI: 10.31154/cogito.v10i1.708.692-705

Yerebakan, M. O., & Hu, B. (2024). Human–Robot Collaboration in Modern Agriculture: A Review of the Current Research Landscape. *Advanced Intelligent Systems*, *6*(7), 2300823. DOI: 10.1002/aisy.202300823

Yu, Z., Feng, G., Liu, H., Peng, H., & Dong, X. (2024). Sustainable market? The impact of downstream market concentration on high-quality agricultural development: Evidence from China's dairy industry. *Frontiers in Sustainable Food Systems*, *8*. Advance online publication. DOI: 10.3389/fsufs.2024.1453115

Zhang, Y., Sun, W., Yang, J., Wu, W., Miao, H., & Zhang, S. (2022). An Approach for Autonomous Feeding Robot Path Planning in Poultry Smart Farm. *Animals (Basel)*, *12*(22), 3089. DOI: 10.3390/ani12223089

Zhao, Z., Yang, Q., Li, R., Yang, J., Liu, Q., Zhu, B., Weng, C., Liu, W., Hu, P., Ma, L., Qiao, J., Xu, M., & Tian, H. (2024). A comprehensive review on the evolution of bio-inspired sensors from aquatic creatures. *Cell Reports. Physical Science*, *5*(7), 102064. DOI: 10.1016/j.xcrp.2024.102064

Compilation of References

Aarif, K. O. M., Alam, A., & Hotak, Y. (2025). Smart sensor technologies shaping the future of precision agriculture: Recent advances and future outlooks. *Journal of Sensors*, *2025*(1), 2460098.

Abdulhussain, S. H., Mahmmod, B. M., Alwhelat, A., Shehada, D., Shihab, Z. I., Mohammed, H. J., & Hussain, A. (2025). A comprehensive review of sensor technologies in IOT: Technical aspects, challenges, and future directions. *Computers*, *14*(8), 342.

Abrol, D. P. (2011). Biochemical basis of plant-pollination interaction. In *Pollination biology: biodiversity conservation and agricultural production* (pp. 413–458). Springer Netherlands.

Abtin, S., Seyedaghamiri, F., Aalidaeijavadi, Z., Farrokhi, A. M., Moshrefi, F., Ziveh, T., & Haghparast, A. (2024). A review on the consequences of molecular and genomic alterations following exposure to electromagnetic fields: Remodeling of neuronal network and cognitive changes. *Brain Research Bulletin*, *217*, 111090.

Adeel, A., Gogate, M., & Hussain, A. (2020). Contextual deep learning-based audio-visual switching for speech enhancement in real-world environments. *Information Fusion*, *59*, 163–170.

Afzal, N., & Rehman, M. ur, Seneviratne, L., & Hussain, I. (2025). The Convergence of AI and animal-inspired robots for ecological conservation. *Ecological Informatics*, *85*, 102950. DOI: 10.1016/j.ecoinf.2024.102950

Agbese, M., Mohanani, R., Khan, A. A., & Abrahamsson, P. (2023). Implementing AI Ethics. *Making Sense of the Ethical Requirements.*, *62*. Advance online publication. DOI: 10.1145/3593434.3593453

Ajay, N., Upadhyaya, A. N., Debnath, P., Rani, P., Selvakumar, P., Manjunath, T. C., & Bhattacharya, S. (2025). AI-Based Disaster Prediction and Early Warning Systems. In Mobo, F. (Ed.), *AI and Emerging Technologies for Emergency Response and Smart Cities* (pp. 55–74). IGI Global Scientific Publishing.

Akman, E. (2025). The impact of AI-based visual designs on students' AI literacy and attitudes toward AI. *Interactive Learning Environments*, •••, 1–19.

Aldad, T. S., Gan, G., Gao, X. B., & Taylor, H. S. (2012). Fetal radiofrequency radiation exposure from 800-1900 mhz-rated cellular telephones affects neurodevelopment and behavior in mice. *Scientific Reports*, *2*, 312.

Alharthi, A., Alqurashi, A., Alharbi, T., Alammar, M., Aldosari, N., Bouchekara, H., & Al Ayidh, A. (2025). Explainable AI for sensor signal interpretation to revolutionize human health monitoring: A review. *IEEE Access : Practical Innovations, Open Solutions*.

Aliane, N. (2025). Drones and AI-Driven Solutions for Wildlife Monitoring. *Drones (Basel)*, *9*(7), 455. DOI: 10.3390/drones9070455

Allen, A. N., Harvey, M., Harrell, L., Jansen, A., Merkens, K. P., Wall, C. C., & Oleson, E. M. (2021). A convolutional neural network for automated detection of humpback whale song in a diverse, long-term passive acoustic dataset. *Frontiers in Marine Science*, *8*, 607321.

American Bar Association. (2025). Bridging species: Legal and ethical implications of AI- assisted animal communication. *Tort Trial & Insurance Practice Law Journal*. https://www.americanbar.org/groups/tort_trial_insurance_practice/resources/brief/202 5-winter/legal-ethical-implications-ai-assisted-animal-communication/

Anastasiou, E., Kasimati, A., Papadopoulos, G., Vatsanidou, A., Gemtou, M., Schwierz, F., Meyer-Aurich, A., & Fountas, S. (2025). Assessing the agroecological impact of digital tools in livestock production: A systematic review. *Smart Agricultural Technology*, *12*, 101480. DOI: 10.1016/j.atech.2025.101480

Andersen, P. D., Jørgensen, B. H., Lading, L., & Rasmussen, B. (2004). Sensor foresight—Technology and market. *Technovation*, *24*(4), 311–320.

Anderson, T. L., & Donath, M. (1990). Animal behavior as a paradigm for developing robot autonomy. *Robotics and Autonomous Systems*, *6*(1), 145–168. DOI: 10.1016/S0921-8890(05)80033-8

Andreassen, T., Surlykke, A., & Hallam, J. (2014). Semi-automatic long-term acoustic surveying: A case study with bats. *Ecological Informatics*, *21*, 13–24.

Aoi, S., Manoonpong, P., Ambe, Y., Matsuno, F., & Wörgötter, F. (2017). Adaptive Control Strategies for Interlimb Coordination in Legged Robots: A Review. *Frontiers in Neurorobotics, 11*. Advance online publication. DOI: 10.3389/fnbot.2017.00039

Arac, A., Zhao, P., Dobkin, B. H., Carmichael, S. T., & Golshani, P. (2019). Deep-Behavior: A Deep Learning Toolbox for Automated Analysis of Animal and Human Behavior Imaging Data. *Frontiers in Systems Neuroscience, 13*, 20. DOI: 10.3389/fnsys.2019.00020 PMID: 31133826

Arkow, P. (1998). Application of ethics to animal welfare. *Applied Animal Behaviour Science, 59*(1-3), 193–200.

Arsawan, I. W. E., Koval, V., Rajiani, I., Rustiarini, N. W., Supartha, W. G., & Suryantini, N. P. S. (2022). Leveraging knowledge sharing and innovation culture into SMEs sustainable competitive advantage. *International Journal of Productivity and Performance Management, 71*(2), 405–428.

Assom, L. (2023). *Decoding communication of non-human species: Unsupervised machine learning to infer syntactical and temporal patterns in fruit-bats vocalizations.*

Ayyaswamy, K., Gobinath, V. M., Kathirvel, N., & Maheswaran, C. P. (2025). AI-Based Automated Disease Detection Based on Symptoms Healthcare for Pets and Birds. In AI and the Revival of Big Data (pp. 311-346). IGI Global Scientific Publishing.

Babiker, A., Alshakhsi, S., Al-Thani, D., Montag, C., & Ali, R. (2025). Attitude towards AI: Potential influence of conspiracy belief, XAI experience and locus of control. *International Journal of Human-Computer Interaction, 41*(13), 7939–7951.

Babu, G. C. N., Gupta, S., Bhambri, P., Leo, L. M., Rao, B. H., & Kumar, S. (2021). A semantic health observation system development based on the IoT sensors. *Turkish Journal of Physiotherapy and Rehabilitation, 32*(3), 1721–1729.

Bakker, K. (2022). *The sounds of life: How digital technology is bringing us closer to the worlds of animals and plants.*

Barkema, H. W., von Keyserlingk, M. A. G., Kastelic, J. P., Lam, T. J. G. M., Luby, C., Roy, J.-P., LeBlanc, S. J., Keefe, G. P., & Kelton, D. F. (2015). *Invited review:* Changes in the dairy industry affecting dairy cattle health and welfare. *Journal of Dairy Science, 98*(11), 7426–7445. DOI: 10.3168/jds.2015-9377

Barker, S. B., & Wolen, A. R. (2008). The benefits of human–companion animal interaction: A review. *Journal of Veterinary Medical Education, 35*(4), 487–495. PMID: 19228898

Barton, M.-C., & Pöppelbuß, J. (2022). Prinzipien für die ethische Nutzung künstlicher Intelligenz. *HMD Praxis der Wirtschaftsinformatik*, *59*(2), 468. DOI: 10.1365/s40702-022-00850-3

Bauer, E., Greiff, S., Graesser, A. C., Scheiter, K., & Sailer, M. (2025). Looking beyond the hype: Understanding the effects of AI on learning. *Educational Psychology Review*, *37*(2), 45.

Beecher, M. D. (2021). Why are no animal communication systems simple languages? *Frontiers in Psychology*, *12*, 602635. PMID: 33815200

Benneth, E. O. (2026). *Introduction: Principles of animal behavior.* In *Principles of Animal Behaviour* (pp. 1–15). CRC Press.

Bergdahl, N., & Sjöberg, J. (2025). Attitudes, perceptions and AI self-efficacy in K-12 education. *Computers and Education: Artificial Intelligence*, *8*, 100358.

Bernardes, R. C., Lima, M. A. P., Guedes, R. N. C., Da Silva, C. B., & Martins, G. F. (2021). Ethoflow: Computer Vision and Artificial Intelligence-Based Software for Automatic Behavior Analysis. *Sensors (Basel)*, *21*(9), 3237. DOI: 10.3390/s21093237 PMID: 34067084

Bhat, S., & Hortal, E. (2021). GAN-Based Data Augmentation For Improving The Classification Of EEG Signals. DOI: 10.1145/3453892.3461338

Bhuiyan, M. R., & Wree, P. (2023). Animal Behavior for Chicken Identification and Monitoring the Health Condition Using Computer Vision: A Systematic Review. *IEEE Access : Practical Innovations, Open Solutions*, *11*, 126601–126610. DOI: 10.1109/ACCESS.2023.3331092

Bianco, M. J., Gerstoft, P., Traer, J., Ozanich, E., Roch, M. A., Gannot, S., & Deledalle, C. A. (2019). Machine learning in acoustics: Theory and applications. *The Journal of the Acoustical Society of America*, *146*(5), 3590–3628. PMID: 31795641

Bonnet, F., Mills, R., Szopek, M., Schönwetter-Fuchs, S., Halloy, J., Bogdan, S., Correia, L., Mondada, F., & Schmickl, T. (2019). Robots mediating interactions between animals for interspecies collective behaviors. *Science Robotics*, *4*(28), eaau7897. Advance online publication. DOI: 10.1126/scirobotics.aau7897

Borotschnig, H. (2025). Emotions in Artificial Intelligence. DOI: 10.48550/ARXIV.2505.01462

Bouji, M., Lecomte, A., Gamez, C., Blazy, K., & Villégier, A. S. (2016). Neurobiological effects of repeated radiofrequency exposures in male senescent rats. *Biogerontology*, *17*(5), 841–857.

Boulanger, M. D., Miyamae, J. A., Martin, T., Hish, G., & Moore, T. Y. (2025). *Computer Vision for Lesser Egyptian Jerboa (* Jaculus jaculus *) Behavioral Analysis and Animal Care Refinement. Animal Behavior and Cognition.* Advance online publication. DOI: 10.1101/2025.07.30.667603

Bradshaw, G. A. (2021). *Elephants on the edge: What animals teach us about humanity*. Yale University Press.

Brady, A. P., Brkljačić, B., McNulty, J., Loewe, C., Coffey, M., & Visvikis, D. & European Society of Radiology, in Conjunction with ESTRO, EFRS, EFOMP and EANM. (2025). A project to define standards for radiology staffing, education and training across the European Union. *Insights Into Imaging*, *16*(1), 56.

Brandao, P. R. (2025). The Impact of Artificial Intelligence on Modern Society. *AI*, *6*(8), 190. DOI: 10.3390/ai6080190

Brandy, S. (2023). Overcoming challenges and unlocking the potential: Empowering Small and Medium Enterprises (SMEs) with data analytics solutions. *International Journal of Information Technology and Computer Science Applications*, *1*(3), 150–160.

Brickson, L., Vollrath, F., & Titus, A. J. (2023). Elephants and Algorithms: A Review of the Current and Future Role of AI in Elephant Monitoring [Review of Elephants and Algorithms: A Review of the Current and Future Role of AI in Elephant Monitoring]. arXiv (Cornell University). Cornell University. /arxiv.2306.13803DOI: 10.48550

Brown, K. (2012). *Little sinners, and other stories*. University of Nebraska Press.

Brugarolas, R., Lee, J., & Martins, N. R. B. (2020). Wearable devices for animal health monitoring. *Frontiers in Veterinary Science*, *7*(63), 1–11. DOI: 10.3389/fvets.2020.00063

Brumm, H., & Slabbekoorn, H. (2005). Acoustic communication in noise. *Advances in the Study of Behavior*, *35*, 151–209.

Buckler, B. (1996). A learning process model to achieve continuous improvement and innovation. *The Learning Organization*, *3*(3), 31–39.

Burnham, R. (2023). Animal Calling Behaviours and What This Can Tell Us about the Effects of Changing Soundscapes. *Acoustics (Basel, Switzerland)*, *5*(3), 631–652. DOI: 10.3390/acoustics5030039

Cakir, G., Bezbradica, M., & Helfert, M. (2019, May). The Shift from financial to non-financial measures during transition into digital retail–a systematic literature review. In *International conference on business information systems* (pp. 189-200). Cham: Springer International Publishing.

Cambraia, A., & Pyrrho, S. (2025). Generative artificial intelligence and the risk of technodigital colonialism. *Frontiers in Political Science*, *7*, 1628139. DOI: 10.3389/fpos.2025.1628139

Carpenter, C. C. (1978). *Ritualistic social behaviors in lizards.* In *Behavior and neurology of lizards* (pp. 253–267).

Cartmill, E. A. (2023). *Using the senses in animal communication.* In *A new companion to linguistic anthropology* (pp. 369–390).

Castro, A., Silva, F., & Santos, V. (2021). Trends of Human-Robot Collaboration in Industry Contexts: Handover, Learning, and Metrics. *Sensors (Basel)*, *21*(12), 4113. DOI: 10.3390/s21124113

Cave, S., & Dihal, K. (2020). *AI narratives: A history of imaginative thinking about intelligent machines*. Oxford University Press.

Chavan, V., Cenaj, A., Shen, S., Bar, A., Binwani, S., Del Becaro, T., Funk, M., Greschner, L., Hung, R., Klein, S., Kleiner, R., Krause, S., Olbrych, S., Parmar, V., Sarafraz, J., Soroko, D., Don, D. W., Zhou, C., Vu, H. T. D., & Fresquet, X. (2025). Feeling Machines: Ethics, Culture, and the Rise of Emotional AI. DOI: 10.48550/ARXIV.2506.12437

Chen, C., Qiu, J., He, L., Wu, Q., & Lu, X. (2024). Research on the application of image recognition technology in live-stock and poultry behavior recognition. In Zhang, X., & Yin, M. (Eds.), *Fifth International Conference on Computer Vision and Data Mining (ICCVDM 2024)* (p. 60). SPIE. https://doi.org/DOI: 10.1117/12.3048224

Chen, D., Chauhan, K., Parsa, R., Liu, Z. A., Liu, F., Mak, E., Eng, L., Hannon, B., Croke, J., Hope, A., Fallah-Rad, N., Wong, P., & Raman, S. (2025). Patient perceptions of empathy in physician and artificial intelligence chatbot responses to patient questions about cancer. *NPJ Digital Medicine*, *8*(1). Advance online publication. DOI: 10.1038/s41746-025-01671-6 PMID: 40360673

Cheng, J., Xie, B., Lin, C., & Ji, L. (2012). A comparative study in birds: Call-type-independent species and individual recognition using four machine-learning methods and two acoustic features. *Bioacoustics*, *21*(2), 157–171.

Chisom, O. N., Biu, P. W., Umoh, A. A., Obaedo, B. O., Adegbite, A. O., & Abatan, A. (2024). Reviewing the role of AI in environmental monitoring and conservation: A data-driven revolution for our planet. *World Journal of Advanced Research and Reviews, 21*(1), 161–171.

Choi, J. I., Yang, E., & Goo, E. H. (2024). The effects of an ethics education program on artificial intelligence among middle school students: Analysis of perception and attitude changes. *Applied Sciences (Basel, Switzerland), 14*(4), 1588.

Choi, S. R., & Lee, M. (2023). Transformer architecture and attention mechanisms in genome data analysis: A comprehensive review. *Biology (Basel), 12*(7), 1033. PMID: 37508462

Chopra, S., Bhambri, P., & Singh, B. (2011). Segmentation of the mammogram images to find breast boundaries. *International Journal of Computer Science and Technology*, *2*(2), 164–167.

Chowdhary, S., Selvakumar, P., Sangeetha, B., Sarvaani, V. V., Anute, N., & Sharma, M. (2025). The Role of Social Capital in Entrepreneurial Development. In Martínez-Falcó, J., Sánchez-García, E., & Marco-Lajara, B. (Eds.), *Support Networks for Entrepreneurs: Social Capital and Strategic Collaborations* (pp. 311–338). IGI Global Scientific Publishing.

Christov-Moore, L., Reggente, N., Vaccaro, A. G., Schoeller, F., Pluimer, B. R., Douglas, P. K., Iacoboni, M., Man, K., Damásio, A. R., & Kaplan, J. (2023). Preventing antisocial robots: A pathway to artificial empathy [Review of Preventing antisocial robots: A pathway to artificial empathy]. Science Robotics, 8(80). American Association for the Advancement of Science. DOI: 10.1126/scirobotics.abq3658

Ciszewska-Mlinarič, M., & Trąpczyński, P. (2019). Foreign market adaptation and performance: The role of institutional distance and organizational capabilities. *Sustainability*, *11*(6), 1793.

Coates, J. M., Santhanalakshmi, K., Dinesh, N., Gandhimathi, S., & Chandel, P. S. Mohit, & Selvakumar, P. (2025). Assessment and Evaluation in Transnational Staff Development. In G. Morris & J. Ye (Eds.), Training and Development in Transnational Higher Education (pp. 387-406). IGI Global Scientific Publishing.

Coatsworth, P., Gonzalez-Macia, L., Collins, A. S. P., Bozkurt, T., & Güder, F. (2023). Continuous monitoring of chemical signals in plants under stress. *Nature Reviews. Chemistry*, *7*(1), 7–25. PMID: 37117825

Cohen, A., & Galily, Y. (2024). Behavioral and Psychosocial Dynamics of Engagement: The Digital Divide in Artificial Intelligence [AI]-Driven Sports Podcasts. *Technologies, 14*(10), 911. DOI: 10.3390/technologies14100911

Corrêa, N. K., Galvão, C., Santos, J. W., Pino, C. D., Pinto, E. P., Barbosa, C., Massmann, D., Mambrini, R., Galvão, L., Terem, E., & Oliveira, N. de. (2023). Worldwide AI ethics: A review of 200 guidelines and recommendations for AI governance [Review of Worldwide AI ethics: A review of 200 guidelines and recommendations for AI governance]. Patterns, 4(10), 100857. Elsevier BV. DOI: 10.1016/j.patter.2023.100857

Corvite, S., Roemmich, K., Rosenberg, T. I., & Andalibi, N. (2023). Data Subjects' Perspectives on Emotion Artificial Intelligence Use in the Workplace: A Relational Ethics Lens. *Proceedings of the ACM on Human-Computer Interaction, 7*, 1. DOI: 10.1145/3579600

Cowie, J., Cowie, S., & Gentry, K. (2023). Artificial intelligence technologies and compassion in healthcare: A systematic scoping review. *Journal of Medical Systems, 47*(1), 1–17. DOI: 10.1007/s10916-023-01931-1

Crawford, K. (2021). *Atlas of AI: Power, politics, and the planetary costs of artificial intelligence*. Yale University Press.

Crist, E. (2019). *Abundant earth: Toward an ecological civilization*. University of Chicago Press.

Crovini, C. (2022). Crisis Management and Risk Management in SMEs: Towards an Integrated Early Warning System. In *Crisis Management for Small and Medium-Sized Enterprises (SMEs) Strategies for External Crises* (pp. 215–233). Springer International Publishing.

Cui, Z., Li, L., Wang, Y., Zhong, Z., & Li, J. (2023). Review of research and control technology of underwater bionic robots. *Intelligent Marine Technology and Systems, 1*(1), 7. DOI: 10.1007/s44295-023-00010-3

Czakon, W. (1999). Relational capability of organizations–theoretical advances. *Journal of Sociology (Melbourne, Vic.), 104*(5), 48–65.

d'Isa, R. (2025). Robotic animals as new tools in rodent neuroscience research: Proposed applications of zooinspired robots for mouse behavioral testing. *Frontiers in Behavioral Neuroscience, 19*, 1545352. DOI: 10.3389/fnbeh.2025.1545352

Damián-Rodríguez, M., & Gómez-Gómez, A. (2024). Artificial Intelligence Alone Will Not Democratise Education: On Educational Inequality, Techno-Solutionism and Inclusive Tools. *Sustainability, 16*(2), 781. DOI: 10.3390/su16020781

Darney, P. E., Fernando, A. V., Krishnan, R. S., Raj, J. R. F., Lavanya, V., & Kumar, V. V. (2025). RhinoGuardNet: An Integrated Deep Learning System for Monitoring, Movement Prediction, and Threat Detection in Javan Rhino Conservation. In *2025 International Conference on Machine Learning and Autonomous Systems (ICMLAS)* (pp. 101-108). IEEE.

Dave, R., Kaunert, C., & Singh, B. (2025). Wildlife and Forest Resource Management With Artificial Intelligence. In Machine Learning and Internet of Things in Fire Ecology (pp. 301-324). IGI Global Scientific Publishing.

Deng, X., Wan, T., Xu, K., Gao, T., Qiao, P., Feng, D., & Dou, Y. (2025, April). Scaling Bioacoustic Signal Pre-training with Million Samples Via Mask-Modeling. In *ICASSP 2025-2025 IEEE International Conference on Acoustics, Speech and Signal Processing (ICASSP)* (pp. 1-5). IEEE.

Devendra, C., & Thomas, D. (2002). Smallholder farming systems in Asia. *Agricultural Systems*, *71*(1), 17–25. DOI: 10.1016/S0308-521X(01)00033-6

Devi, V. B., Privadharshini, K., & Haritha, M. (2025). Smart Irrigation System by Detecting Plant Squeals Using IoT. In *2025 International Conference on Computing and Communication Technologies (ICCCT)* (pp. 1-8). IEEE.

Dewberry, N. K., AlHmoud, I., Benton, K., Suarez, D., Chen, Y.-P., Karkaria, V., Tsai, Y.-K., Brock, M., Alazzawi, N., Chowdhury, S., Chen, W., Cao, J., & Gokaraju, B. (2025). A real-time VR-enabled digital twin framework for multi-user interaction in Industry 4.0. *Manufacturing Letters*, *44*, 1486–1497. DOI: 10.1016/j.mfglet.2025.06.168

Dhivya, P., & Abhishek, P., HariHaran, M., & Midhun, J. (2025). Deep Learning-Based Acoustic Classification of Animal Species Using Convolutional Neural Networks. In *2025 International Conference on Inventive Computation Technologies (ICICT)* (pp. 159-165). IEEE.

Díaz-Boladeras, M. (2023). Bond formation with pet-robots: An integrative approach. *Current Psychology (New Brunswick, N.J.)*, *42*(4), 2591–2608. DOI: 10.1007/s12144-022-03792-7

Dignum, V. (2018). Ethics in artificial intelligence: Introduction to the special issue. *Ethics and Information Technology*, *20*(1), 1. DOI: 10.1007/s10676-018-9450-z

Ding, L., Zhang, C., Yue, Y., Yao, C., Li, Z., Hu, Y., Yang, B., Ma, W., Yu, L., Gao, R., & Li, Q. (2025). Wearable Sensors-Based Intelligent Sensing and Application of Animal Behaviors: A Comprehensive Review. *Sensors (Basel)*, *25*(14), 4515. DOI: 10.3390/s25144515

Diwan, A., Sunil, R., Mer, P., Mahadeva, R., & Patole, S. P. (2024). Advancements in Emotion Classification via Facial and Body Gesture Analysis: A Survey. *Expert Systems: International Journal of Knowledge Engineering and Neural Networks*. Advance online publication. DOI: 10.1111/exsy.13759

Donaldson, S., & Kymlicka, W. (2011). *Zoopolis: A political theory of animal rights*. Oxford University Press.

Dwivedi, D. N., Mahanty, G., & nath Dwivedi, V. (2024). Intelligent conservation: a comprehensive study on AI-enhanced environmental monitoring and preservation. In The Convergence of Self-Sustaining Systems With AI and IoT (pp. 215-226). IGI Global.

Elbasi, E., Mostafa, N., AlArnaout, Z., Zreikat, A. I., Cina, E., Varghese, G., Shdefat, A., Topcu, A. E., Abdelbaki, W., Mathew, S., & Zaki, C. (2023). Artificial Intelligence Technology in the Agricultural Sector: A Systematic Literature Review. *IEEE Access : Practical Innovations, Open Solutions*, *11*, 171–202. DOI: 10.1109/ACCESS.2022.3232485

Elendu, C., Amaechi, D. C., Elendu, T. C., Jingwa, K. A., Okoye, O. K., Okah, M. J., Ladele, J. A., Farah, A. H., & Alimi, H. A. (2023). Ethical implications of AI and robotics in healthcare: A review [Review of Ethical implications of AI and robotics in healthcare: A review]. Medicine, 102(50). Wolters Kluwer. DOI: 10.1097/md.0000000000036671

Elgazzar, M. A. G., & Dawood, M. E. T. (2025). Mastering the Phases of Design Strategy: A Comprehensive Guide to Navigating from Conceptualization to Launch in the Digital Product Life Cycle. *Journal of Art. Design and Music*, *4*(2), 1.

Elshaekh, D. N., Mohamed, H. A., Menam, L. Y. A. E., Sharshar, K. A., & Kayed, S. I. (2025). Multiband printed rectenna for radio frequency energy harvesting (RF-EH). *Discover Electronics*, *2*(1), 39.

Essien, D., & Neethirajan, S. (2025). Multimodal AI systems for enhanced laying hen welfare assessment and productivity optimization. *Smart Agricultural Technology*, *12*, 101564. DOI: 10.1016/j.atech.2025.101564

Ezanno, P., Picault, S., Beaunée, G., Bailly, X., Muñoz, F., Duboz, R., & Guégan, J. F. (2021). Research perspectives on animal health in the era of artificial intelligence. *Veterinary Research*, *52*, 1–15. PMID: 33676570

Fascista, A. (2022). Toward Integrated Large-Scale Environmental Monitoring Using WSN/UAV/Crowdsensing: A Review of Applications, Signal Processing, and Future Perspectives. *Sensors (Basel)*, *22*(5), 1824. DOI: 10.3390/s22051824

Fei, M., Yuan, Y., Xie, Y., Ren, H., Liu, I., He, Y., Ren, F., Yu, F. R., & Ni, S. (2024). Generative Technology for Human Emotion Recognition: A Scope Review [Review of Generative Technology for Human Emotion Recognition: A Scope Review]. DOI: 10.2139/ssrn.4889990

Fernandes, A. F. A., Dórea, J. R. R., & Rosa, G. J. D. M. (2020). Image Analysis and Computer Vision Applications in Animal Sciences: An Overview. *Frontiers in Veterinary Science*, *7*, 551269. DOI: 10.3389/fvets.2020.551269 PMID: 33195522

Fernández-Rodicio, E., Castro-González, Á., Gamboa-Montero, J. J., Carrasco-Martínez, S., & Salichs, M. A. (2024). Creating Expressive Social Robots That Convey Symbolic and Spontaneous Communication. *Sensors (Basel)*, *24*(11), 3671. DOI: 10.3390/s24113671

Ferrante, G. S., Rodrigues, F. M., Andrade, F. R. H., Goularte, R., & Meneguette, R. I. (2021). Understanding the state of the Art in Animal detection and classification using computer vision technologies. *2021 IEEE International Conference on Big Data (Big Data)*, 3056–3065. https://doi.org/DOI: 10.1109/BigData52589.2021.9672049

Figueiredo, A. F., Boy, J., & Guggenberger, G. (2021). Common mycorrhizae network: A review of theories and mechanisms behind underground interactions. *Frontiers in Fungal Biology*, *2*, 735299. PMID: 37744156

Fiske, A., Henningsen, P., & Buyx, A. (2019). Your Robot Therapist Will See You Now: Ethical Implications of Embodied Artificial Intelligence in Psychiatry, Psychology, and Psychotherapy [Review of Your Robot Therapist Will See You Now: Ethical Implications of Embodied Artificial Intelligence in Psychiatry, Psychology, and Psychotherapy]. Journal of Medical Internet Research, 21(5). JMIR Publications. DOI: 10.2196/13216

Fitzpatrick, J., Browning, H., & Brakel, S. (2022). Ethical challenges in precision livestock farming. *Journal of Agricultural & Environmental Ethics*, *35*(2), 1–19. DOI: 10.1007/s10806-022-09898-3

Floreano, D., & Mattiussi, C. (2008). *Bio-inspired artificial intelligence: Theories, methods, and technologies*. MIT Press.

Floridi, L., Cowls, J., Beltrametti, M., Chatila, R., Chazerand, P., Dignum, V., & Vayena, E. (2018). AI4People—An ethical framework for a good AI society: Opportunities, risks, principles, and recommendations. *Minds and Machines*, *28*(4), 689–707. DOI: 10.1007/s11023-018-9482-5

Fogelson, D. M., Rutledge, C., & Zimbro, K. S. (2022). The Impact of Robotic Companion Pets on Depression and Loneliness for Older Adults with Dementia During the COVID-19 Pandemic. *Journal of Holistic Nursing*, *40*(4), 397–409. DOI: 10.1177/08980101211064605

Freeberg, T. M., Baker, M. C., Bloomfield, L. L., Charrier, I., Gammon, D. E., Hailman, J. P., & Sturdy, C. B. (2007). *Complexities in vocal communication. Ecology and behavior of chickadees and titmice: an integrated approach.* Oxford University Press.

Fröhlich, M., & van Schaik, C. P. (2018). The function of primate multimodal communication. *Animal Cognition*, *21*, 619–629. PMID: 29876698

Fu, Y., & Weng, Z. (2024). Navigating the Ethical Terrain of AI in Education: A Systematic Review on Framing Responsible Human-Centered AI Practices [Review of Navigating the Ethical Terrain of AI in Education: A Systematic Review on Framing Responsible Human-Centered AI Practices]. Computers and Education Artificial Intelligence, 100306. Elsevier BV. DOI: 10.1016/j.caeai.2024.100306

Gao, Z., Shi, Q., Fukuda, T., Li, C., & Huang, Q. (2019). An overview of biomimetic robots with animal behaviors. *Neurocomputing*, *332*, 339–350. DOI: 10.1016/j.neucom.2018.12.071

Garlisi, D., Pagano, A., Giuliano, F., Croce, D., & Tinnirello, I. (2025). Interference Analysis of LoRaWAN and Sigfox in Large-Scale Urban IoT Networks. *IEEE Access : Practical Innovations, Open Solutions.*

Garstang, M. (2010). Elephant infrasounds: Long-range communication. [). Elsevier.]. *Handbook of Behavioral Neuroscience*, *19*, 57–67.

Gaworski, M., & Kic, P. (2024). Assessment of Production Technologies on Dairy Farms in Terms of Animal Welfare. *Applied Sciences (Basel, Switzerland)*, *14*(14), 6086. DOI: 10.3390/app14146086

Geary, W. L., Bode, M., Doherty, T. S., Fulton, E. A., Nimmo, D. G., Tulloch, A. I., & Ritchie, E. G. (2020). A guide to ecosystem models and their environmental applications. *Nature Ecology & Evolution*, *4*(11), 1459–1471.

Geldenhuys, C. M., & Niesler, T. R. (2025). Learning to rumble: Automated elephant call and sub-call classification, detection and endpointing using deep architectures. *Bioacoustics*, *34*(3), 307–354.

Ghafoor, A., Symeonidis, I., Rydberg, A., Lindahl, C., & Abbasi, A. Q. (2025). Towards Empowering Stakeholders Through Decentralized Trust and Secure Livestock Data Sharing. *Cryptography*, *9*(3), 52. DOI: 10.3390/cryptography9030052

Ghatoray, S. K., & Li, Y. (2025). Automated UX Insights from User Research Videos by Integrating Facial Emotion and Text Sentiment. DOI: 10.48550/ARXIV.2503.22510

Ghavi Hossein-Zadeh, N. (2025). Artificial intelligence in veterinary and animal science: Applications, challenges, and future prospects. *Computers and Electronics in Agriculture*, *235*, 110395. DOI: 10.1016/j.compag.2025.110395

Ghotbi, N. (2022). The Ethics of Emotional Artificial Intelligence: A Mixed Method Analysis. *Asian Bioethics Review*, *15*(4), 417. DOI: 10.1007/s41649-022-00237-y PMID: 37808444

Giri, G., Iza, K. C., Khatiwada, P., Adhikari, S. K., & Shakya, S. (2025). CNN-Based Bird Sound Detection: A Comparative Performance Study. *IACSIT International Journal of Engineering and Technology*, *2*(2), 176–187.

Goel, P. K., Saifi, S., Goel, N., Aeron, S., Yadav, S., & Ansari, S. A. (2025). Ethical Considerations in AI: Applications for Wildlife Conservation. In AI and Machine Learning Techniques for Wildlife Conservation (pp. 247-266). IGI Global Scientific Publishing.

Gökçek-Saraç, Ç. (2020). Effects of 2.1 GHz Electromagnetic Radiation on Locomotor Activity, Recognition Memory, and Anxiety-Related Behavior in Rats. *Neurophysiology*, *52*, 261–266. DOI: 10.1007/s11062-021-09881-w

Gordon, R., & Seckbach, J. (Eds.). (2016). *Biocommunication: Sign-mediated interactions between cells and organisms* (Vol. 1). World Scientific.

Gorgeva, E., Robertson, J., Voss, S., & Hoogewerff, J. (2025). The potential of bioacoustics for surveying carrion insects. *The Australian Journal of Forensic Sciences*, *57*(1), 122–141.

Gove, R., & Uzdzinski, J. (2013). A performance-based system maturity assessment framework. *Procedia Computer Science*, *16*, 688–697.

Goyal, A., Bhakhar, R., & Singh, S. (2025). Limitations and Challenges of AI in Wildlife Conservation. In AI and Machine Learning Techniques for Wildlife Conservation (pp. 363-394). IGI Global Scientific Publishing.

Guan, H., Dong, L., & Zhao, A. (2022). Ethical Risk Factors and Mechanisms in Artificial Intelligence Decision Making. *Behavioral Sciences (Basel, Switzerland)*, *12*(9), 343. DOI: 10.3390/bs12090343 PMID: 36135147

Guarnido-Lopez, P., Pi, Y., Tao, J., Mendes, E. D. M., & Tedeschi, L. O. (2024). Computer vision algorithms to help decision-making in cattle production. *Animal Frontiers : the Review Magazine of Animal Agriculture*, *14*(6), 11–22. DOI: 10.1093/af/vfae028 PMID: 39764526

Guiochet, J., Machin, M., & Waeselynck, H. (2017). Safety-critical advanced robots: A survey. *Robotics and Autonomous Systems*, *94*, 43–52. DOI: 10.1016/j.robot.2017.04.004

Gupta, S., Vijarania, P., Gupta, S., & Vijarania, M. (2025). Artificial Intelligence in the Current Era: Navigating the Impacts on Society and Wildlife Conservation. In AI and Machine Learning Techniques for Wildlife Conservation (pp. 65-96). IGI Global Scientific Publishing.

Gupta, S., Hasan, W., Singh, S., Kumar, D., Ansari, M. J., & Nisar, S. (2024). *Agriculture 4.0: Smart Farming with IoT and Artificial Intelligence* (1st ed.). CRC Press., DOI: 10.1201/9781003570219

Gupta, T., & Bansal, S. (2023). AI-Driven Emotional Recognition in Digital Ads: A Novel Approach to Consumer Engagement. *Journal of Marketing & Supply Chain Management*, *2*(3), 1. DOI: 10.47363/jmscm/2023(2)131

Gur, T., & Maaravi, Y. (2025). The algorithm of friendship: Literature review and integrative model of relationships between humans and artificial intelligence (AI). *Behaviour & Information Technology*, *1*. Advance online publication. DOI: 10.1080/0144929x.2025.2502467

Gustafson, K., Sandstrom, T., & Townsend, L. (2018). The bush war to save the rhino: Improving counter-poaching through intelligence. *Small Wars & Insurgencies*, *29*(2), 269–290.

Hammouch, H., & Raghav, A. (2025). AI and ML Applications in Wildlife Conservation and Forest Management: A Comprehensive Review. Machine Learning and Internet of Things in Fire Ecology, 49-68.

Han, J., Siegford, J., Colbry, D., Lesiyon, R., Bosgraaf, A., Chen, C., Norton, T., & Steibel, J. P. (2023). Evaluation of computer vision for detecting agonistic behavior of pigs in a single-space feeding stall through blocked cross-validation strategies. *Computers and Electronics in Agriculture*, *204*, 107520. DOI: 10.1016/j.compag.2022.107520

Harleen, B. (2016). A prediction technique in data mining for diabetes mellitus. *Journal of Management Science and Technology*, *4*(1), 1–12.

Heijden, M. G. V. D. (2016). Underground networking. *Science*, *352*(6283), 290–291. PMID: 27081054

Hermann, E., Hermann, G., & Tremblay, J. C. (2021). Ethical Artificial Intelligence in Chemical Research and Development: A Dual Advantage for Sustainability. *Science and Engineering Ethics*, *27*(4). Advance online publication. DOI: 10.1007/s11948-021-00325-6 PMID: 34231042

Hess, K. P., Entin, E. E., Hess, S. M., Hutchins, S. G., Kemple, W. G., Kleinman, D. L., ... & Serfaty, D. (2000). Building adaptive organizations: a bridge from basic research to operational exercises.

Hirsch, A. M., Bauer, W. D., Bird, D. M., Cullimore, J., Tyler, B., & Yoder, J. I. (2003). Molecular signals and receptors: Controlling rhizosphere interactions between plants and other organisms. *Ecology*, *84*(4), 858–868.

Hirsch, A. M., & Fujishige, N. A. (2011). Molecular signals and receptors: Communication between nitrogen-fixing bacteria and their plant hosts. In *Biocommunication of plants* (pp. 255–280). Springer.

Ho, T. M., Mantello, P., & Ho, M. (2023). An analytical framework for studying attitude towards emotional AI: The three-pronged approach [Review of An analytical framework for studying attitude towards emotional AI: The three-pronged approach]. MethodsX, 10, 102149. Elsevier BV. DOI: 10.1016/j.mex.2023.102149

Höfling, T. T. A., Gerdes, A. B. M., Föhl, U., & Alpers, G. W. (2020). Read My Face: Automatic Facial Coding Versus Psychophysiological Indicators of Emotional Valence and Arousal. *Frontiers in Psychology*, *11*. Advance online publication. DOI: 10.3389/fpsyg.2020.01388 PMID: 32636788

Hogenboom, J. A., Pellegrino, L., Sandrucci, A., Rosi, V., & D'Incecco, P. (2019). *Invited review:* Hygienic quality, composition, and technological performance of raw milk obtained by robotic milking of cows. *Journal of Dairy Science*, *102*(9), 7640–7654. DOI: 10.3168/jds.2018-16013

Holloway, L., Bear, C., & Wilkinson, K. (2014). Re-capturing bovine life: Robot–cow relationships, freedom and control in dairy farming. *Journal of Rural Studies*, *33*, 131–140. DOI: 10.1016/j.jrurstud.2013.01.006

Holzinger, A., Fister, I., Fister, I., Kaul, H.-P., & Asseng, S. (2024). Human-Centered AI in Smart Farming: Toward Agriculture 5.0. *IEEE Access : Practical Innovations, Open Solutions*, *12*, 62199–62214. DOI: 10.1109/ACCESS.2024.3395532

Horsevad, N., Kwa, H. L., & Bouffanais, R. (2022). Beyond Bio-Inspired Robotics: How Multi-Robot Systems Can Support Research on Collective Animal Behavior. *Frontiers in Robotics and AI, 9*. Advance online publication. DOI: 10.3389/frobt.2022.865414

Horvat, D., Ivanišević, R., & Gluščević, L. (2024). Risks associated with robotic process automation. *Journal of Process Management and New Technologies, 12*(1–2), 72–82. DOI: 10.5937/jpmnt12-50617

Hossain, M. S., & Muhammad, G. (2018). Emotion recognition using deep learning approach from audio–visual emotional big data. *Information Fusion, 49*, 69. DOI: 10.1016/j.inffus.2018.09.008

Hossain, M., Samiun, M., Tiwari, A., Johora, F. T., Hosain, M. S., Saleh, M. A., & Vanu, N. (2024). Data Analytics in Wildlife Management Business Insights and Technological Solutions. In *2024 International Conference on Progressive Innovations in Intelligent Systems and Data Science (ICPIDS)* (pp. 539-544). IEEE.

Hsu, C. Y., & Li, W. (2023). Explainable GeoAI: Can saliency maps help interpret AI's learning process? *International Journal of Geographical Information Science, 37*(5), 963–987.

Huang, C., Zhang, Z., Mao, B., & Yao, X. (2022). An Overview of Artificial Intelligence Ethics. *IEEE Transactions on Artificial Intelligence, 4*(4), 799. DOI: 10.1109/tai.2022.3194503

Hubballi, R. B., Selvakumar, P., Seenivasan, R., & Basava Aradhya, S. G., Dinesh, N., & Seelam, P. K. (2025). Overview of Current AI Technologies in Education. In L. Kyei-Blankson & E. Ntuli (Eds.), Transformative AI Practices for Personalized Learning Strategies (pp. 1-26). IGI Global Scientific Publishing.

Hughey, L. F., Hein, A. M., Strandburg-Peshkin, A., & Jensen, F. H. (2018). Challenges and solutions for studying collective animal behaviour in the wild. *Philosophical Transactions of the Royal Society B: Biological Sciences, 373*(1746), 20170005. https://doi.org/DOI: 10.1098/rstb.2017.0005

Hutchison, D., Kanade, T., Kittler, J., Kleinberg, J. M., Mattern, F., Mitchell, J. C., Naor, M., Nierstrasz, O., & Rangan, P. C., Steffen, B., Sudan, M., Terzopoulos, D., Tygar, D., Vardi, M. Y., Weikum, G., Branson, S., Wah, C., Schroff, F., Babenko, B., … Belongie, S. (2010). Visual Recognition with Humans in the Loop. In K. Daniilidis, P. Maragos, & N. Paragios (Eds.), *Computer Vision – ECCV 2010* (Vol. 6314, pp. 438–451). Springer Berlin Heidelberg. https://doi.org/DOI: 10.1007/978-3-642-15561-1_32

Ihde, D. (1990). *Technology and the lifeworld: From garden to earth.* Indiana University Press.

Irving, C. J. (2023). *An investigation into multi-component warning colouration* (Doctoral dissertation, Macquarie University).

Jacuzzi, G., & Olden, J. D. (2025). Few-shot transfer learning enables robust acoustic monitoring of wildlife communities at the landscape scale. *Ecological Informatics, 80*, 103294.

Janik, V. M. (2014). Cetacean vocal learning and communication. *Current Opinion in Neurobiology, 28*, 60–65. PMID: 25057816

Janssen, J. H., Tacken, P., de Vries, G.-J., van den Broek, E. L., Westerink, J. H. D. M., Haselager, P., & IJsselsteijn, W. A. (2013). Machines Outperform Laypersons in Recognizing Emotions Elicited by Autobiographical Recollection. *Human-Computer Interaction, 28*(6), 479. DOI: 10.1080/07370024.2012.755421

Järvenpää, A. M., Kunttu, I., Jussila, J., & Mäntyneva, M. (2021, April). Data-driven decision-making in circular economy SMEs in Finland. In *The International Research & Innovation Forum* (pp. 371–382). Springer International Publishing.

Jawad, M., Talreja, K., Bhutto, S. A., & Faizan, K. (2024). Investigating how AI Personalization Algorithms Influence Self-Perception, Group Identity, and Social Interactions Online. *Review of Applied Management and Social Sciences, 7*(4), 533–550.

Jayanthi, K., Chitradevi, D., & Muthaiyah, S. (2025). Early Prediction of Harmful Environmental Changes Based on Bioacoustics Analysis. In *Technological Advancements for Deep Sea Ecosystem Conservation and Exploration* (pp. 129–148). IGI Global Scientific Publishing.

Jiang, T., Sun, Z., Fu, S., & Lv, Y. (2024). Human-AI interaction research agenda: A user-centered perspective. *Data and Information Management, 8*(4), 100078. DOI: 10.1016/j.dim.2024.100078

Jianhua, L. (1998). Communication and the evolution of ecosystems. *World Futures: Journal of General Evolution, 52*(1), 35–73.

Jobin, A., Ienca, M., & Vayena, E. (2019). The global landscape of AI ethics guidelines. *Nature Machine Intelligence, 1*(9), 389. DOI: 10.1038/s42256-019-0088-2

Jung, H., Park, S., Joe, S., Woo, S., Choi, W., & Bae, W. (2025). AI-Driven Control Strategies for Biomimetic Robotics: Trends, Challenges, and Future Directions. *Biomimetics, 10*(7), 460. DOI: 10.3390/biomimetics10070460

Kabra, M., Robie, A. A., Rivera-Alba, M., Branson, S., & Branson, K. (2013). JAABA: Interactive machine learning for automatic annotation of animal behavior. *Nature Methods*, *10*(1), 64–67. DOI: 10.1038/nmeth.2281 PMID: 23202433

Kalinaki, K., Acaru, S. F., Kugonza, J., & Nsubuga, R. (2024). Towards an Intelligent Tomorrow: Machine Learning Enabling Sustainable Development. In Methodologies, Frameworks, and Applications of Machine Learning (pp. 66-89). IGI Global.

Kalita, A. J., Deka, J., Devi, C., & Kalita, P. (2025). Importance of AI-Tech in Managing a Protected Area: Considering Forest Health Dynamics. In Artificial Intelligence and Animal Ecology (pp. 193-220). CRC Press.

Kamal, R. (2024). *Internet of Things Architecture and Design Principles* (Fifth). MacGraw Hill Education (India) PVT LTD.

Karthik, P. C., Bharathi, S. S., Sivakumar, A., Thangavelu, C., Nalini, S., & Lalitha, S. D. (2025). Emerging Artificial Intelligence Tools, Advancements, and Opportunities for Protecting Wild Animals. In AI and Machine Learning Techniques for Wildlife Conservation (pp. 213-246). IGI Global Scientific Publishing.

Katsis, L. K., Rhinehart, T. A., Dorgay, E., Sanchez, E. E., Snaddon, J. L., Doncaster, C. P., & Kitzes, J. (2025). A comparison of statistical methods for deriving occupancy estimates from machine learning outputs. *Scientific Reports*, *15*(1), 1–13.

Kemmerzell, N., Schreiner, A. F., Khalid, H., Schalk, M., & Bordoli, L. (2025). Towards a Better Understanding of Evaluating Trustworthiness in AI Systems [Review of Towards a Better Understanding of Evaluating Trustworthiness in AI Systems]. ACM Computing Surveys. Association for Computing Machinery. DOI: 10.1145/3721976

Kershenbaum, A., Akçay, Ç., Babu-Saheer, L., Barnhill, A., Best, P., Cauzinille, J., & Dunn, J. C. (2025). Automatic detection for bioacoustic research: A practical guide from and for biologists and computer scientists. *Biological Reviews of the Cambridge Philosophical Society*, *100*(2), 620–646.

Kershenbaum, A., Blumstein, D. T., Roch, M. A., Akçay, Ç., Backus, G., Bee, M. A., & Yuan, C. (2016). Acoustic sequences in non-human animals: A tutorial review and prospectus. *Biological Reviews of the Cambridge Philosophical Society*, *91*(1), 13–52. DOI: 10.1111/brv.12160

Kim, E. H., MyungJin, L., & Shin, J. (2025). MMER-LMF: Multi-Modal Emotion Recognition in Lightweight Modality Fusion. *Electronics (Basel)*, *14*(11), 2139. DOI: 10.3390/electronics14112139

Kim, J. H., Sohn, U. D., Kim, H. G., & Kim, H. R. (2018). Exposure to 835 MHz RF-EMF decreases the expression of calcium channels, inhibits apoptosis, but induces autophagy in the mouse hippocampus. *The Korean Journal of Physiology & Pharmacology; Official Journal of the Korean Physiological Society and the Korean Society of Pharmacology, 22*(3), 277.

Kim, S., Park, J., & Lee, J. (2024). AI-Based Smart Monitoring Framework for Livestock Farms. *Applied Sciences (Basel, Switzerland), 15*(10), 5638. DOI: 10.3390/app151005638

Kimura, K., Fukuyama, I., & Fukuyama, K. (2025). Deep learning-based detector of invasive alien frogs, Polypedates leucomystax and Rhinella marina, on an island at invasion front. *Biological Invasions, 27*(3), 1–11.

Kling-Eveillard, F., Allain, C., Boivin, X., Courboulay, V., Créach, P., Philibert, A., Ramonet, Y., & Hostiou, N. (2020). Farmers' representations of the effects of precision livestock farming on human-animal relationships. *Livestock Science, 238*, 104057. DOI: 10.1016/j.livsci.2020.104057

Koch, W. (2016). *Tracking and sensor data fusion.* Springer-Verlag Berlin An.

Kóczi, D., & Sárosi, J. (2025). Safety Engineering for Humanoid Robots in Everyday Life—Scoping Review. *Electronics (Basel), 14*(23), 4734. DOI: 10.3390/electronics14234734

Koger, B., Deshpande, A., Kerby, J. T., Graving, J. M., Costelloe, B. R., & Couzin, I. D. (2023). Quantifying the movement, behaviour and environmental context of group-living animals using drones and computer vision. *Journal of Animal Ecology, 92*(7), 1357–1371. DOI: 10.1111/1365-2656.13904 PMID: 36945122

Kotha, S. S., Akter, N., Abhi, S. H., Das, S. K., & Islam, Md. R., Ali, Md. F., Ahamed, Md. H., Islam, Md. M., Sarker, S. K., Badal, Md. F. R., Das, P., Tasneem, Z., & Hasan, Md. M. (2024). Next generation legged robot locomotion: A review on control techniques. *Heliyon, 10*(18), e37237. DOI: 10.1016/j.heliyon.2024.e37237

Kotte, K. R., Attaluri, V., & Selvakumar, P. (2025). Revolutionising Retail Banking with AI and Virtual Technologies AI-Driven Financial Training. In Nair, A. J., Manohar, S., Limbu, Y. B., & Huhmann, B. A. (Eds.), *Intersecting Natural Language Processing and FinTech Innovations in Service Marketing* (pp. 421–440). IGI Global Scientific Publishing.

Kovac, M., Bonnet, F., & Floreano, D. (2021). Robotic animals for scientific research and conservation. *Annual Review of Control, Robotics, and Autonomous Systems, 4*(1), 185–209. DOI: 10.1146/annurev-control-062220-092241

Kowar, S., Paul, T., Mukhopadhayay, N. D., & Barua, B. (2025). From Data to Decisions: Exploring AI's Role in Tackling Environmental Issues. In Cases on AI-Driven Solutions to Environmental Challenges (pp. 1-22). IGI Global Scientific Publishing.

Kshirsagar, P. R.. (2022). Machine learning algorithm for leaf disease detection. *AIP Conference Proceedings*, *2393*(1).

Kumar, R., Sankat, M., & Panda, A. (2025). Innovations in Biodiversity Monitoring: AGI Drones at the Forefront. In Artificial General Intelligence-Based Drones for Climate Change (pp. 211-248). IGI Global Scientific Publishing.

Kumar, V. (2025). The complex interaction between plants and acoustic signals: Friends or foes? *Naturwissenschaften*, *112*(3), 43. PMID: 40445373

Kvsn, R. R., Montgomery, J., Garg, S., & Charleston, M. (2020). Bioacoustics data analysis: A taxonomy, survey and open challenges. *IEEE Access : Practical Innovations, Open Solutions*, *8*, 57684–57708.

Lamanna, M., Bovo, M., & Cavallini, D. (2025). Wearable Collar Technologies for Dairy Cows: A Systematized Review of the Current Applications and Future Innovations in Precision Livestock Farming. *Animals (Basel)*, *15*(3), 458. DOI: 10.3390/ani15030458

Landowska, A. (2018). Towards New Mappings between Emotion Representation Models. *Applied Sciences (Basel, Switzerland)*, *8*(2), 274. DOI: 10.3390/app8020274

Laru, J., Celik, I., Jokela, I., & Mäkitalo, K. (2025). The antecedents of pre-service teachers' AI literacy: Perceptions about own AI driven applications, attitude towards AI and knowledge in machine learning. *European Journal of Teacher Education*, •••, 1–23.

Latif, S., Rana, R., Khalifa, S., Jurdak, R., & Epps, J. (2019). Direct Modelling of Speech Emotion from Raw Speech. *Interspeech*, *2022*. Advance online publication. DOI: 10.21437/interspeech.2019-3252

LeDoux, J. E. (2012). Rethinking the Emotional Brain. *Neuron*, *73*(4), 653. DOI: 10.1016/j.neuron.2012.02.004 PMID: 22365542

Lee, Y.-G., Han, K., Chung, C., & Ji, I. (2024). Effects of Smart Farming on the Productivity of Korean Dairy Farms: A Case Study of Robotic Milking Systems. *Sustainability*, *16*(22), 9991. DOI: 10.3390/su16229991

Lettieri, G., Handjaras, G., Ricciardi, E., Leo, A., Papale, P., Betta, M., Pietrini, P., & Cecchetti, L. (2019). Emotionotopy in the human right temporo-parietal cortex. *Nature Communications*, *10*(1). Advance online publication. DOI: 10.1038/s41467-019-13599-z PMID: 31804504

Li, G. (2024). Advancements, Challenges, and Future Prospects in Vision-Driven Animal Behavior Recognition Technology. In Y. Wang (Ed.), *Proceedings of the 2024 International Conference on Artificial Intelligence and Communication (ICAIC 2024)* (Vol. 185, pp. 522–532). Atlantis Press International BV. https://doi.org/DOI: 10.2991/978-94-6463-512-6_55

Liang, H., Lee, S.-C., Bae, W., Kim, J., & Seo, S. (2023). Towards UAVs in Construction: Advancements, Challenges, and Future Directions for Monitoring and Inspection. *Drones (Basel)*, *7*(3), 202. DOI: 10.3390/drones7030202

Liang, Y., Liu, B., Ji, J., & Li, X. (2019). Network Representations of Facial and Bodily Expressions: Evidence From Multivariate Connectivity Pattern Classification. *Frontiers in Neuroscience*, *13*. Advance online publication. DOI: 10.3389/fnins.2019.01111 PMID: 31736683

Liao, Y., Gao, Y., Wang, F., Zhang, L., Xu, Z., & Wu, Y. (2025). Emotion recognition with multiple physiological parameters based on ensemble learning. *Scientific Reports*, *15*(1). Advance online publication. DOI: 10.1038/s41598-025-96616-0 PMID: 40473884

Limami, F., Hdioud, B., & Thami, R. O. H. (2024). Contextual emotion detection in images using deep learning. *Frontiers in Artificial Intelligence*, *7*. Advance online publication. DOI: 10.3389/frai.2024.1386753 PMID: 38952408

Lin, C., Wang, J., & Ji, L. (2023). An AI-based Wild Animal Detection System and Its Application. *Biodiversity Information Science and Standards*, *7*, e112456. DOI: 10.3897/biss.7.112456

Lin, J. C. (2025). Health and safety practices and policies concerning human exposure to RF/microwave radiation. *Frontiers in Public Health*, *13*, 1619781.

Liu, F. (2024). Artificial Intelligence in Emotion Quantification : A Prospective Overview. *CAAI Artificial Intelligence Research*, *9150040*. Advance online publication. DOI: 10.26599/air.2024.9150040

Liu, M., Schwab, J., & Heß, U. (2023). Language and face in interactions: Emotion perception, social meanings, and communicative intentions. *Frontiers in Psychology*, *14*. Advance online publication. DOI: 10.3389/fpsyg.2023.1146494 PMID: 37205062

Li, W., Hu, D., & Yang, L. (2023). Actuation Mechanisms and Applications for Soft Robots: A Comprehensive Review. *Applied Sciences (Basel, Switzerland)*, *13*(16), 9255. DOI: 10.3390/app13169255

Locke, J. L. (2021). The indexical voice: Communication of personal states and traits in humans and other primates. *Frontiers in Psychology*, *12*, 651108. PMID: 33935911

London, A. J. (2019). Artificial intelligence and black-box medical decisions: Accuracy versus explainability. *The Hastings Center Report*, *49*(1), 15–21. PMID: 30790315

Luphade, N. (2023). Bioacoustic classification of Hainan gibbon call types using deep learning. *Journal of Bioacoustics Research*, *15*(2), 112–123.

Lu, X. (2022). Deep Learning Based Emotion Recognition and Visualization of Figural Representation. *Frontiers in Psychology*, *12*. Advance online publication. DOI: 10.3389/fpsyg.2021.818833 PMID: 35069400

Machová, K., Szabóová, M., Paralič, J., & Mičko, J. (2023). Detection of emotion by text analysis using machine learning. *Frontiers in Psychology*, *14*. Advance online publication. DOI: 10.3389/fpsyg.2023.1190326 PMID: 37799520

Mahajan, R., & Zaveri, M. A. (2021). Harnessing Emotive Features for Emotion Recognition from Text. *International Journal of Advanced Computer Science and Applications*, *12*(7). Advance online publication. DOI: 10.14569/ijacsa.2021.0120719

Mahato, S., & Neethirajan, S. (2025). Integrating Artificial Intelligence in dairy farm management – biometric facial recognition for cows. *Information Processing in Agriculture*, *12*(3), 312–325. DOI: 10.1016/j.inpa.2024.10.001

Maheswaran, K., Selvakumar, P., Shaikh, M., Shashank, G., Gupta, S., & Manjunath, T. C. (2025). The Evolution of Smart Urban Mobility with Emerging Technologies. In Ponnusamy, S., Assaf, M., Youssef, B., Jeon, G., & Kalyanaraman, S. (Eds.), *Leveraging VANETs and Blockchain Technology for Urban Mobility* (pp. 1–28). IGI Global Scientific Publishing.

Mahmud, S. A., Ji, M., Nasim, I., Agarwal, V., Mishra, A., & Kasera, S. (2024). *Co-existence of Heterogeneous Wireless Networks in 2.4 GHz and 5 GHz Spectrum* (No. INL/RPT-24-80712-Rev000). Idaho National Laboratory (INL), Idaho Falls, ID (United States).

Mallick, R., Flathmann, C., Lancaster, C., Hauptman, A. I., McNeese, N. J., & Freeman, G. (2023). The pursuit of happiness: The power and influence of AI teammate emotion in human-AI teamwork. *Behaviour & Information Technology, 43*(14), 3436. DOI: 10.1080/0144929x.2023.2277909

Manikandan, V., & Neethirajan, S. (2025). Decoding Poultry Welfare from Sound—A Machine Learning Framework for Non-Invasive Acoustic Monitoring. *Sensors (Basel), 25*(9), 2912.

Manjunatha, G. R., & Raveesha, H. R. (2025). AI for Biodiversity Conservation: Harnessing Predictive Analytics and Habitat Monitoring to Protect Endangered Species and Ecosystems. In AI and Ecological Change for Sustainable Development (pp. 33-64). IGI Global Scientific Publishing.

Mansoor, S., Iqbal, S., Popescu, S. M., Kim, S. L., Chung, Y. S., & Baek, J.-H. (2025). Integration of smart sensors and IOT in precision agriculture: Trends, challenges and future prospectives. *Frontiers in Plant Science, 16*, 1587869. DOI: 10.3389/fpls.2025.1587869

Mantello, P., & Ho, T. M. (2023). Emotional AI and the future of wellbeing in the post-pandemic workplace. *AI & Society, 39*(4), 1883. DOI: 10.1007/s00146-023-01639-8 PMID: 36776535

Maroto-Gómez, M., Álvarez-Arias, S., Rodríguez-Huelves, J., Segura-Bencomo, A., & Malfaz, M. (2025). A Robot Companion with Adaptive Object Preferences and Emotional Responses Enhances Naturalness in Human–Robot Interaction. *Electronics (Basel), 14*(18), 3711. DOI: 10.3390/electronics14183711

Martínez Balvanera, S., Mac Aodha, O., Weldy, M. J., Pringle, H., Browning, E., & Jones, K. E. (2025). Whombat: An open-source audio annotation tool for machine learning assisted bioacoustics. *Methods in Ecology and Evolution, 16*(1), 19–28.

Martínez-Vega, J., & García-Martínez, A. (2025). AI Algorithms in the Agrifood Industry: Application Potential in the Spanish Agrifood Context. *Applied Sciences (Basel, Switzerland), 15*(4), 2096. DOI: 10.3390/app15042096

Martin, T., Gasselin, P., Hostiou, N., Feron, G., Laurens, L., Purseigle, F., & Ollivier, G. (2022). Robots and transformations of work in farm: A systematic review of the literature and a research agenda. *Agronomy for Sustainable Development, 42*(4), 66. DOI: 10.1007/s13593-022-00796-2

Masvingise, K., & Mujuru, N. M. (2025). AI-Powered Solutions for Sustainable Tourism Practices Through Wildlife Conservation Initiatives. In AI Technologies for Personalized and Sustainable Tourism (pp. 113-148). IGI Global.

Megalingam, R. K., Vadivel, S. R. R., Kotaprolu, S. S., Nithul, B., Kumar, D. V., & Rudravaram, G. (2025). Cleaning Robots: A Review of Sensor Technologies and Intelligent Control Strategies for Cleaning. *Journal of Field Robotics*, *42*(5), 2234–2259. DOI: 10.1002/rob.22515

Melnick, R. L., Moskowitz, J. M., Héroux, P., Mallery-Blythe, E., McCredden, J. E., & Herbert, M., & International Commission on the Biological Effects of Electromagnetic Fields (ICBE-EMF). (2025). The WHO-commissioned systematic reviews on health effects of radiofrequency radiation provide no assurance of safety. *Environmental Health : A Global Access Science Source*, *24*(1), 70.

Menaker, T., Zamansky, A., Van Der Linden, D., Kaplun, D., Sinitica, A., Karl, S., & Huber, L. (2020). Towards a Methodology for Data-Driven Automatic Analysis of Animal Behavioral Patterns. *Proceedings of the Seventh International Conference on Animal-Computer Interaction*, 1–6. https://doi.org/DOI: 10.1145/3446002.3446126

Mienye, I. D., Swart, T. G., & Obaido, G. (2024). Recurrent neural networks: A comprehensive review. *Information (Basel)*, *15*(9), 517.

Migdał, P., Plotnik, M., Bieńkowski, P., Berbeć, E., Latarowski, K., Białecka, N., & Murawska, A. (2025). The Influence of an Electromagnetic Field at a Radiofrequency of 900 MHz on the Behavior of a Honey Bee. *Agriculture*, *15*(12), 1266.

Miklósi, Á. (2015). *Dog behaviour, evolution, and cognition*. Oxford University Press.

Miles, A., Fields, N. L., Bennett, M., Xu, L., Magruder, K., Stringfellow, M. K., Sesay, B. J., & Handique, S. (2024). Robotic Animal Use among Older Adults Enrolled in Palliative or Hospice Care: A Scoping Review and Framework for Future Research. *Robotics (Basel, Switzerland)*, *13*(6), 92. DOI: 10.3390/robotics13060092

Miranda-de la Lama, G. C., & Mota-Rojas, D. (2020). Association between Attitude and Empathy with the Quality of Human-Livestock Interactions. *Animals (Basel)*, *10*(8), 1304. DOI: 10.3390/ani10081304

Mishra, R. C., Ghosh, R., & Bae, H. (2016). Plant acoustics: Searching for a sound mechanism for signaling in plants. *Journal of Experimental Botany*, *67*(15), 4483–4494. PMID: 27342223

Mohanta, H. C.. (2022). An optimized PI controller-based SEPIC converter for microgrid-interactive hybrid renewable power sources. *Wireless Communications and Mobile Computing*, *2022*, 6574825.

Mohmmad, S., Chauhan, G. S., Rao, B. S., Sravani, K., & Sridhar, K. (2025). Detection of Animals based on their Species by Implementing Bio-Acoustic Classification for Forest Wildlife Monitoring. In *2025 International Conference on Intelligent Computing and Control Systems (ICICCS)* (pp. 1244-1249). IEEE.

Monadizadeh, S., Kibert, C. J., Li, J., Woo, J., Asutosh, A., Roostaie, S., & Kouhirostami, M. (2021). A review of protocols and guidelines addressing the exposure of occupants to electromagnetic field radiation (EMFr) in buildings. *Journal of Green Building*, *16*(2), 55–81.

Morcatty, T. Q., Su, S., Siriwat, P., Andersson, A. A., Atoussi, S., Feddema, K., & Roberts, D. L. (2024). Navigating ethical challenges in online wildlife trade research. *Conservation Biology*, *38*(5), e14341. PMID: 39248761

Nagesh, K. M., & Piovene, C. (2011). *Emission and function of volatile organic compounds in response to abiotic stress.* In *Abiotic stress in plants: Mechanisms and adaptations* (p. 367).

Nakade, D., & Dhadse, S. (2024). The plant bioacoustics: A system of plant-sound relationship. *Plant Science*, *7*(1), 1–8.

Nalayini, C. M., Kalpana, V., Hemamalini, S., & Sathyamoorthy, K. (2025). A YOLOv8-based AI System for Real-Time Endemic Species Threat Detection and. *Journal of Innovative Image Processing*, *7*(1), 50–73.

Narimisaei, J., Naeim, M., Imannezhad, S., Samian, P., & Sobhani, M. (2024). Exploring emotional intelligence in artificial intelligence systems: a comprehensive analysis of emotion recognition and response mechanisms [Review of Exploring emotional intelligence in artificial intelligence systems: a comprehensive analysis of emotion recognition and response mechanisms]. Annals of Medicine and Surgery, 86(8), 4657. Wolters Kluwer. DOI: 10.1097/ms9.0000000000002315

Nartey, J. (2025). The Ethics of Emotion Recognition Technology: Implications for Privacy and Consent. DOI: 10.2139/ssrn.5200140

National Toxicology Program. Toxicology and carcinogenesis studies in Sprague Dawley (Hsd:Sprague Dawley SD) rats exposed to whole-body radio frequency radiation at a frequency (900 MHz) and modulations (GSM and CDMA) used by cell phones. Natl Toxicol Program Tech Rep Ser. 2018 Nov;(595):NTP-TR-595. doi: . PMID: 33562898; PMCID: PMC8039879.DOI: 10.22427/NTP-TR-595

Naveen, S., Dhamotharan, R., Sahebzathi, S., Krishnan, R. S., Kumar, V. V., & Sundaravadivel, P. (2025). ElephantGuardNet: A Comprehensive Solution for Elephant Conservation and Human-Elephant Conflict Reduction. In 2025 4th International Conference on Sentiment Analysis and Deep Learning (ICSADL) (pp. 1470-1477). IEEE.

Nazat, S., Alayed, W., Li, L., & Abdallah, M. (2025). Ensemble Learning Framework for Anomaly Detection in Autonomous Driving Systems. *Sensors (Basel)*, *25*(16), 5105. DOI: 10.3390/s25165105

Nazir, S., & Kaleem, M. (2021). Advances in image acquisition and processing technologies transforming animal ecological studies. *Ecological Informatics*, *61*, 101212. DOI: 10.1016/j.ecoinf.2021.101212

Neethirajan, S. (2023a). Artificial Intelligence and Sensor Technologies in Dairy Livestock Export: Charting a Digital Transformation [Review of Artificial Intelligence and Sensor Technologies in Dairy Livestock Export: Charting a Digital Transformation]. Sensors, 23(16), 7045. Multidisciplinary Digital Publishing Institute. DOI: 10.3390/s23167045

Neethirajan, S., Reimert, I., & Kemp, B. (2021). Measuring Farm Animal Emotions—Sensor-Based Approaches [Review of Measuring Farm Animal Emotions—Sensor-Based Approaches]. Sensors, 21(2), 553. Multidisciplinary Digital Publishing Institute. DOI: 10.3390/s21020553

Neethirajan, S. (2023b). AI in Sustainable Pig Farming: IoT Insights into Stress and Gait. *Agriculture*, *13*(9), 1706. DOI: 10.3390/agriculture13091706

Neethirajan, S., Scott, S., Mancini, C., Boivin, X., & Strand, E. (2024). Human-computer interactions with farm animals—Enhancing welfare through precision livestock farming and artificial intelligence. *Frontiers in Veterinary Science*, *11*, 1490851. DOI: 10.3389/fvets.2024.1490851

Nessler, J. N. (2025). Editorial: Artificial intelligence in animal behaviour, veterinary behaviour and neurology. *Frontiers in Veterinary Science*, *12*, 1607623. DOI: 10.3389/fvets.2025.1607623

Nguyen Chi, T., Vu, T. T., Nguyen, H. T., & Clink, D. J. (2025). Circadian rhythms and the use of transfer learning for critically endangered crested argus Rheinardia ocellata in the central highlands of Vietnam: the implications for conservation. *Philosophical Transactions B, 380*(1928).

Nikshya, J. E., Raj, J. R. F., Krishnan, R. S., Rajkumar, G. V., Settu, I. J., & Kavitha, S. (2025). Deep Learning Models for Acoustic Monitoring of the Himalayan Monal and Other Elusive Bird. In *2025 4th International Conference on Sentiment Analysis and Deep Learning (ICSADL)* (pp. 1392-1399). IEEE.

Nilsson, S. R., Goodwin, N. L., Choong, J. J., Hwang, S., Wright, H. R., Norville, Z. C., Tong, X., Lin, D., Bentzley, B. S., Eshel, N., McLaughlin, R. J., & Golden, S. A. (2020). Simple Behavioral Analysis (SimBA) – an open source toolkit for computer classification of complex social behaviors in experimental animals. *Animal Behavior and Cognition*. Advance online publication. DOI: 10.1101/2020.04.19.049452

Nisha Angeline, M., Selvakumar, P., Gupta, M. K., Nazar, N., Elumalai, D., & Bhattacharya, S. (2025). Emerging Materials for Energy Storage and Environmental Applications. In K. Muduli, B. Moharana, S. Ales, & D. Biswal (Eds.), Using Computational Intelligence for Sustainable Manufacturing of Advanced Materials (pp. 53-72). IGI Global Scientific Publishing.

Ododo, E. P., Offiong, A. E., Odiong, U. O., & Edet, E. E. (2024). UNMASKING AI: HOW MEDIA SHAPES STUDENTS' PERCEPTIONS AND ATTITUDES TOWARDS TECHNOLOGY. *The American Journal of Interdisciplinary Innovations and Research*, *6*(09), 19–29.

Ojeda-Castelo, J. J., Capobianco-Uriarte, M. D. L. M., Piedra-Fernandez, J. A., & Ayala, R. (2022). A survey on intelligent gesture recognition techniques. *IEEE Access : Practical Innovations, Open Solutions*, *10*, 87135–87156.

Okuyama, J., Nakajima, K., Matsui, K., Nakamura, Y., Kondo, K., Koizumi, T., & Arai, N. (2015). Application of a computer vision technique to animal-borne video data: Extraction of head movement to understand sea turtles' visual assessment of surroundings. *Animal Biotelemetry*, *3*(1), 35. DOI: 10.1186/s40317-015-0079-y

Oladele, O. K. (2025). *AI in Wildlife Conservation: Species Monitoring*. Poaching Prevention, and Habitat Restoration.

Olojede, H. T., & Polo, E. P. (2025). "In Praise of Normative Science: Arts and Humanities in the Age of Artificial Intelligence," pages 1-9. In *International Journal ofSocial Sciences and Humanities: Africa Research Corps Network (ARCN) Journals*. ISSN: 2360-9877. Volume 11, Issue 2, (January, 2025). DOI: . Available at https://arcnjournals.org/images/27265774111101121-1.pdfDOI: 27265774111101121-1

Oritsegbemi, O. (2023). Human Intelligence versus AI: Implications for Emotional Aspects of Human Communication. *Journal of Advanced Research in Social Sciences*, *6*(2), 76. DOI: 10.33422/jarss.v6i2.1005

Oyewole, J. A., Awolaju, B. A., & Ayomide, P. A. O. (2025). Exploring interspecies communication: A sociolinguistic analysis of animal signals and their human interpretations. *Biological Sciences*, *5*(2), 898–906.

Özentürk, U., Chen, Z., Jamone, L., & Versace, E. (2024). Robotics for poultry farming: Challenges and opportunities. *Computers and Electronics in Agriculture*, *226*, 109411. DOI: 10.1016/j.compag.2024.109411

OZGER, Z. B., CIHAN, P., & GOKCE, E. (2024). A Systematic Review of IoT Technology and Applications in Animals. *Kafkas Üniversitesi Veteriner Fakültesi Dergisi*, *30*, 411–431.

Pant, A., Hoda, R., Spiegler, S. V., Tantithamthavorn, C., & Turhan, B. (2023). Ethics in the Age of AI: An Analysis of AI Practitioners' Awareness and Challenges. *ACM Transactions on Software Engineering and Methodology*, *33*(3), 1. DOI: 10.1145/3635715

Pantic, I. V., Pantic, J. P., Valjarevic, S., Corridon, P. R., & Topalovic, N. (2025). AI approaches using random forests for signal analysis. *Chemico-Biological Interactions*, •••, 111624. PMID: 40582691

Pavlovic, V. I., Sharma, R., & Huang, T. S. (1997). Visual interpretation of hand gestures for human-computer interaction: A review. *IEEE Transactions on Pattern Analysis and Machine Intelligence*, *19*(7), 677–695.

Pedrazzi, L., Naik, H., Sandbrook, C., Lurgi, M., Fürtbauer, I., & King, A. J. (2025). Advancing animal behaviour research using drone technology. *Animal Behaviour*, *222*, 123147. DOI: 10.1016/j.anbehav.2025.123147

Pepperberg, I. M. (2019). *The Alex studies: Cognitive and communicative abilities of grey parrots*. Harvard University Press.

Perez, M., & Toler-Franklin, C. (2023). *CNN-Based Action Recognition and Pose Estimation for Classifying Animal Behavior from Videos: A Survey* (Version 1). arXiv. https://doi.org/DOI: 10.48550/ARXIV.2301.06187

Pflanzer, M., Traylor, Z., Lyons, J. B., Dubljević, V., & Nam, C. S. (2022). Ethics in human–AI teaming: Principles and perspectives. *AI and Ethics*, *3*(3), 917. DOI: 10.1007/s43681-022-00214-z

Pinto, A., Solovov, A., Simões, A. C., & Menezes, P. (2025). User Acceptance in Human–Robot Interaction: Exploring the Role of Anthropomorphic Mechanisms in Manufacturing Environments—A Systematic Literature Review. *Journal of Intelligent & Robotic Systems*, *111*(2), 46. DOI: 10.1007/s10846-025-02248-x

Pise, A. A., Alqahtani, M. A., Verma, P., Purushothama, K., Karras, D. A., Prathibha, S., & Halifa, A. (2022). Methods for Facial Expression Recognition with Applications in Challenging Situations. *Computational Intelligence and Neuroscience, 1*. Advance online publication. DOI: 10.1155/2022/9261438 PMID: 35665283

Podder, T., Bhattacharya, D., Majumder, P., & Bălaş, V. E. (2023). A feature boosted deep learning method for automatic facial expression recognition. *PeerJ. Computer Science, 9*. Advance online publication. DOI: 10.7717/peerj-cs.1216 PMID: 37346544

Polo, E. P., Emmanuel, H., & Obeka, O. B. (2024). Examining the epistemological status of AI-aided research in the information age: Research integrity of Margaret Lawrence University in Delta State. *International Journal of Social Sciences and Humanities: Africa Research Corps Network (ARCN). Journals, 11*(1), 197–207. https://www.arcnjournals.org/images/27265774111101125.pdf

Portelange, N. (2020). Implementing an IoT-based information system for the data recording in a phase I clinical center.

Potluri, S., Satpathy, S., & Basa, S. S. (Eds.). (2025). *AI in Agriculture for Sustainable and Economic Management*. Taylor & Francis., DOI: 10.1201/9781003451648

Pradhan, T., Sarkar, D., & Nanda, C. (2025). Transformative solutions through ai and digital platforms in human-elephant conflict management. Rais journal for social sciences, 9(1), 221-230.

Pradhan, N., Sille, R., & Sagar, S. (2022). Artificial intelligence empowered models for UAV communications. In *Unmanned aerial vehicle cellular communications* (pp. 95–113). Springer International Publishing.

Pratap, A. R., & Annamalai, S. (2025). Sensor Technologies for Environmental Data Collection. *Environmental Monitoring Using Artificial Intelligence*, 133-166.

Praveena, K., Kapali, B. S. C., & Patrer, D. (2020). Emotion recognition with AI: Techniques and applications. *World Journal of Advanced Research and Reviews, 8*(2), 344. DOI: 10.30574/wjarr.2020.8.2.0362

Prunier, A., Mounier, L., Le Neindre, P., Leterrier, C., Mormède, P., Paulmier, V., & Guatteo, R. (2013). Identifying and monitoring pain in farm animals: A review. *Animal, 7*(6), 998–1010. PMID: 23254122

Puri, V., Nayyar, A., & Raja, L. (2019). Agriculture drones in India: A review. *Computers and Electronics in Agriculture, 162*, 164–177. DOI: 10.1016/j.compag.2019.04.001

Quansah, E., Hartz, D. E., & Salipante, P. (2022). Adaptive practices in SMEs: Leveraging dynamic capabilities for strategic adaptation. *Journal of Small Business and Enterprise Development*, *29*(7), 1130–1148.

Rahman, M. A., Emon, M. M. H., Khan, T., Hamid, A. B. A., & Yaakub, N. I. (2025). Eco-Intelligence: AI's Contribution to a Sustainable World. In AI and Green Technology Applications in Society (pp. 31-62). IGI Global Scientific Publishing.

Rani, S.. (2025). nBLAST-JC: An accelerated deployment of nucleotide BLAST on high-performance cluster. *Heliyon*, *11*(10), e41866. Advance online publication. DOI: 10.1016/j.heliyon.2025.e41866

Regan, T. (2004). *The case for animal rights*. University of California Press.

Resnik, D. B., & Hosseini, M. (2024). The ethics of using artificial intelligence in scientific research: New guidance needed for a new tool. *AI and Ethics*. Advance online publication. DOI: 10.1007/s43681-024-00493-8 PMID: 40337745

Robinson, D. G., & Draguhn, A. (2021). Plants have neither synapses nor a nervous system. *Journal of Plant Physiology*, *263*, 153467. PMID: 34247030

Robinson, P., & el Kaliouby, R. (2009). Computation of emotions in man and machines. *Philosophical Transactions of the Royal Society of London. Series B, Biological Sciences*, *364*(1535), 3441. DOI: 10.1098/rstb.2009.0198 PMID: 19884138

Roemmich, K., Schaub, F., & Andalibi, N. (2023). Emotion AI at Work. *Implications for Workplace Surveillance, Emotional Labor, and Emotional Privacy.*, *1*. Advance online publication. DOI: 10.1145/3544548.3580950

Rogers, L. J., & Kaplan, G. T. (2002). *Songs, roars, and rituals: Communication in birds, mammals, and other animals*. Harvard University Press.

Romano, D., Porfiri, M., Zahadat, P., & Schmickl, T. (2024). Animal–robot interaction—An emerging field at the intersection of biology and robotics. *Bioinspiration & Biomimetics*, *19*(2), 020201. DOI: 10.1088/1748-3190/ad2086

Romeo, Z., & Testolin, A. (2025). Artificial intelligence can emulate human normative judgments on emotional visual scenes. *Royal Society Open Science*, *12*(7). Advance online publication. DOI: 10.1098/rsos.250128 PMID: 40740716

Rudroff, T. (2024). Artificial Intelligence as a Replacement for Animal Experiments in Neurology: Potential, Progress, and Challenges. *Neurology International*, *16*(4), 805–820. DOI: 10.3390/neurolint16040060

Russell, S., & Norvig, P. (2020). *Artificial intelligence: A modern approach* (4th ed.). Pearson.

Saarela, M., & Podgorelec, V. (2024). Recent Applications of Explainable AI (XAI): A Systematic Literature Review. *Applied Sciences (Basel, Switzerland)*, *14*(19), 8884. DOI: 10.3390/app14198884

Saghir, R. B., Campos, I., Gouvêa, T. S., & Sonntag, D. (2025). EcoScape Analyzer: A Tool for Performing Soundscape Analysis With Flexible Pipeline for Biodiversity Assessment. In *Companion Proceedings of the 30th International Conference on Intelligent User Interfaces* (pp. 137-140).

Sakthivel, M., Jaganathan, A., Mohanraj, M., & Selvakumar, P. Mohit, & Manjunath, T. (2025). Brand Loyalty through AI and Personalisation AI-Generated Content for Financial Marketing. In A. J Nair, S. Manohar, Y. B. Limbu, & B. A. Huhmann (Eds.), Intersecting Natural Language Processing and FinTech Innovations in Service Marketing (pp. 77-96). IGI Global Scientific Publishing.

Salma, Z., Hijón-Neira, R., Pizarro, C., & Abdul Moqeet, A. (2025). Effectiveness of Robot-Mediated Learning in Fostering Children's Social and Cognitive Development. *Applied Sciences (Basel, Switzerland)*, *15*(7), 3567. DOI: 10.3390/app15073567

Salmerón, J. L., & Celma, A. R. (2021). Synthetic Emotions for Empathic Building. *Mathematics*, *9*(7), 701. DOI: 10.3390/math9070701

Samijayani, O. N., Sabrina, A., Thamrin, R. Z., Astharini, D., & Syahriar, A. (2024). Performance of Interconnected Hybrid ZigBee-Optic for Extended Wireless Sensor Networks. *J. ICT Res. Appl*, *18*, 69–80.

Santhia, R. K., & Jayalakshmi, S. L. (2025). Advancing Environmental Surveillance: A Comprehensive Review of Sound Event Classification Techniques and Applications. In *2025 8th International Conference on Trends in Electronics and Informatics (ICOEI)* (pp. 1329-1337). IEEE.

Sardana, G. D. (2009). Evaluating the business performance of an SME: A conceptual framework. *International Journal of Globalisation and Small Business*, *3*(2), 137–159.

Sarma, H. Joshi, S., Lahiri, D., Ray, R. R., & Davoodbasha, M. (Eds.). (2023). *Microbial biofilms: Challenges and advances in metabolomic study.*

Sathiya, D., Palanisamy, P., Nandhini, E., Jayamurugan, V., Kirthikcharan, K., & Manoj, N. (2025). Deep Sound Search Leveraging Pre-trained CNNs and Faiss for Animal Vocalization Analysis. In *International Conference on Sustainability Innovation in Computing and Engineering (ICSICE 2024)* (pp. 410-423). Atlantis Press.

Savage-Rumbaugh, S., Shanker, S. G., & Taylor, T. J. (1993). *Apes, language, and the human mind.* Oxford University Press.

Schneider, F. D., Brose, U., Rall, B. C., & Guill, C. (2016). Animal diversity and ecosystem functioning in dynamic food webs. *Nature Communications*, *7*(1), 12718. PMID: 27703157

Schneider, S., Taylor, G. W., Linquist, S., & Kremer, S. C. (2019). Past, present and future approaches using computer vision for animal re-identification from camera trap data. *Methods in Ecology and Evolution*, *10*(4), 461–470. DOI: 10.1111/2041-210X.13133

Schuh, G., Patzwald, M., & Cardoso, M. C. I. (2019, August). Resilient technology strategy in volatile environments: Derivation of requirements to enable long-term strategic positioning in times of volatility, uncertainty, complexity and ambiguity. In *2019 Portland International Conference on Management of Engineering and Technology (PICMET)* (pp. 1-15). IEEE.

Schuhmann, C., Kaczmarczyk, R., Rabby, G., Friedrich, F., Kraus, M., Kalyan, K., Nadi, K., Nguyen, H., Kersting, K., & Auer, S. (2025). EmoNet-Face: An Expert-Annotated Benchmark for Synthetic Emotion Recognition. DOI: 10.48550/ARXIV.2505.20033

Sebeok, T. A. (1990). *Essays in zoosemiotics*. Springer.

Sethi, S. S., & Jain, K. (2024). AI technologies for social emotional learning: Recent research and future directions. *Journal of Research in Innovative Teaching & Learning*, *17*(2), 213. DOI: 10.1108/jrit-03-2024-0073

Sethumadhavan, R. P. P., S., Selvakumar, P., Kadu, S. G., Chandel, P. S., & Gupta, S. (2025). Environmental Sustainability and Employee Participation. In C. Şeker, E. Örücü, & A. Kaya (Eds.), *Green Leadership Strategies for Increased Employee Engagement and Performance* (pp. 275-298). IGI Global Scientific Publishing.

ShabanaFathima, M., Shanmugapriya, I., Pavithira, L., Mihirkumar, B. S., Rajeswari, J., Revathi, R., & Vengatesh, T. (2025). Deep Learning Approaches for Monitoring and Preserving Ecological Biodiversity: Challenges and Innovations. *Cuestiones de Fisioterapia*, *54*(3), 844–858.

Shao, C., Nah, S., Makady, H., & McNealy, J. (2025). Understanding user attitudes towards AI-enabled technologies: An integrated model of Self-Efficacy, TAM, and AI Ethics. *International Journal of Human-Computer Interaction*, *41*(5), 3053–3065.

Sharma, M. K., Prasad, S. V., Anusha, G., Das, A., & Sambathkumar, M. (2024). Tech-Driven Solutions for Environmental Conservation by AI Collaboration Processes. In Environmental Applications of Carbon-Based Materials (pp. 1-29). IGI Global.

Sharma, R., Bhambri, P., & Sohal, A. K. (2020). Energy bio-inspired for MANET. *International Journal of Recent Technology and Engineering*, *8*(6), 5580–5585.

Sharma, S., Sato, K., & Gautam, B. P. (2023). A methodological literature review of acoustic wildlife monitoring using artificial intelligence tools and techniques. *Sustainability*, *15*(9), 7128.

Shojaeifard, M. B., Jarideh, S., Owjfard, M., Nematollahii, S., Talaei-Khozani, T., & Malekzadeh, M. (2018). Electromagnetic Fields of Mobile Phone Jammer Exposure on Blood Factors in Rats. *Journal of Biomedical Physics & Engineering*, *8*(4), 403–408.

Siddiqi, N., Moin, F., & Al-Kindi, M. (2024). The Impact of Mobile Phone Electromagnetic Waves on the Neurons and Blood Brain Barrier Integrity in the Chick Embryo. *Biomedical & Pharmacology Journal*, *17*(3).

Sienkiewicz, Z. J., Bartram, R., Haylock, R. G., & Saunders, R. D. (2001, January). Single, brief exposure to a 50 Hz magnetic field does not affect the performance of an object recognition task in adult mice. *Bioelectromagnetics*, *22*(1), 19–26. DOI: 10.1002/1521-186x(200101)22:1<19:aid-bem3>3.0.co;2-0

Sienkiewicz, Z. J., Haylock, R. G., Bartrum, R., & Saunders, R. D. (1998). 50 Hz magnetic field effects on the performance of a spatial learning task by mice. *Bioelectromagnetics*, *19*(8), 486–493. DOI: 10.1002/(sici)1521-186x(1998)19:8<486:aid-bem6>3.0.co;2-k

Silva, D. E., Chen, C., & Zhu, Y. (2024). Facets of algorithmic literacy: Information, experience, and individual factors predict attitudes toward algorithmic systems. *New Media & Society*, *26*(5), 2992–3017.

Singer, P. (1975). *Animal liberation*. HarperCollins.

Singh, A. P.. (2021). Sketching of EV network: A complete roadmap. In Singh, H. (Eds.), *Sustainable development through engineering innovations* (pp. 431–442). Springer., DOI: 10.1007/978-981-15-9554-7_37

Singh, S. K., & Shahi, S. (2024). AI for Earth: Enhancing Environmental Sustain Ability with Technology. *Research Journal of Medical Sciences*, *18*, 219–225.

Stahl, B. C., Timmermans, J., & Mittelstadt, B. (2022). The ethics of computing: A survey of professional bodies and academic institutions. *Computer*, *55*(6), 65–75. DOI: 10.1109/MC.2022.3164232

Stern, U., He, R., & Yang, C.-H. (2015). Analyzing animal behavior via classifying each video frame using convolutional neural networks. *Scientific Reports, 5*(1), 14351. DOI: 10.1038/srep14351 PMID: 26394695

Stix, C. (2022). Artificial intelligence by any other name: A brief history of the conceptualization of "trustworthy artificial intelligence.". *Discover Artificial Intelligence, 2*(1). Advance online publication. DOI: 10.1007/s44163-022-00041-5

Stokes, J. E., Mullan, S., Takahashi, T., Monte, F., & Main, D. C. (2020). Economic and welfare impacts of providing good life opportunities to farm animals. *Animals (Basel), 10*(4), 610. PMID: 32252335

Stowell, D. (2022). *Computational bioacoustics with deep learning.*

Stowell, D. (2022). Computational bioacoustics with deep learning: A review and roadmap. *PeerJ, 10*, e13152. PMID: 35341043

Straulino, E., Scarpazza, C., & Sartori, L. (2023). What is missing in the study of emotion expression? [Review of What is missing in the study of emotion expression?]. Frontiers in Psychology, 14. Frontiers Media. DOI: 10.3389/fpsyg.2023.1158136

Straw, I., & Callison-Burch, C. (2020). Artificial Intelligence in mental health and the biases of language based models [Review of Artificial Intelligence in mental health and the biases of language based models]. PLoS ONE, 15(12). Public Library of Science. DOI: 10.1371/journal.pone.0240376

Subasi, A., Qaisar, S. M., & Nisar, H. (2024). *Artificial intelligence and multimodal signal processing in human-machine interaction*. Elsevier.

Sun, J. J., Zhou, H., Zhao, L., Yuan, L., Seybold, B., Hendon, D., Schroff, F., Ross, D. A., Adam, H., Hu, B., & Liu, T. (2024). Video Foundation Models for Animal Behavior Analysis. *Animal Behavior and Cognition*. Advance online publication. DOI: 10.1101/2024.07.30.605655

Swift, S., Downie, J. A., Whitehead, N. A., Barnard, A. M., Salmond, G. P., & Williams, P. (2001). Quorum sensing as a population-density–dependent determinant of bacterial physiology. *Philosophical Transactions of the Royal Society of London. Series B, Biological Sciences, 357*(1420), 593–607. PMID: 11450110

Symeonidou, N., Leiponen, A., Autio, E., & Bruneel, J. (2022). The origins of capabilities: Resource allocation strategies, capability development, and the performance of new firms. *Journal of Business Venturing, 37*(4), 106208.

Tamborini, M. (2025). The ethics of bioinspired animal-robot interaction: A relational meta-ethical approach. *Journal of Responsible Technology*, *22*, 100116. DOI: 10.1016/j.jrt.2025.100116

Tang, J., Zhang, Y., Yang, L., Chen, Q., Tan, L., Zuo, S., & Zhu, G. (2015). Exposure to 900 MHz electromagnetic fields activates the mkp-1/ERK pathway and causes blood-brain barrier damage and cognitive impairment in rats. *Brain Research*, *1601*, 92–101.

Tan, Y., & Zheng, Z. (2013). Research Advance in Swarm Robotics. *Defence Technology*, *9*(1), 18–39. DOI: 10.1016/j.dt.2013.03.001

Thirunagalingam, A., & Selvakumar, P. (2025). Application of Automation and Robotics in the Oil and Gas Industry. In Abdullayev, V., & Khang, A. (Eds.), *Revolutionizing AI and Robotics in the Oil and Gas Industry* (pp. 229–244). IGI Global Scientific Publishing.

Trigueiros, P., Ribeiro, F., & Reis, L. P. (2015). Hand gesture recognition system based in computer vision and machine learning. *Developments in medical image processing and computational vision*, 355-377.

Tulsulkar, G., Mishra, N., Thalmann, N. M., Lim, H. E., Lee, M. P., & Cheng, S. K. (2021). Can a humanoid social robot stimulate the interactivity of cognitively impaired elderly? A thorough study based on computer vision methods. *The Visual Computer*, *37*(12), 3019–3038. DOI: 10.1007/s00371-021-02242-y

Tzafestas, S. G. (2018). Roboethics: Fundamental Concepts and Future Prospects. *Information (Basel)*, *9*(6), 148. DOI: 10.3390/info9060148

Ullah, F., Saqib, S., & Xiong, Y. C. (2025). Integrating artificial intelligence in biodiversity conservation: Bridging classical and modern approaches. *Biodiversity and Conservation*, *34*(1), 45–65.

Ullah, S., Ou, J., Xie, Y., & Tian, W. (2024). Facial expression recognition (FER) survey: A vision, architectural elements, and future directions. *PeerJ. Computer Science*, *10*. Advance online publication. DOI: 10.7717/peerj-cs.2024 PMID: 38855254

Valdez-Juárez, L. E., & Castillo-Vergara, M. (2021). Technological capabilities, open innovation, and eco-innovation: Dynamic capabilities to increase corporate performance of SMEs. *Journal of Open Innovation*, *7*(1), 8.

Valente, D., Wang, H., Andrews, P., Mitra, P. P., Saar, S., Tchernichovski, O., & Benjamini, Y. (2007). Characterizing animal behavior through audio and video signal processing. *IEEE MultiMedia*, *14*(4), 32–41.

Velagaleti, S. B. (2024). Empathetic Algorithms: The Role of AI in Understanding and Enhancing Human Emotional Intelligence. *Deleted Journal*, *20*, 2051. DOI: 10.52783/jes.1806

Vicci, H. (2024). Emotional Intelligence in Artificial Intelligence: A Review and Evaluation Study [Review of Emotional Intelligence in Artificial Intelligence: A Review and Evaluation Study]. SSRN Electronic Journal. RELX Group (Netherlands). DOI: 10.2139/ssrn.4818285

Virant-Doberlet, M., Stritih-Peljhan, N., Žunič-Kosi, A., & Polajnar, J. (2023). Functional diversity of vibrational signaling systems in insects. *Annual Review of Entomology*, *68*(1), 191–210. PMID: 36198397

Vistorte, A. O. R., Deroncele-Acosta, Á., Ayala, J. L. M., Barrasa, Á., López-Granero, C., & Martí-González, M. (2024). Integrating artificial intelligence to assess emotions in learning environments: A systematic literature review. *Frontiers in Psychology*, *15*. Advance online publication. DOI: 10.3389/fpsyg.2024.1387089 PMID: 38966729

Vogg, R., Lüddecke, T., Henrich, J., Dey, S., Nuske, M., Hassler, V., Murphy, D., Fischer, J., Ostner, J., Schülke, O., Kappeler, P. M., Fichtel, C., Gail, A., Treue, S., Scherberger, H., Wörgötter, F., & Ecker, A. S. (2025a). Computer vision for primate behavior analysis in the wild. *Nature Methods*, *22*(6), 1154–1166. DOI: 10.1038/s41592-025-02653-y PMID: 40211003

Voysey, I., Baillie, L., Williams, J., & Herrmann, M. (2025). Co-Designing Educational Robotic Pet Concepts with Animal Welfare Educators and Children. *International Journal of Social Robotics*. Advance online publication. DOI: 10.1007/s12369-025-01328-0

Waddington, S. R., Li, X., Dixon, J., Hyman, G., & de Vicente, M. C. (2010). Getting the focus right: Production constraints for six major food crops in Asian and African farming systems. *Food Security*, *2*(1), 27–48. DOI: 10.1007/s12571-010-0053-8

Waelen, R. (2024). Philosophical Lessons for Emotion Recognition Technology. *Minds and Machines*, *34*(1). Advance online publication. DOI: 10.1007/s11023-024-09671-3

Wagner, M., & Vashisth, D. (2025). The Influence of Inaccurate GPU Power Measurements for Machine Learning Workloads in Industrial Applications. In *International Conference on Human-Computer Interaction* (pp. 233-247). Springer, Cham.

Wall, J., Wittemyer, G., Klinkenberg, B., & Douglas-Hamilton, I. (2014). Novel opportunities for wildlife conservation and research with real-time monitoring. *Ecological Applications*, *24*(4), 593–601. PMID: 24988762

Wang, S., Gao, J. Z., Lin, H., Shitole, M., Reza, L., & Zhou, S. (2019, April). Dynamic human behavior pattern detection and classification. In *2019 IEEE Fifth International Conference on Big Data Computing Service and Applications (BigDataService)* (pp. 159-166). IEEE.

Wang, Y., & Liu, W. (2023). Emotional Simulation of Artificial Intelligence and Its Ethical Reflection. *Academic Journal of Humanities & Social Sciences*, *6*(5). Advance online publication. DOI: 10.25236/ajhss.2023.060503

Waworundeng, J. (2024). IoT-based Environmental Monitoring with Data Analysis of Temperature, Humidity, and Air Quality. *CogITo Smart Journal*, *10*(1), 271–284. DOI: 10.31154/cogito.v10i1.708.692-705

Weaven, S., Quach, S., Thaichon, P., Frazer, L., Billot, K., & Grace, D. (2021). Surviving an economic downturn: Dynamic capabilities of SMEs. *Journal of Business Research*, *128*, 109–123.

Weinstein, B. G. (2018). A computer vision for animal ecology. *Journal of Animal Ecology*, *87*(3), 533–545. DOI: 10.1111/1365-2656.12780 PMID: 29111567

Williams, B., van Merriënboer, B., Dumoulin, V., Hamer, J., Fleishman, A. B., McKown, M., . . . Denton, T. (2025). Using tropical reef, bird and unrelated sounds for superior transfer learning in marine bioacoustics. *Philosophical Transactions B*, *380*(1928).

Williams, S. M. (2024). *Tracking and predicting bee behaviour and movement using machine learning* (Doctoral dissertation, Bangor University).

Williams, C., Schallmo, D., Lang, K., & Boardman, L. (2019). *The International Society for Professional Innovation Management*. ISPIM.

Woelders, H., de Wit, A., Lourens, A., Stockhofe, N., Engel, B., Hulsegge, I., & Zwamborn, P. (2017). Study of potential health effects of electromagnetic fields of telephony and Wi-Fi, using chicken embryo development as animal model. *Bioelectromagnetics*, *38*(3), 186–203.

Wurtz, K., Camerlink, I., D'Eath, R. B., Fernández, A. P., Norton, T., Steibel, J., & Siegford, J. (2019a). Recording behaviour of indoor-housed farm animals automatically using machine vision technology: A systematic review. *PLoS One*, *14*(12), e0226669. DOI: 10.1371/journal.pone.0226669 PMID: 31869364

Wu, X., Wang, H., Yan, Z., Tang, X., Xu, P., Siok, W.-T., Li, P., Gao, J.-H., Lyu, B., & Qin, L. (2025). AI shares emotion with humans across languages and cultures. DOI: 10.48550/ARXIV.2506.13978

Yalabik, B., & Fairchild, R. J. (2011). Customer, regulatory, and competitive pressure as drivers of environmental innovation. *International Journal of Production Economics, 131*(2), 519–527.

Yazdani, A., & Darbani, S. (2023).. . *The Impact of AI on Trends, Design, and Consumer Behavior.*, *1*(4), 4. DOI: 10.61838/kman.aitech.1.4.2

Yerebakan, M. O., & Hu, B. (2024). Human–Robot Collaboration in Modern Agriculture: A Review of the Current Research Landscape. *Advanced Intelligent Systems, 6*(7), 2300823. DOI: 10.1002/aisy.202300823

Yim, I. H. Y., & Wegerif, R. (2024). Teachers' perceptions, attitudes, and acceptance of artificial intelligence (AI) educational learning tools: An exploratory study on AI literacy for young students. *Future in Educational Research, 2*(4), 318–345.

Yi, Q., Xie, C., Guan, D., & Yuan, W. (2025). Man2Marine: Marine mammal sound classification in small samples by transfer learning from human sound data. *Pattern Recognition Letters, 190*, 185–191.

Yu, Z., Feng, G., Liu, H., Peng, H., & Dong, X. (2024). Sustainable market? The impact of downstream market concentration on high-quality agricultural development: Evidence from China's dairy industry. *Frontiers in Sustainable Food Systems, 8*. Advance online publication. DOI: 10.3389/fsufs.2024.1453115

Zandi, A., Hosseinirad, S., Kashani Zadeh, H., Tavakolian, K., Cho, B. K., Vasefi, F., & Tavakolian, P. (2025). A systematic review of multi-mode analytics for enhanced plant stress evaluation. *Frontiers in Plant Science, 16*, 1545025. PMID: 40370362

Zeller Zigaitis, W. L., & Robinson, A. C. (2023). Sensemaking in the wild: A review of practitioner collected geospatial data and its synthesis within protected areas for poaching mitigation. *Annals of GIS, 29*(3), 319–335.

Zhai, C., Wibowo, S., & Li, L. D. (2024). The effects of over-reliance on AI dialogue systems on students' cognitive abilities: a systematic review [Review of The effects of over-reliance on AI dialogue systems on students' cognitive abilities: a systematic review]. Smart Learning Environments, 11(1). Springer Nature. DOI: 10.1186/s40561-024-00316-7

Zhang, Y., Li, Z., Gao, Y., & Zhang, C. (2015). Effects of fetal microwave radiation exposure on offspring behavior in mice. *Journal of Radiation Research, 56*(2), 261–268.

Zhang, Y., Sun, W., Yang, J., Wu, W., Miao, H., & Zhang, S. (2022). An Approach for Autonomous Feeding Robot Path Planning in Poultry Smart Farm. *Animals (Basel)*, *12*(22), 3089. DOI: 10.3390/ani12223089

Zhao, J., Yue, H., Wang, K., Sun, Y., Zhao, L., Wang, H., & Peng, R. (2025). Disruption of the linear relationship between recognition memory and local functional connectivity of left perirhinal cortex in rat after acute 2.856 GHz microwave exposure. *Brain Research*, •••, 149816.

Zhao, Z., Yang, Q., Li, R., Yang, J., Liu, Q., Zhu, B., Weng, C., Liu, W., Hu, P., Ma, L., Qiao, J., Xu, M., & Tian, H. (2024). A comprehensive review on the evolution of bio-inspired sensors from aquatic creatures. *Cell Reports. Physical Science*, *5*(7), 102064. DOI: 10.1016/j.xcrp.2024.102064

About the Contributors

Walid Ben Ameur earned his PhD from the University of Carthage. He currently serves as a teacher-researcher at the Department of Life Sciences, Faculty of Sciences of Gabès, Tunisia, and is a member of the Laboratory of Ecology of Terrestrial Fauna at the University of Gabès, Tunisia. Walid Ben Ameur has authored more than 20 peer-reviewed scientific papers published in international journals. His work has made useful contributions to the field of environmental chemistry, with a particular focus on persistent organic pollutants. He has developed analytical techniques for assessing emerging pollutants and evaluating their effects on both the environment and human health. His research spans a diverse range of samples, including those from environmental, biological, human, and food sources. Dr. Ben Ameur has mentored three Master's students and regularly serves as a reviewer for respected journals such as Science of the Total Environment, Chemosphere, Environmental Science and Pollution Research, Marine Pollution Bulletin, Environmental Pollution, Ecotoxicology and Environmental Safety, and Chemistry Africa.

Adnène Arbi is an Associate Professor at the National Institute of Applied Sciences and Technology (INSAT). His current research activities are in the field of Applied Mathematics and Artificial Intelligence (AI). Major research projects include modelling of mathematical neural networks and model predictive control. He has published more than 34 papers in peer-reviewed international journals since 2014. He edited two books on Neural Networks, Iot, AI, and Data-driven control published in 2025-2026. He is a member of the editorial board of the journals: Journal of Algorithms & Computational Technology, Innovative Applied Mathematics and Computational Sciences, Frontiers in Applied Mathematics and Statistics and Karshi Multidisciplinary International Scientific Journal. He served as a reviewer for more than 100 papers for international journals such as the IMA Journal of Mathematical Control and Information, Mathematics and Computers in Simulation, Chaos, Solitons & Fractals, Journal of Intelligent & Fuzzy Systems,

Computational and Applied Mathematics. American Association for Sciences and Technology, Mathematical Methods in the Applied Sciences, Neural Networks, Asian Journal of Control, Neural processing letters, International Journal of Statistical Distributions and Applications, IEEE Transactions on Neural Networks and Learning Systems, IMA Journal of Mathematical Control and Information, JCAS: International Journal of Control, Automation and Systems. He is an Expert/ Consultant for the Science for Africa Foundation (2023). He has co-organized two international workshops and conferences since 2022. He is a member of the Laboratory of Engineering Mathematics at Tunisia Polytechnic School, University of Carthage, Tunisia. Besides, he is an international member of laboratoire Artificial Intelligence Research and Applications (AIRA), Hassan I University, Settat, Morocco. H-index: 25; Total citations: 1375; Expertise: Time scales, Dynamical Systems, Time delay systems, Neural Networks, Model Predictive Control.

***

Pillarisetty Alekhya working as Assistant Professor in the department of Electronics and Communication Engineering in Mallareddy Technical campus, working since August 2025. My teaching experience is 3 years (Nov 2022- Present) Education: I have completed my M.E. in Embedded Systems from Stanley College of Engineering and Technology For Women (Osmania University, 2021) and a B.Tech in ECE from Megha Institute of Engineering and Technology For Women (JNTU Hyderabad, 2019). My technical skills include C and Python programming languages, and tools such as Matlab, MS-Office, Keil micro vision IDE, Xilinx ISE 14.5, Anaconda Navigator, and NS-2. Subjects taught by me: Electronic Circuit Analysis, Computer Organization and Architecture, Data Communications and Networks, Introduction to Internet of Things, Artificial Intelligence, Python Programming

Pankaj Bhambri is working in the Department of Information Technology of Guru Nanak Dev Engineering College, Ludhiana.

Mahender Reddy Bobbala is an Assistant Professor in the Department of Computer Science and Engineering (Cyber Security) at CVR College of Engineering, an autonomous institution affiliated with JNTUH, located in Vastunagar, Mangalpalli, Ibrahimpatnam, Hyderabad, Telangana – 501510. With over 12 years of teaching experience, he has been actively involved in academic instruction and mentoring in the field of computer science. His areas of interest include Cyber Security and Machine Learning, and he has contributed to the academic community through publications in the domain of machine learning. His dedication to quality education and research reflects his commitment to shaping future professionals in emerging technologies.

Sreedevi E. With a passion for software engineering and machine learning, I embarked on an academic journey that culminated in an M.Tech in Computer Science. Driven by a thirst for knowledge, I pursued a Ph.D. in Software Engineering, specializing in software defect detection through machine learning algorithms. My research delved into the intersection of cutting-edge technology and software development. Throughout my academic endeavours, I honed my expertise in software engineering and cultivated a keen interest in machine learning applications.Eager to contribute to the evolving tech landscape, my journey is marked by a commitment to advancing the realms of software engineering and machine learning.

Soumen Ghosh is currently pursuing a Bachelor of Technology (B.Tech) in Computer Science and Engineering (CSE) from the Global Institute of Management & Technology, located in Krishnanagar, India. With a fervent passion for technology and problem-solving, I engaged in coding competitions, hackathons, and workshops to further hone their skills.

Jayshree Gorde is currently serving as a Medical Administrator and is a PhD Scholar at Symbiosis International (Deemed University), Pune. She possesses extensive industry experience in healthcare management, having held administrative positions at several renowned institutions, including Sahyadri Hospitals, Lokmanya Hospital, Apollo Hospitals, and Jupiter Hospital. Her work focuses on bridging the gap between clinical administration and operational excellence.

Rupam Hazra, a passionate and determined student currently pursuing a Bachelor of Technology (BTech) in Computer Science and Engineering at Krishnanagar GIMT. I have a strong interest in the fields of Artificial Intelligence, Cybersecurity, and emerging technologies. With a dedication to academic excellence and innovation, I have authored 16 research publications in globally recognized platforms, focusing on real-world applications and technological advancement.

Kartigeyan J. is an Associate Professor in the Department of Electrical and Electronics Engineering at J.B. Institute of Engineering and Technology, Hyderabad, India (500 075). He is actively involved in teaching, academic administration, and research, contributing to the development of undergraduate and postgraduate engineering education. His professional interests include advancing technical knowledge in electrical and electronics engineering and mentoring students toward academic and professional excellence.

Sanjeev Kumar is a Professor in Lovely Professional University Punjab India. He received Undergraduate in Hotel Management from Osmania University, Postgraduate in Tourism Management from Madurai Kamaraj University. He completed Doctor of Philosophy in Hospitality Management in 2019 from Amity University Jaipur India. His research areas lies in food Service Industry, his research focuses on Alcoholic beverages, Event management and Sustainable Management Practices, Metaverse, AI Machine Learning and Artificial Intelligence. He published more than 120 research papers, articles and chapters in Scopus Indexed, UGC Approved and peer reviewed Journals and books. Dr. Sanjeev Kumar participated and acted as resource person in various National and International conferences, seminars, session chair, research workshops and industry talks and his work has been widely cited. Dr. Sanjeev also edited more than 20 books with various International Publishers such as IGI Global, Cambridge Scholars, Emerald, CABI, Bentham Science, Taylor & Francis. Dr Sanjeev is an active researcher and has published numerous research articles in prestigious National and International journals and books.

Suresh Kumar is working as a Associate Professor in the Department of Computer Science and Engineering of Maharishi Markandeshwar (Deemed to be University), Mullana, Haryana, India.

T.C. Manjunath was born on 6th February 1967. He did his B.E. from RV College of Engg. in 1989 (B'lore Univ), M.E. in Robotics & Automation from Govt.'s LD College of Engg., (Gujarat Univ., Ahmedabad) in 1992 with FCD-First Rank & PhD in Systems & Control Engg. from IIT Bombay in the year 2007, got best research scholar award for the entire IITB. He is currently working as the Dean Research (R & D) and as Professor in Dept. of Computer Science & Engg. with specialization in IoT, Cyber Security, Blockchain Technology in Rajarajeswari College of Engineering, Bengaluru, Karnataka. He has got a teaching (academic), research & administrative experience of more than 35+ years (10 years as Principal & 8 years as HoD, 1 year as Dean Research & rest as faculty) and has published more than 1500+ research papers in various national & international journals & conferences, with 175+ patents, 100+ book-chapters, 25 awards, 100s of Keynote talks and also published 25 text books in various field of science, engineering & technology, completed projects worth Rs. 9 Crores till date & guided 10 PhD scholars in VTU. He is a member of 30 Professional Society Chapters and is a fellow of IE, IETE, AMIEE, Sr. Member of IEEE and a chartered engineer.

Geetha Manoharan is currently working in Telangana as an assistant professor at SR University. She is the university-level PhD program coordinator and has also been given the additional responsibility of In Charge Director of Publications and Patents under the Research Division at SR University.

Selvakumar P. is currently an Associate Professor in the Department of Science and Humanities at Nehru Institute of Technology, Coimbatore, Tamil Nadu, India. Prior to joining Nehru Institute of Technology, he held positions as an Assistant Professor in the Department of Science and Humanities various engineering colleges in Coimbatore, Tamil Nadu, India, spanning from 2011 to 2023. Dr.P. Selvakumar's academic and research interests lie primarily in the field of phytochemistry. He has made significant contributions to his field, with an impressive record that includes 30 journal publications, 12 patents, and 20 conference papers. Additionally, he has organized 5 seminars and conferences, further demonstrating his commitment to academic and scientific exchange. Throughout his career, Dr. P.Selvakumar has been recognized for his dedication and excellence in teaching, having been honored with several awards, including the Best Teacher Awards. He holds a prominent position in the academic community, contributing actively to research.

Sirikonda Shwetha is an Assistant Professor in the Department of Computer Science and Engineering at Sumathi Reddy Institute of Technology for Women, with more than 12 years of experience in academia. Her teaching and research interests include machine learning, deep learning, medical image analysis, and optimization techniques. She is actively engaged in developing intelligent systems with practical relevance.

Bottu Giridharan Sivakumar is a faculty member in the Department of Mechanical Engineering at the Academy of Maritime Education and Training (AMET), Deemed to be University, located at 135 East Coast Road, Kanathur, Tamil Nadu, India. He is an academic and researcher with expertise in mechanical engineering, actively involved in teaching, research, and academic administration. Dr. Sivakumar's professional interests include advancing engineering education, conducting applied and interdisciplinary research, and mentoring undergraduate and postgraduate students. He contributes to curriculum development, research publications, and institutional activities, with a focus on fostering technical competence, innovation, and professional ethics among future engineers.

Naresh Kumar Sripada is a Professor in the Department of Computer Science and Engineering at SR University, with over 21 years of academic experience. His expertise includes deep learning, machine learning, medical image analysis, and optimization techniques, and his research focuses on developing efficient intelligent systems for real-world applications

Index

A

B

C

D

E

F

G

H

I

L

M

www.ingramcontent.com/pod-product-compliance
Lightning Source LLC
LaVergne TN
LVHW081533300926
849161LV00006B/1404
9798337354835